Tea Chemistry and Health Benefits

Xiaoting Zhai
Editor

Tea Chemistry and Health Benefits

Flavour and Processing Dynamics

Springer

Editor
Xiaoting Zhai
State Key Laboratory of Tea Plant
Germplasm Innovation and Resource
Utilization, School of Tea Science
Anhui Agricultural University
Hefei, China

ISBN 978-3-032-11827-1 ISBN 978-3-032-11828-8 (eBook)
https://doi.org/10.1007/978-3-032-11828-8

This Springer imprint is published by the registered company Springer Nature Switzerland AG
The registered company address is: Gewerbestrasse 11, 6330 Cham, Switzerland

Preface

This book of *Tea Chemistry and Health Benefits: Flavour and Processing Dynamics* compiles eleven chapters written by a young board of tea researchers from China including Dr. Xiaoting Zhai from Anhui Agricultural University, Dr. Shanshan Hu from Anhui Agricultural University, Dr. Zisheng Han from Anhui Agricultural University, Dr. Yujie Wang from Zhejiang University, Dr. Piaopiao Long from Anhui Agricultural University, Ph.D. candidate Mr. Dongzhu Huang from Anhui Agricultural University, and Ph.D. candidate Miss. Jing Wang from Hunan Agricultural University. It reports the latest developments in relevant and interesting topics such as *Camellia sinensis*. tea, a traditional worldwide beverage: an introduction, flavor characteristics in six types of teas including color, aroma, and taste, the effects of processing on the quality of color, aroma and taste, technologies for the flavor quality improvement in tea-beverage processing, as well as health benefits of tea including antioxidant, anti-inflammatory effect, and regulation of gut microflora by tea.

Chapter 1 summarizes the worldwide distribution of tea plant, native and cultivated tea plant, fresh leaf harvesting and tea plant cultivation procedures, regional differences, classification of teas and reasons for regional diversification, hot tea infusions, the current situation of new-style ready-to-drink tea beverages, flavor chemical profiles of tea products, and the overview of health benefits of tea products. Chapter 2 focuses on three types of color compounds including catechins and their oxidation products, flavones, flavonols, and their glycosides, amino acid contributing compounds. The common pigment substances of plant including chlorophyll a, chlorophyll b, carotenoids, xanthophylls, and anthocyanins are also discussed. Chapter 3 brings new insights on the pigments dynamics in the production of different teas. It describes the formation mechanisms of colorants during tea production including enzymatic browning, microbe-induced transformation, and maillard products in thermal processing. Chapter 4 deals with the classic aroma characteristics of six types of tea (green tea, black tea, oolong tea, white tea, yellow tea, dark tea) and their key volatile contributors, and constructs a map correlating tea aroma perception with chemical composition. Chapter 5 discusses aroma formation and changes induced by different processing techniques. Chapter 6 gives an insight on taste characteristics of six major tea types, focusing on key nonvolatile compounds and their sensory impacts. The respective taste receptors and compound interactions are also discussed. Chapter 7 gives an overview on the effects of key processing steps on taste quality and metabolites in different tea types, and some novel processing methods are also described. Chapter 8 focuses on the key flavor compounds in tea beverages and the critical factors affecting their flavor qualities. It further explored the impacts of major processing technologies, including extraction, clarification, and sterilization, on the flavor development and overall quality, with a special emphasis on flavor improvement methods. Additionally, global innovations and advancements in tea beverage processing were discussed, highlighting their potential for improving sensory attributes and extending shelf life. Chapter 9 deals with the biological mechanisms of oxidative stress and its health impacts, with an in-depth analysis of the antioxidant effects of tea bioactive compounds such as polyphenols, alkaloids, amino acids, and flavonoids. Chapter 10 gives an overview on the biological mechanisms of inflammation and its pathological consequences, as well as the anti-inflammatory effects and molecular pathways of tea bioactive compounds. Chapter 11 focuses on the pivotal role of gut microbiota in human health and disease prediction. The new findings underscore tea's potential in preventive health and disease intervention, offering modern scientific evidence for the health benefits of traditional tea consumption.

In summary, this book presents the combined efforts of six professionals developing the research in tea production and health benefits with a variety of backgrounds and expertise. The editor wishes to thank the production staff and all the contributors for sharing their experience and for making this book possible.

Hefei, China Xiaoting Zhai

Competing Interests This book of Tea Chemistry and Health Benefits: Flavour and Processing Dynamics compiles eleven chapters written by a young board of tea researchers from China including Dr. Xiaoting Zhai from Anhui Agricultural University, Dr. Shanshan Hu from Anhui Agricultural University, Dr. Zisheng Han from Anhui Agricultural University, Dr. Yujie Wang from Zhejiang University, Dr. Piaopiao Long from Anhui Agricultural University, Ph.D. candidate Mr. Dongzhu Huang from Anhui Agricultural University, and Ph.D. candidate Miss. Jing Wang from Hunan Agricultural University. The authors have no competing interests to declare that are relevant to the content of this manuscript.

Contents

List of Contributors

Zisheng Han State Key Laboratory of Tea Plant Germplasm Innovation and Resource Utilization, School of Tea Science, Anhui Agricultural University, Hefei, China;
International Joint Laboratory on Tea Chemistry and Health Effects of Ministry of Education, Anhui Agricultural University, Hefei, China

Shanshan Hu State Key Laboratory of Tea Plant Germplasm Innovation and Resource Utilization, School of Tea Science, Anhui Agricultural University, Hefei, China;
International Joint Laboratory on Tea Chemistry and Health Effects of Ministry of Education, Anhui Agricultural University, Hefei, China

Dongzhu Huang State Key Laboratory of Tea Plant Germplasm Innovation and Resource Utilization, School of Tea Science, Anhui Agricultural University, Hefei, China;
International Joint Laboratory on Tea Chemistry and Health Effects of Ministry of Education, Anhui Agricultural University, Hefei, China

Piaopiao Long State Key Laboratory of Tea Plant Germplasm Innovation and Resource Utilization, School of Tea Science, Anhui Agricultural University, Hefei, China;
International Joint Laboratory on Tea Chemistry and Health Effects of Ministry of Education, Anhui Agricultural University, Hefei, China

Jing Wang Key Laboratory of Tea Science of Ministry of Education, Hunan Agricultural University, Changsha, 410128 China

Yujie Wang College of Biosystems Engineering and Food Science, Zhejiang University, Hangzhou, China

Xiaoting Zhai State Key Laboratory of Tea Plant Germplasm Innovation and Resource Utilization, School of Tea Science, Anhui Agricultural University, Hefei, China;
International Joint Laboratory on Tea Chemistry and Health Effects of Ministry of Education, Anhui Agricultural University, Hefei, China

Chapter 1
Camellia Sinensis. Tea, A Traditional Worldwide Beverage. An Introduction

Xiaoting Zhai

Abstract This chapter discusses the traditional use of the valuable plant *Camellia sinensis*, which is spread globally and prominently found in various countries across Asia, Africa, and South America, including China, India, Sri Lanka, Kenya, Brazil, Argentina, and so on. This plant has been used for centuries due to its various purported properties, primarily as a beverage (infusion) known by names such as tea, Cha (茶), Tu (荼), and Ming (茗). The global interest in these infusions is tied to the widespread distribution of the products, though the origin of *Camellia sinensis* is the southwest of China. On one hand, the plant and its infusions are renowned for their pharmacological effects against common ailments. On the other hand, various manufacturing processes, such as withering, fixation, shaping, yellowing, fermentation, aeration, and drying, significantly alter the sensory and chemical properties of the products. Moreover, new-style ready-to-drink tea beverages based on high-quality tea and diverse ingredients are expected for better flavor. Innovating modern technologies is useful for the flavor quality improvement in new-style tea beverage processing. Therefore, a thorough analysis of tea chemistry and health benefits focusing on flavour and processing dynamics is recommended.

Keywords *Camellia sinensis* · Tea infusion · Manufacturing process · Flavor quality · Pharmacological effect · Cultural heritage

X. Zhai (✉)
State Key Laboratory of Tea Plant Germplasm Innovation and Resource Utilization, School of Tea Science, Anhui Agricultural University, Hefei, China
e-mail: xiaotingzhai@ahau.edu.cn

International Joint Laboratory On Tea Chemistry and Health Effects of Ministry of Education, Anhui Agricultural University, Hefei, China

X. Zhai (ed.), *Tea Chemistry and Health Benefits*,
https://doi.org/10.1007/978-3-032-11828-8_1

1.1 Tea Infusions and Regional Products. A Worldwide Perspective

Nowadays, tea is recognized as one of the most popular non-alcoholic beverages worldwide, largely due to its unique flavor and numerous prominent health benefits. The history of tea consumption spans nearly 5,000 years (Wei et al., 2018; Xia et al., 2019). People around the world also regard tea-drinking as a way for socialization. Commercial tea products are primarily derived from the fresh buds, tender leaves, mature leaves, and stems of *Camellia* species which belong to the Theaceae family of angiosperms, specifically *Camellia sinensi*s (L.) O. Kuntze, *C. sinensis* var. *assamica* (Masters) Kitamura, and *C. sinensis* var. *Cambodiensis* (Meng et al., 2019). Additionally, leaves and stems of *Ilex Paraguariensis* A. St.-Hil., *Aspalathus linearis* (Burm. f.) R. Dahlgren, and *Ilex Kudingcha* CJ Tseng are used to produce other herbal teas in certain countries, such as *Mate* tea, rooibos tea, and Kuding tea, respectively (Zhang et al., 2019a).

The tea plant *C. sinensis* is an endemic species in southwestern China, and now is cultivated all over the world, particularly in tropical and sub-tropical regions including Asia, Africa, South America. In 2022, the item "Traditional tea processing techniques and associated social practices in China" was added to the intangible cultural heritage list of the United Nations Educational, Scientific, and Cultural Organization (https://www.unesco.org/zh/intangible-cultural-heritage) (UNESCO). The item involves knowledge, skills, and practices related to tea garden management, tea leaf harvesting, tea hand-making as well as the consumption and sharing of tea. Since ancient time, Chinese people have been growing, picking, processing, and drinking tea. Tea producers have developed green tea, yellow tea, black tea, white tea, oolong tea, dark tea, and flower-scented teas based on different core techniques such as fixation, yellowing, pile-fermentation, withering, shaping, fermentation, and scenting. There are over 2,000 types of tea products and different traditional customs that are still present in the daily life, ceremonies, and festive activities of Chinese people.

The Food and Agriculture Organization of the United Nations (FAOSTAT) database indicates that global production of tea leaves has increased by approximately 59% in production in the last decade, reaching 32 million tons across 4.8 million hectares in 50 tea-growing countries (https://www.fao.org/faostat/) (Fig. 1.1). Moreover, China, India, Kenya, Sri Lanka, Türkiye, Vietnam, Indonesia, Japan, Argentina, Bangladesh were top 10 producers of tea leaves production (Fig. 1.2). The combination of genetic variation, environmental factors, and diverse tea processing methods has led to a vast array of tea products with varying flavors, such as umami, bitter, astringent, and sweet, to fulfill global consumer request. Evidently, tea plant is the most widely consumed plant source for tea globally.

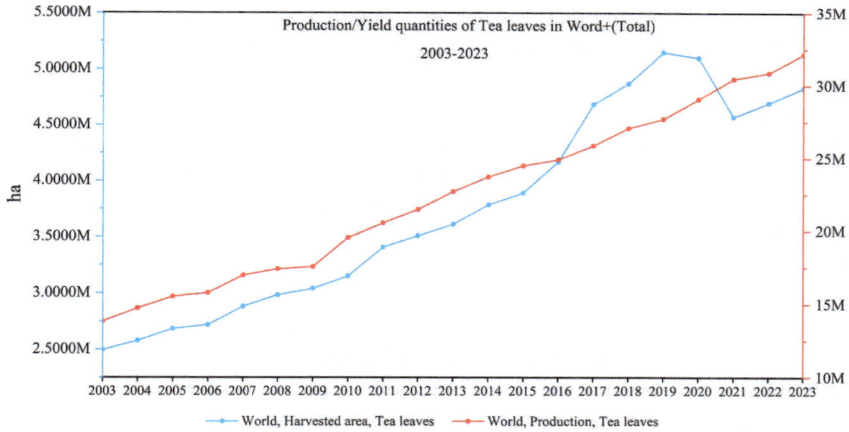

Fig. 1.1 Production and yield quantities of tea leaves in world (2003–2023) from FAO database

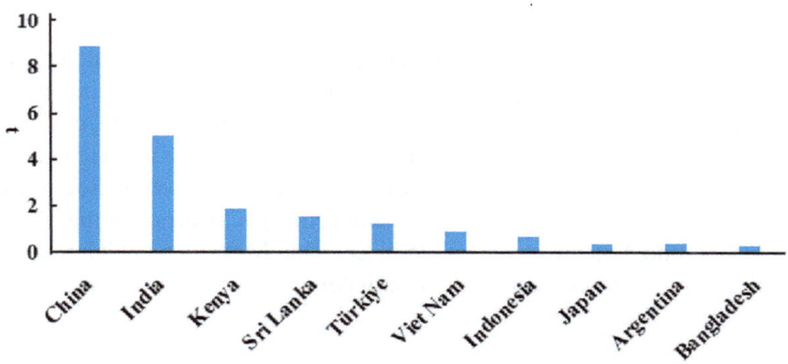

Fig. 1.2 Production of tea leaves: top 10 producers (Average 2003–2023) from FAO database

1.2 Cultivation of C. Sinensis, A Valuable Plant

1.2.1 Worldwide Distribution of Tea Plant

As above mentioned, *C. sinensis* is diffused in tropical and sub-tropical regions, and widely present in major countries of Asia, Africa, and South America: China, India, Sri Lanka, Kenya, Brazil, Argentina, and so on. The commercial and scientific interest in tea plant is tied to its widespread distribution all over the world, though the origin of *C. sinensis* is only in Southwest China.

After being spread and artificially cultivated, the adaptability range of tea plants has far exceeded their original area of growth. Currently, tea plants are distributed worldwide, ranging from the Carpathians outside Ukraine at a northern latitude of 49°N to Natal, South Africa at a southern latitude of 33°S. The most concentrated area of tea plant cultivating and production occurs between 6° and 32°N. Tea plant is primarily found in subtropical and tropical regions, with a vertical distribution ranging from below sea level to an altitude of 2300 m (Java Island, Indonesia). Tea is widely produced in five continents. According to data from FAOSTAT (https://www.fao.org/faostat/), Asia's tea production accounts for about 87.3% of the world's total production in 2023; Africa follows with 11.5%; and other continents account for only 1.2%. Currently, 50 countries worldwide cultivate tea leaves and produce dried tea, among which 18 countries in Africa, 16 in Asia, 7 in South America, 4 in Europe, 2 in Oceania, and 2 in North America.

Based on tea production distribution and climate conditions, the world's tea planting areas can be divided into the following six regions: East Asia, Southeast Asia, South Asia, West Asia, Europe, East Africa, and South America. East Asia, with China and Japan as the main tea-producing countries, produced approximately 51% of the world's total tea production in 2023, making it a major tea-producing region in the world. South Asia, comprising India, Sri Lanka, and Bangladesh as the main tea-producing countries, accounted for about 26% of the world's total tea production. In Southeast Asia, countries such as Indonesia, Vietnam, Myanmar, and Malaysia contributed to 6% of the world's total tea production in 2023. In West Asia, countries such as Georgia, Azerbaijan, Turkey, and Iran contributed to approximately 4% of the world's tea production. In East Africa, countries such as Kenya, Malawi, Uganda, Tanzania, Mozambique, South Africa's Natal region, central Zaire, Rwanda, Cameroon, and Mauritius in the Indian Ocean are the main tea producers, accounting for approximately 11% of the world's total tea production. South America started growing tea trees in the early twentieth century. Countries in the region that produce tea include Argentina, Brazil, Peru, Ecuador, Mexico, and Colombia. South American tea production makes up approximately 1% of the global production.

1.2.2 Native Versus Cultivated Tea Plant. An Introduction to Domestication and Biodiversity

For over 100 years, the origin and evolution of cultivated tea plants have attracted the attention of numerous researchers (Baildon, 1877; Higdon & Frei, 2003; Huang et al., 2014; Meegahakumbura et al., 2016; Roberts et al., 1957; Zhang et al., 2018a, 2018b). Despite some significant progress, the understanding of the domestication history of tea plants still lags far behind that of other crops such as wheat (*Triticum aestivum*) and maize (*Zea mays*), remaining many controversies and mysteries. This section reviews and summarizes the achievements related to the origin and evolution of cultivated tea plants, particularly recent researches on tea genetic diversity.

Native Tea Plant Resources

Any cultivated plant has its native ancestors, and searching for the native types or wild relatives of cultivated plants is the best and most common way to trace their evolutionary footsteps. Tea plant is one of the earliest cultivated woody crops in China, with a long history of cultivation and a vast cultivation area. Therefore, for a long time, almost no one would doubt that the cultivation of tea plant originated in China. However, since the 1930s, when the British discovered a suspected wild tea plant in the Assam region of northeastern India, some have proposed the hypothesis that Assam, India is the original home of the tea plant.

The book "Chinese Ancient Tea Plant" and a series of literatures (Chen et al., 1979; Guo, 1990; He et al., 2009; Li et al., 2017; Tian et al., 2008; Wang & Yu, 2002; Yu, 1989; Yu 2016) have revealed that many native tea plants are found in not only Southwestern China including Yunnan, Sichuan, Chongqing, Guizhou, but also in Guangxi, Guangdong, Fujian, Hunan, Hubei, Jiangxi, and other regions. In 1958, British scholar J. H. Sealy recorded 5 species and 2 varieties of tea plants in book "*A Revision of the Genus Camellia*", and they were tea [*Camellia sinensis* (L.) O. Kuntze], *Dali* tea [*C. taliensis* (W. W. Smith) Melch.] *Irrawaddy* tea (*C. irrawadiensis* Burua), slender-stemmed tea (*C. gracilipes* Merrill ex Sealy), and hairy-stemmed tea (*C. pubicosta* Merrill). Among them, tea is further divided into Chinese variety [*C. sinensis* var. *sinensis* (C. B. Clarke) Ridley] and Assam variety [*C. sinensis* var. *assamica* (Masters) Kitamura]. Afterwards, book "A Systematic Study of the Genus Camellia" increased tea plant group to 17 species (Zhang, 1981), while "*Flora of China*" published in 1998 increased tea plant group to 32 species. In this book, the type locality of all these plants is in China, especially in southwestern China. The above work greatly promoted our understanding of the cultivation of native tea plant relatives, and became an important basis for studying the origin of cultivated tea (Zhang et al., 2018a, 2018b).

Afterwards, the tea plant group was reclassified based on the distance between the gynoecium and the bracts. Therefore, the short-stamen hairy-ribbed tea (*C. pubicosta* Merr.) was distinguished from the tea group, while the remaining tea group plants were categorized into 12 species and 10 varieties. Among these, the *Laotian* tea (*C. sealyama* Ming) was identified as a new species (Min, 1992, 2000). According to this classification method, cultivated tea plants fall into two varieties under *C. sinensis* (L.) O. Kuntze: Assam variety (also known as large-leaf tea) and Chinese variety (also known as small-leaf tea). Additionally, there are several other species or varieties cultivated in low quantities, such as *Dali* tea, large-bract tea (*C. grandibracteata*), *Dehong* tea (*C. sinensis* var. *dehungensis* (Chang & B. H. Chen)), white wool tea (*C. sinensis* var. *pubilimba* Chang), and hairy-leaf tea (*C. ptilophylla* Chang). It is noteworthy that a unique type of tea cultivated in the Indochina Peninsula and South Asia was identified as *Camellia sinensis* (L.) O. Kuntze var. *lasiocalyx* (G. Watt) and it is called *Cambod* tea (Das & Ghosh, 2016). Some species, such as large-bract tea and *Dali* tea, are cultivated near villages in low quantities. Recently, hairy-leaf tea has been cultivated on a large scale. Except for these groups, some valuable fresh leaves of native tea plants including *Da Chang* tea (*C. tachangensis* F. C. Zhang), thick-stem tea (*C. crassicolumna* Chang), and Guangxi tea (*C. kwangsiensis* Chang) are collected by to produce tea on a small scale. In recent years, due to the pursuit of tea leaves made from native tea plants, various tea leaves are produced from wild or semi-wild arbor type tea plants, causing significant damage to tea tree resources (Li et al., 2015; Zhao et al., 2014).

Cultivated Tea Plant

Since the end of twentieth century, scholars from China, India, Japan, and other countries have utilized various molecular marker technologies to analyze the genetic diversity and genetic structure of tea plants in different regions and cultivars. With the development of molecular marker technology, important information has been obtained that helps further understand the domestication origin and spread of tea plants.

(1) *Interbreeding between tea varieties, gene flow, as well as hybridization and gene flow between cultivated tea and wild relatives are very common due to human intervention*

For example, *Cambodian* tea is likely derived from a hybridization between *Camellia sinensis* var. *assamica* and *Camellia sinensis* var. *Sinensis* (Meegahakumbura et al., 2016), while cultivated tea varieties in India are highly mixed. Raina et al. (2012) studied the genetic diversity of 1,644 cultivated tea samples in India and found a rich variety of hybridizations, and their genetic types were not consistent with the morphology-based classification. Many cultivated teas in India are hybrids between *Camellia sinensis* var. *assamica* from India and *Camellia sinensis* var. *sinensis* from China (Meegahakumbura et al., 2016). The history of cultivated tea in Africa spans less than 200 years. The populations exhibit a wide range of types and significant genetic differences between varieties. These include small-leaf teas from China, large-leaf teas from India, and hybrids between them. It is common for hybrids between large-leaf teas to have a maternal parent that is also a large-leaf tea. Genetic analysis of the populations of *C. irrawadiensis* Burua and large-leaf teas in China has shown that the genetic material of *C. irrawadiensis* Burua has infiltrated into large-leaf teas. This suggests the possible involvement of *C. irrawadiensis* Burua in the domestication of large-leaf teas (Li & Qin, 2015; Liu et al., 2010; Wambulwa et al., 2016a, 2016b).

(2) *Cultivated tea plants are categorized into three genetic lineages representing three different types: small-leaf tea, Chinese large-leaf tea (mostly found in southern and western Yunnan, China), and Indian large-leaf tea*

Meegahakumbura et al. (2016) applied simple sequence repeats (SSR) markers to analyze the genetic composition and structure of 392 tea germplasm resources from China and India. These samples were consisted by a significant number of native tea plants and samples of tea plant varieties from main tea-producing regions in China and India. The results showed that cultivated tea trees could be categorized into three genetic lineages representing three different types of tea trees: small-leaf tea, Chinese large-leaf tea mostly found in southern and western Yunnan, China, and Indian large-leaf tea. Although the latter two belong to large-leaf tea, they are from different genetic lineages, suggesting that small-leaf tea, Chinese large-leaf tea, and Indian large-leaf tea likely originated independently. It is suggested that Chinese large-leaf tea may originate in southwest Yunnan, while Indian large-leaf tea may originate in the Assam region of India. Chloroplast DNA fragments analysis also indicated the cultivation of a lineage of tea plant in Assam that is distinct from Chinese large-leaf tea (Meegahakumbura et al., 2016).

(3) *Cultivated tea plants have many regional types, and the genetic compositions of local varieties significantly differ from those in other regions*

Except for the Cambodian tea from the Central-Southern Peninsula and the Assam tea from India mentioned above, cultivated tea plant in Japan also exhibits significant differences Taniguchi et al. (2014) used SSR markers to analyze the genetic compositions of 788 tea germplasm resources collected from various Asian countries. The results showed that these teas can be divided into three categories. Among them, the varieties cultivated in Japan constitute a separate category, but with low genetic diversity. Other samples were separated into two categories: Assam tea and small-leaf tea. Varieties from countries in the Central-Southern Peninsula, Bangladesh, Sri Lanka, and parts of India are considered as Assam tea, while many varieties from China, Korea, and India are classified as small-leaf tea. Although Japanese tea is originated from China, long-term artificial selection has led to genetic differences between Japanese cultivated tea and Chinese small-leaf tea. In fact, the genetic gap between them is even wider than that between Assam tea and small-leaf tea (Matsumoto et al., 2002). According to the analysis by Katoh et al. (2003), the genetic types of tea plants cultivated in Myanmar are also very diverse, including not only Assam tea and small-leaf tea, but also some from *Dali* tea (*C. irrawadiensis* Burua).

(4) *Cultivated tea in china has rich genetic diversity and cultivation types*

In addition to the small-leaf tea and large-leaf tea mentioned above, local residents have also cultivated *Dali* tea, *Da Bao* tea, and *Dehong* tea in the southwestern region of Yunnan (Zhao et al., 2014). Yao et al. (2012) analyzed the genetic structure of 450 tea plant samples collected from various tea-cultivating regions in China. The results revealed obvious regional differences and those samples were divided into five categories based on genetic structure. The cultivated tea plants from the provinces along the lower reaches of the Yangtze River are genetically highly consistent with samples from Fujian, forming one category. The cultivated tea varieties in Guangdong and Guangxi, Guizhou and Chongqing, as well as Hubei and Sichuan, have highly similar genetic compositions and were divided into three categories, respectively.

Local varieties in Yunnan are separated into a unique category. Tea plants in Yunnan, Guizhou, and Guangxi provinces exhibit high genetic diversity. These samples can be differentiated into five categories by genetic composition, which didn't correspond to taxonomic species or subspecies. Although the local varieties grown in Hubei, Sichuan, Chongqing, and Guizhou provinces belong to small-leaf tea, their genetic composition is not consistent with that of the small-leaf tea from the lower reaches of the Yangtze River (Yao et al., 2012).

1.2.3 *Tea Plant Cultivation. Harvesting and Cultivation Procedures, and Regional Differences*

The tea plant needs subtropical environments based on the following conditions:

(a) *Light* Tea plant need sufficient sunlight for photosynthesis, and suitable light conditions are crucial for the growth of tea plant.
(b) *Temperature* The optimal temperature for tea plant is 15–30 °C, the highest temperature in summer should not exceed 30 °C, and the lowest temperature in winter should not be lower than 5 °C.
(c) *Soil* Tea plant is suitable to cultivation in acidic soil. pH value should be around 4.5–5.5. The soil should be rich in organic matter.
(d) *Moisture* Tea plant need sufficient moisture, but too much moisture can lead to root diseases, so the water supply needs to be controlled.
(e) *Altitude* Tea plant is suitable for planting at altitudes of 500–1500 m, high altitudes are usually sunny and provide suitable temperature for the growth of tea plant.

These conditions work together to affect the growth of the tea plant as well as the yield and quality of tea leaves (Manivel, 2022a). The fresh leaves are commonly picked up from March to September, and then produced into different kinds of tea. The environmental features of the world's tea producing regions are summarized in details as follows.

Harvesting Time of Fresh Leaf

Tea plants have a strong adaptability to the environment and can be cultivated in areas with significant differences in annual average temperatures as well as in regions with considerable variations in precipitation (Zhang et al., 2019a, 2019b). Tea plants are native to subtropical regions and prefer a mild and humid climate. However, most tea-cultivating areas in the world are located in subtropical and tropical climate zones. Therefore, under different climatic conditions, the growth of tea plants also varies. In the tea-cultivating areas between 16°S and 20°N, tea plants can be cultivated and harvested throughout the year. In tea-cultivating areas above 20°N, tea plants have a distinct growth rest period throughout the annual cycle. Generally, tea plants with a temperature difference of <10 °C between January and July can grow throughout the year, such as tea plants in Bogor, Indonesia, Colombo, and Sri Lanka. Tea-cultivating areas with a temperature difference of 10–15 °C from January to July are long-seasonal producing region, such as Guangzhou of China, while those with a temperature difference of 15–25 °C are short-seasonal producing region, such as Hankou of China and Georgia.

Precipitation

The amount of precipitation in tea-cultivating regions around the world varies greatly, with some receiving over 6000 mm annually and others as little as 700 mm. The annual precipitation in some main tea-cultivating regions, for example, Northeastern India, ranges from 1300 to 4000 mm. Besides, the annual precipitation in Sri Lanka ranges from 1800 to 6200 mm. Shizuoka Prefecture in Japan rains around 2500 mm, while Kyoto Prefecture rains only 1500 mm. In regions with little temperature difference between January and July, the growth of tea plants and the yield of tea leaves are often constrained by the amount of precipitation. When the precipitation is abundant and evenly distributed precipitation, tea plants grow vigorously and the monthly yield of fresh tea leaves is also high. For example, in Java, Indonesia, the precipitation is abundant and evenly distributed throughout the year, which results in tea harvesting proportions ranging from 7 to 10% per month. On the other hand, regions like Assam in Northeast India and Yunnan and Guangdong in China, the very little precipitation in winter and early spring leads to slow growth of tea plant and yield of tea leaves (Nowogrodzki, 2019).

Soil Acidity

The soil conditions in tea-producing areas around the world are generally acidic, with red and yellow soils. The soil texture varies greatly, ranging from heavy clay to extremely loose sand. The soil acidity ranges from pH 4 to 6.5. The fertility of the soil varies from rich organic forest soil to barren land with little organic matter. In Japan, most tea-producing areas are located on sandy loam and alluvial soil along rivers. These soils exhibit significant differences in texture but are characterized by high fertility (Manivel, 2022b). In Sri Lanka, tea-producing areas are predominantly found on red soil, which is further divided into forest, grassland, as well as lowland types. In the mountainous tea gardens of India, most of the soil is grayed red soil and gray-brown soil derived from weathered slate as well as alluvial soil derived from weathered sandstone in the plains.

1.3 Consumption Behaviors and Tea Products

The tea plant has been used for centuries due to its various purported properties, primarily as a beverage (infusion) known by names such as tea, Cha (茶), Tu (荼), and Ming (茗). Fresh tea leaves contain varying amounts of umami and bitter-astringent compounds, such as theanine, caffeine, catechins, and flavonol glycosides, but release little aroma. However, abundant flavor compounds are formed during manufacturing process.

1.3.1 Classification of Teas and Reasons for Regional Diversification

According to the international standard ("ISO publishes standard on classification of tea types," 2023, ISO 20715:2023) tea can be categorized into six types including green tea, oolong tea, black tea, white tea, yellow tea, and dark tea based on the different processing steps (Fig. 1.3). Specifically speaking

(a) *Green Tea* The production of green tea involves fixation, rolling, and drying. The production of oolong tea involves withering, shaking, rolling, and drying.
(b) *Black Tea* The production of black tea involves withering, rolling, fermentation, and drying.
(c) *White Tea* The production of white tea involves withering and drying.
(d) *Yellow Tea* The production of yellow tea involves fixation, rolling, yellowing, and drying.

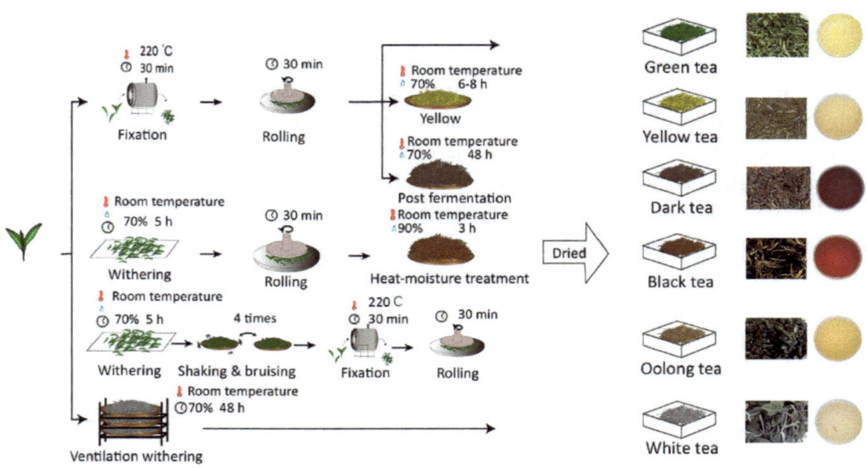

Fig. 1.3 The manufacturing processes of six types of tea. Modified according to Wang et al. (2019)

(e) *Oolong Tea* The production of oolong tea involves withering, shaking, rolling, and drying.
(f) *Dark Tea* The production of dark tea involves fixation, postfermentation, shaping, and drying.

Meanwhile, depending on the fermentation condition and degree, green tea is unfermented tea, white and yellow teas are lightly-fermented tea, oolong tea is semi-fermented tea, black tea is fully-fermented tea, and most Chinese dark teas that undergo a microbial fermentation at relative elevated temperature and humidity are recognized as post-fermented tea (Shitandi & Anakalo, 2013; Shitandi et al., 2013; Teranishi et al., 1999).

Commonly, young tea leaves and buds plucked in early spring and under sunny conditions can be more easily produced into higher grade tea than leaves plucked in summer or autumn because of their mellow, umami taste, and flowery aroma (Ji et al., 2018; Joshi & Gulati, 2015; Jumtee et al., 2011). The enzyme mediated reactions start from leaf plucking, which leads to flavor formation.

Unfermented Green Tea

In China, most of the tea crops harvested are then manufactured into green tea. A moderate fixation of the leaves helps to deactivate the enzymes (e. g., polyphenol oxidase, peroxidase) from both inside the leaves and microflora during heating on the one hand, and also causes the emission of the green odor note on the other side (Del Rio et al., 2004). Raw pu-erh tea is similarly produced to green tea. White tea, oolong tea, and black tea are fermented teas with different involvement of polyphenol oxidase.

Lightly-, Semi-, and Full-Fermented Tea

The production of white tea only contains withering and drying, while for oolong tea and black tea, the withered leaves are firstly rolled and crushed, which disrupts cellular compartmentation and leads to contact between phenolic compounds and polyphenol oxidases to form oxidative oligomers including theaflavins, theasinensins, and thearubigins. Additionally, black tea needs a full-fermentation step for about 5 h at room temperature and high moisture, leading to a series of oxidation, condensation, and degradation reactions of tea polyphenols (Harbowy et al., 2010; Kongpichitchoke et al., 2016). Furthermore, a unique yellowing step during yellow tea production results in.

Post-fermented Dark Tea

Post-fermented tea, i.e., pu-erh tea (ripe) and fu brick tea, is commonly produced from sun-dried green tea at higher temperature and humidity (Lv et al., 2013). Drying, the last process for all teas, cannot only remove water content important for transportation and storage, but also helps to form Maillard reaction products during heating treatment, and thus, generate the special sensory quality such as a chestnut-like odor note (Zhang et al., 2020a, 2020b).

Regional Diversification

Among the six types of tea, black tea originally from China is today the most highly consumed tea in the world, accounting for 75% of global tea consumption, particularly in Europe and North America. However green tea, comprising 15% of global consumption, is more popular in Asia, especially in China (Zhang et al., 2019a). Oolong tea is well-known in China and Southeast Asia due to its flowery/fruity aroma and brisk/smooth taste. According to the China Tea Marketing Association's 2023 report, oolong tea accounted for 10.0% of total tea consumption in China. Other types of tea with smaller production are almost exclusively produced and consumed in China, with consumption ratios of dark, white, and yellow tea being 13.7%, 3.0%, and 0.7% of total consumption in 2023, respectively.

1.3.2 Hot Tea Infusions. Preparation Procedures

The Brewing Process of Hot Tea Infusion in China

Tea is a traditional beverage in China which contains various varieties, origins, and processing methods. Even the same type of tea leaves present significantly different flavors in the tea infusions due to different brewing methods. The changes of brewing conditions such as tea leaves amount, water quality and temperature, brewing duration, and tea ware affect the substances extraction yield and the flavor perception of the tea infusion.

Tea brewing and drinking utensils include tea leaf-water separation utensils, covered bowl, tea pot, and tea-sipping cup. Auxiliary tools include fair mug, filter, tea tongs, tea spoon, tea stick, tea holder, tea towel, tea basin, tea tray, etc. The brewing process is mainly carried out according to the steps of preparing utensils, adding tea leaves, pouring water, filtering tea infusion, and tasting infusion (Fig. 1.4) (Huang & Wang, 2015).

Each type of tea has its own unique characteristics that are influenced by the brewing conditions. Corresponding brewing methods are often applied to highlight the advantages of each type of tea. For example, when brewing traditional white tea White Peony, the optimal conditions are 3.05 g of tea leaves added into 152.5 mL

Fig. 1.4 Tea brewing and drinking utensils. From left to right, they are tea spoon, tea holder, covered bowl, filter, fair mug, and tea-sipping cup, respectively. Photo Xiaoting Zhai

boiling water (tea leave-to-water ratio of 1:50) and brewed for 5.03 min. The tea infusion evokes pure and fresh aroma, mellow taste, and orange-yellow color. The content of catechins and free amino acids is significantly influenced by brewing conditions. In details, the amount of tea leaves has the greatest impact on the extraction yield of free amino acids, while the brewing temperature has the greatest impact on the extraction yield of catechins. When brewing *Tieguanyin* oolong tea, 8.4 g of tea leaves is added 130 mL boiling water into covered bowl (tea leave-to-water ratio of 1:15.5). The leaves can be brewed for 30 s for the 1st infusion, 10 s for the 2nd infusion, 20 s for the 3rd infusion, and 30 s for the 4th infusion. This brewing method can preserve the aroma for over 7 infusions, and even maintain the flavor for more than 10 infusions. In conclusion, tea infusions can fully showcase their advantages under suitable brewing methods. The tea leave-to-water ratio and brewing duration can be adjusted according to personal preference (Yu, 2023).

When tasting a cup of tea, the main factors deciding the acceptance of the infusion are its color, aroma, and taste. First of all, you can evaluate the type, intensity, purity, and persistence of the tea aroma. The ideal aroma characteristics should be clear and long-lasting, pure and rich. For example, green tea mainly contains chestnut-like aroma, along with clear, fresh, floral, and nutty aroma. Black tea evokes sweet, honey-like, floral, fruity, and fresh aroma. Another important factor is the color. The color of green tea is mainly bright green, while that of black tea is mainly bright red. The last but not the least factor is the taste. For instance, green tea tastes mainly bitter and umami, while black tea tastes sweet, mellow, and fresh (Gan et al., 2024).

Chinese Traditional Sensory Evaluation of Tea

The sensory evaluation is an important and professional step for evaluating tea quality that mainly focuses on the evaluation of color, aroma, taste, and overall quality of the tea infusion by a trained panel or a consumer panel. The aim of sensory analysis is to get reproducible and reliable results for tea manufacturers and consumers, thus it is of necessity of appropriate vocabularies such as odor and taste descriptors. With the expanding demand for tea products and processes, Chinese national standard has deeply explored suitable and comprehensive methodology for traditional sensory evaluation of tea.

The latest version of GB/T 23776-2018 presents different grades of green tea, black tea, oolong tea, yellow tea, dark tea, and white tea in terms of color, aroma, and taste of the tea infusion as well as the appearance of tea leaves after brewed (https://openstd.samr.gov.cn/bzgk/gb/newGbInfo?hcno=DE8871908 72380558CCAF17BE9D850E8, GB/T 23776-2018). For instance, green tea infusion with high quality appears green and bright color, fresh, chestnut-like or flowery aroma impression, and mellowish taste. An excellent example of sensory assessment of six types of tea processed from fresh tea leaves plucked from the same cultivar and growing area is given in Table 1.1 (Feng et al., 2019a, 2019b; Wang et al., 2019). However, expert panelists with rich sensory evaluation experience are very important because the terminologies of odor and taste descriptors such as "mellowish" are difficult to distinguish for consumers, even for normal sensory panel. Therefore, the combination of the abovementioned traditional sensory evaluation of tea and quantitative descriptive analysis (QDA) for a comprehensive sensory analysis is rather significant for tea quality assessment (Yang et al., 2021; Zhai et al., 2022).

The Brewing Process of Hot Tea Infusion in Other Regions

Specific tea brewing and drinking utensils occur in other countries. For example, Uruguayans like to drink Mate tea. The utensils for Mate tea include a gourd-shaped ladle with a silver straw. One end of the straw contains a small-holed filter, while the other end contains a mouthpiece. Mate tea is placed in the ladle, brewed with boiling water. When cooled slightly, Mate tea can be slowly sipped through the straw. The green tea brewing process in Morocco is also interesting. Briefly, tea leaves are placed in a large belly, long-spouted silver or copper tea pot. A little boiling water is poured into the pot, to brew for a while and then poured out. Then, fresh mint leaves and sugar (tea to sugar ratio is 1:10) are added, and brewed for a few minutes with new boiling water.

Due to differences in cultural traditions, religious beliefs, ideological consciousness, preferences, and taboos, tea drinking customs vary between different countries. The following are the main tea drinking customs in other countries and regions except for China.

(1) *Addition of Milk or Sugar in Tea Infusion* This style is widely favored in many countries including UK, Netherlands, France, New Zealand, Australia, India, Sri Lanka, Iran, Egypt, United Arab Emirates, Argentina, Venezuela, Tunisia, Canada, etc. Pakistanis also add dried fruits like almonds and apricots into their milk tea (Smith, 2016).

Table 1.1 Sensory assessment of six types of tea processed from fresh tea leaves of a single *Camellia sinensis* cultivar according to the method of GB/T 23776-2018

Dry leaves						
Tea infusion	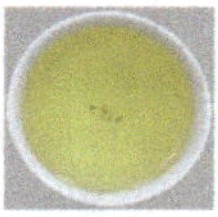			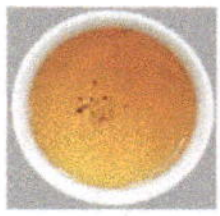	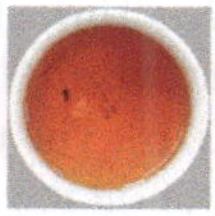	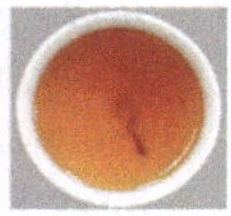
	Green tea	White tea	Yellow tea	Oolong tea	Black tea	Dark tea
Color	Green with yellow tint, brighter	Orange yellow, bright	Yellow green, bright	Orange yellow, brighter	Orange red, bright	Darker orange red, cloudy, brighter
Aroma	Clean, fresh, chest-nut-like	Sweet floral, honey-like	Chest-nut-like, slightly sweet and cooked, fresh	Floral, slightly fermented	Fruity	Slightly stale and rough
Taste	Mellowish	Mellowish, thicker body, slightly astringent	Mellowish	Thicker body, mellowish	Sweet mellow, thicker body	Slight mellowish, astringent

Modified according to Feng et al. (2019a, 2019b), Wang et al. (2019)

(2) *Addition of Sign Plant Ingredients in Tea Infusion* Italian, Turkish, Libyan, and Algerian mainly drink black tea or green tea. They like to add fresh mint, rose, saffron, lemon, and other ingredients to increase the taste quality (Pellissery & Madaan, 2016).

(3) *Addition of Various Ingredients in Tea Infusion* Russian people love to add various ingredients together to tea infusion, such as sugar, dairy, fruit juice, and jam (Ritchie, 2018). When drinking black tea, people often mix in strawberry and rose. Nowadays, black tea is commonly consumed with sugar, milk, lemon, or alcohol. Green tea is enjoyed with cream or milk instead of sugar. Mauritanians like to add sugar and fresh mint leaves to green tea, cook them with boiling water until it evokes strong aroma and dark color. The tea infusion presents a rich, sweet, slight bitter taste.

(4) *Sencha* The tradition of drinking tea in Japan and Korea has a long history. Typically, Gyokuro tea and Sencha are preferred in these two countries and often consumed without the addition of sugar or milk (Kim & Lim, 2019).

(5) *Addition of Mike and Salt in Tea Infusion* In Mongolia, brick tea is mainly consumed, with a minority favoring black tea. They enjoy salty-milk tea. Regarding the production, brick tea is crushed and grinded in a wooden mortar, placed into a special iron pot, boiled the leaves. Then, milk and salt are added

in the infusion, the ratio of milk to water is 1:4 to 1:5. Finally, the infusion is boiled it again before consumption.

(6) *Tea Salad* Tea salad is commonly served in Myanmar. It is made by crushing tender tea leaves, adding sesame oil and salt, mixing with peanuts, sesame seeds, soybeans, dried shrimp, coconut slices, and garlic, then deep-frying until crispy. Finally, boiling water is added, resulting in a refreshing and delicious drink that helps to eliminate fatigue.

(7) *Boiling Tea* Turkish have a special process of brewing and drinking tea. Specifically, they use a large and a small copper kettle for brewing tea (Haddad & Fidan, 2016). The tea leaves are placed in the small kettle. Water is boiled in the large kettle on the stove, poured into the small kettle. When the water reaches 80% of volume, the small kettle is then simmered over the fire for a while. The final tea infusion is shared.

1.3.3 New-Style Ready-to-Drink Tea Beverages. Current Situation

With the continuous expansion of Chinese tea industry, the consumption of tea has not grown synchronously, leading to the overcapacity problem. To address this issue, stimulating the potential of the traditional tea consumption market has become a key problem that urgently needs to be solved (Yang, 2023). In recent years, the new tea beverage industry has achieved rapid development through multidimensional marketing innovation, becoming a new medium and new window to activate the consumption market and traditional tea culture. It also meets the diverse needs of tea consumers in the new era.

The new-style tea beverage commonly refers to ready-to-drink tea beverages made from high-quality tea leaves and products as the main raw materials, combined with fresh milk or dairy products (such as skimmed milk, condensed milk, cheese), fruits or fruit juice, sugar, spices, grains, herbs, legumes, alcohol, carbon dioxide and other raw materials (Yin, 2021). All the raw materials are extracted and blended on site to prepare the beverage. It can be specifically categorized into freshly brewed tea, tea bag, bottled tea, and instant tea, representing the innovative expression of traditional tea drinking methods (Yu et al., 2014). According to data from iiMedia Research (https://www.iimedia.cn/c400/102348.html), the Chinese market of new-style tea beverage reached 354.72 billion RMB in 2024, with an increasing rate of 6.4%, and is expected to exceed 400 billion RMB in 2028.

1.4 An Introduction to Chemical Profiles of Tea Products

Although being differentiated by their origin and manufacturing technique, tea consists mainly of hydrolyzable tannins, flavan-3-ols and their derivatives, phenolic acids, purine alkaloids, and terpenoids. High concentrations of polyphenols, flavonoids, alkaloids, and theanine, which are known as important substances for the growth and metabolism of tea tree, the quality properties of tea infusion, as well as bioactivities (Meng et al., 2019).

The drinking quality of the tea infusion is influenced by a number of factors, including the tea plant cultivar, the geographical position of tea plantation, the harvest conditions of the tea leaves, and the processing and storage conditions. Today, the commercial tea products are mainly manufactured from the fresh tender leaves and buds plucked from two most widely cultivated *Camellia* species (Theaceae), *Camellia sinensis* (L.) O. Kuntze and *C. sinensis* var. *assamica* (Masters) Kitamura (Meng et al., 2019).

1.4.1 Tea Flavor

A major factor for the popularity of tea beverage is its pleasant flavor, which is mainly derived from the tea leaves. Whereas fresh tea leaves exhibit little aroma, the aroma develops during processing and brewing. The tea-derived compounds contributing to the pleasant flavor of tea, odorants particularly, have been increasingly studied in detail since the 1930s and include, e.g., green/grassy smelling hexanal and 1-hexanol, flowery smelling methyl jasmonate and linalool, rose-like smelling geraniol, violet-like smelling *β*-ionone, and honey-like/fruity smelling *(E)-β*-damascenone (Yang et al., 2013). Sometimes, however, improper stored or processed tea is tainted with off-flavors that may be transferred into the final tea product and lead to consumers rejecting the tea (Hara et al., 1987).

Although it is important to ensure tea leaves and buds are tender, free from insect, mold, and mechanical damage, and shaped with a certain uniform size, however, the final determiner of the drinking quality of tea infusion is its flavor (Chaturvedula & Prakash, 2013). The perception of flavor is the result of a multitude of the brain's integration and interpretation of sensations from odorants detected during inhalation (orthonasal) or the chewing and swallowing process (retronasal) and taste receptors on the tongue (Wang et al., 2019). Thus, the perception is commonly affected by three main ingredients of flavor: the mental interpretation specific to the tea drinker (psychology), odor-active volatile compounds (aroma), and nonvolatile components (taste).

1.4.2 Tea Color

The color of tea infusion has been demonstrated to influence the perception of tea and resulting affective or hedonic judgements of tea flavor. In China, the colors of tea infusions as it relates to tea types are generally assessed by the application of Chinese National Standard methods to evaluate the quality of tea (Wang et al., 2019). The desired color of green tea is greenish (chlorophyll A) or yellowish green (chlorophyll B) without any trace of red or brown color (Chaturvedula & Prakash, 2013). The yellow color in green tea infusion is mainly determined by the water soluble flavonols (i.e., kaempferol, quercetin, and isoquercetin), flavones (i.e., apigenin, isovitexin, and vitexin) as well as their glycosides. Due to highly colored products such as theaflavins and thearubigins formed by the enzymatic oxidation and condensation of catechins during the full fermentation process, the red color is the main color of black tea (Chaturvedula & Prakash, 2013). For oolong tea, the infusion color is generally from dark greenish to reddish-brown color depending on the degree of fermentation (Chaturvedula & Prakash, 2013). Commonly, consumers are attracted by the desired infusion color and psychologically consider good flavor quality of the infusion before smelling and tasting, while unexpected color may possibly be a sign of poor-quality tea.

1.4.3 Tea Aroma

As the main taste aspects of tea infusions are limited to the bitter, astringency, and sometimes umami dimensions (Scharbert et al., 2004), and arguably the visual dimensions relating to tea infusion color, the greatest source of variability in tea flavor may be related to odor-active volatiles which are capable of interacting with odor receptors expressed by olfactory sensory cells embedded in the olfactory mucosa located in the nasal passages, and thus, induce an aroma impression (Belitz et al., 2004). However, not every volatile substance can evoke a typical odor note. A few odorants can be recognized by multiple types of distinct olfactory receptors and cause different activation levels. Moreover, one receptor can specifically be activated by more than one type of compound or functional group (Belitz et al., 2004). In general, natural products may normally contain between 3 and 40 odorants, resulting in the variability of food solely in terms of aroma (Belitz et al., 2004).

Up to now, more than 700 volatile compounds have been identified in tea leaves and infusions (Chaturvedula & Prakash, 2013). However, only around 184 compounds are aroma-active and can contribute to the overall aroma of tea. They are summarized as aldehydes, alcohols, ketones, phenols, pyrazines, sulfur-containing compounds, furans, furanones, acids, esters, lactones, indoles, pyrroles, oxidation products, benzenes, and terpenes (Chen et al., 2019, 2020; Flaig et al., 2020a, 2020b; Gong et al., 2017; Ho et al., 2015; Kang et al., 2019; Pang et al., 2019; Schuh & Schieberle, 2006; Wang et al., 2020; Zhu et al., 2015, 2018, 2021). Among them, alcohols, ketones, oxides are responsible for the flowery odor note. Aldehydes mainly contribute to the fatty, green and malty flavor, phenols impart smoky, phenolic and clove-like flavor to the tea, sulfur-containing compounds contribute to the sulfury smell of tea, and some aldehydes and alcohols are responsible for the citrus-like and fruity odor note in tea. Besides, benzenes mainly emit stale and musty odor notes while pyrazines mostly emit roasty and earthy odor notes.

1.4.4 Tea Taste

The taste of tea beverage is influenced by the nonvolatile components of the dry tea leaves, because these are directly brewed into tea infusion and then assessed by taste and touch receptors on the tongue and in the oral cavity (Lawless & Heymann, 2010). Phytochemical studies indicate that all types of dry tea products manufactured globally are composed primarily of polyphenols (18–36%, mainly catechins), protein (20–30%), and soluble sugars (20–25%) (Kapoor et al., 2013). Among them, some macronutrients including galloylated catechins and protein have not been reported to be identified by recognized human taste receptors. Tea also contains a small amount of free amino acids (3–4%), caffeine (2 – 4%), and organic acids (2%) (Kapoor et al., 2013). Kaneko et al. investigated the primary descriptive taste dimensions of green tea to be astringency, bitterness, and umami taste, with little sweetness aftertaste, and no saltiness and sourness observed by panelists (Kaneko et al., 2006). Scharbert and Hofmann reported that the main taste qualities of black tea are astringency and bitterness, with little sweetness, and no umami taste, saltiness and sourness (Scharbert & Hofmann, 2005). This is not surprising, as tea infusions contain almost no salts, but only trace amounts of sour-tasting organic acids, soluble carbohydrates, and sweet-tasting amino acids, which are unlikely to be tasted at trace levels and high taste thresholds (Scharbert & Hofmann, 2005). In contrast, non-galloylated catechins, sodium gallate, and gallic acid have been found to be potent contributors to sweet aftertaste, a term applied to special and positive flavor quality of tea infusion (Cao et al., 2019; Zhang et al., 2016). Approximately 70% of the umami taste intensity of green tea mainly derive from amino acids such as L-glutamic acid, L-glutamate (Nakagawa, 1975), aspartic acid, and particularly *L*-theanine which is the most abundant type of amino acids in unfermented teas and fresh tea leaves (Kaneko et al., 2006). In the study, Kaneko discovered that the interplay between L-glutamate and umami enhancers including L-theanine, succinic acid, gallic acid, and theogallin raises the umami intensity of Matcha beverage with increasing concentration of these umami-tasting modifiers.

1.5 The Health Benefits of Tea Products

The tea plant is of great economic importance and has been cropped for the use as medicine or refreshing beverage over centuries. Due to satisfactory sensory experience of pleasant flavor and numerous reported health benefits such as antioxidative (Damiani et al., 2014), antitumor and antimutagenicity (Li et al., 2013), antidiabetic (Butacnum et al., 2017), hypolipidemic and antiobesity properties (Uchiyama et al., 2011), and protection of the nervous system (Mo et al., 2013), the tea infusion prepared from the leaves of tea plant is worldwide the most popular beverage after water.

References

Baildon, S. (1877). *Tea in Assam: A pamphlet on the origin, culture, and manufacture of tea in Assam*. W. Newman & Company.

Belitz, H. D., Grosch, W., & Schieberle, P. (2004). *Food chemistry*. Springer.

Butacnum, A., Chongsuwat, R., & Bumrungpert, A. (2017). Black tea consumption improves postprandial glycemic control in normal and pre-diabetic subjects: A randomized, double-blind, placebo-controlled crossover study. *Asia Pacific Journal of Clinical Nutrition, 26*(1), 59–64. https://doi.org/10.6133/apjcn.112015.08

Cao, Q. Q., Zou, C., Zhang, Y. H., Du, Q. Z., Yin, J. F., Shi, J., Xue, S., & Xu, Y. Q. (2019). Improving the taste of autumn green tea with tannase. *Food Chemistry, 277*, 432–437. https://doi.org/10.1016/j.foodchem.2018.10.146

Chaturvedula, V. S. P., & Prakash, I. (2013). Tea-aroma, taste, color and bioactive constituents. *Journal of Medicinal Plant Research, 5*(11), 2110–2124. https://doi.org/10.1002/cmdc.201100121

Chen, X., Chen, G., Zhang, F., Liu, Z., Zhan, C., Yang, W., & Zhan, Z. (1979). Big tea tree in China. *Journal of Hunan Agricultural College* (03), 55–63+89–92. https://doi.org/10.13331/j.cnki.jhau.1979.03.011

Chen, X., Chen, D., Jiang, H., Sun, H., Zhang, C., Zhao, H., Li, X., Yan, F., Chen, C., & Xu, Z. (2019). Aroma characterization of Hanzhong black tea (*Camellia sinensis*) using solid phase extraction coupled with gas chromatography-mass spectrometry and olfactometry and sensory analysis. *Food Chemisty, 274*, 130–136. https://doi.org/10.1016/j.foodchem.2018.08.124

Chen, X. H., Sun, H. Y., Qu, D., Yan, F., Jin, W. A., Jiang, H., Chen, C., Zhang, Y. F., Li, C. Y., & Xu, Z. M. (2020). Identification and characterization of key aroma compounds in Chinese high altitude and northernmost black tea (*Camellia sinensis*) using distillation extraction and sensory analysis methods. *Flavour and Fragrance Journal, 35*(6), 666–673. https://doi.org/10.1002/ffj.3605

Damiani, E., Bacchetti, T., Padella, L., Tiano, L., & Carloni, P. (2014). Antioxidant activity of different white teas: Comparison of hot and cold tea infusions. *Journal of Food Composition and Analysis, 33*(1), 59–66. https://doi.org/10.1016/j.jfca.2013.09.010

Das, A. P., & Ghosh, C. (2016). Camellia sinensis var. lasiocalyx (G. Watt) A.P. Das & C. Ghosh—New combination name for the Cambod variety of tea. *Pleione, 10*, 167–168.

Del Rio, D., Stewart, A. J., Mullen, W., Burns, J., Lean, M. E., Brighenti, F., & Crozier, A. (2004). HPLC-MS^n analysis of phenolic compounds and purine alkaloids in green and black tea. *Journal of Agricultural and Food Chemistry, 52*(10), 2807–2815. https://doi.org/10.1021/jf0354848

Feng, Z., Li, M., Li, Y., Wan, X., & Yang, X. (2019a). Characterization of the orchid-like aroma contributors in selected premium tea leaves. *Food Research International, 108841*. https://doi.org/10.1016/j.foodres.2019.108841

Feng, Z., Li, Y., Li, M., Wang, Y., Zhang, L., Wan, X., & Yang, X. (2019b). Tea aroma formation from six model manufacturing processes. *Food Chemistry, 285*, 347–354. https://doi.org/10.1016/j.foodchem.2019.01.174

Flaig, M., Qi, S., Wei, G., Yang, X., & Schieberle, P. (2020a). Characterization of the key odorants in a high-grade Chinese green tea beverage (*Camellia sinensis; Jingshan cha*) by means of the sensomics approach and elucidation of odorant changes in tea leaves caused by the tea manufacturing process. *Journal of Agricultural and Food Chemistry, 68*(18), 5168–5179. https://doi.org/10.1021/acs.jafc.0c01300

Flaig, M., Qi, S. C., Wei, G., Yang, X., & Schieberle, P. (2020b). Characterisation of the key aroma compounds in a Longjing green tea infusion (*Camellia sinensis*) by the sensomics approach and their quantitative changes during processing of the tea leaves. *European Food Research and Technology, 246*(12), 2411–2425. https://doi.org/10.1007/s00217-020-03584-y

Gan, S., Chen, Y., Zhao, L., Zhao, X., Qiu, T., Zhai, X., & Dai, Q. (2024). Characterization of the aroma-active compounds in Xiaokeng green tea by three pretreatment methods combined with gas chromatography-olfactometry (GC-O). *Food Research International, 187*, Article 114359. https://doi.org/10.1016/j.foodres.2024.114359

Gong, X. W., Han, Y., Zhu, J. C., Hong, L., Zhu, D. L., Liu, J. H., Zhang, X., Niu, Y. W., & Xiao, Z. B. (2017). Identification of the aroma-active compounds in Longjing tea characterized by odor activity value, gas chromatography-olfactometry, and aroma recombination. *International Journal of Food Properties, 20*(S1), 1107–1121. https://doi.org/10.1080/10942912.2017.1336719

Guo, Y. (1990). Preliminary investigations on wild tea trees in Hainan Island. *Guangdong Tea* (02), 36–40.

Haddad, J., & Fidan, S. (2016). Turkish tea culture: An intangible heritage. *International Journal of Humanities and Social Science, 4*(1), 62–67.

Hara, T., Kubota, E., & Horita, H. (1987). Off-flavor components in stored packaged green tea *(In Japanese). Nippon Nōgeikagaku Kaishi, 61*(4), 471–473.

Harbowy, M. E., Balentine, D. A., Davies, A. P., & Cai, Y. (2010). Tea chemistry. *Critical Reviews in Plant Sciences, 16*(5), 415–480. https://doi.org/10.1080/07352689709701956

He, Q., Wang, P., Jiang, H., & Wu, G. (2009). Investigation on wild tea resources in some areas of Yunnan Province. *Modern Agricultural Science and Technology, 24*, 80–81.

Higdon, J. V., & Frei, B. (2003). Tea catechins and polyphenols: Health effects, metabolism, and antioxidant functions. *Taylor & Francis*. https://doi.org/10.1080/10408690390826464

Ho, C. T., Zheng, X., & Li, S. (2015). Tea aroma formation. *Food Science and Human Wellness, 4*(1), 9–27. https://doi.org/10.1016/j.fshw.2015.04.001

Huang, Q., & Wang, L. (2015). The history and culture of Chinese tea. *International Journal of Culture and History, 1*(1), 13–16.

Huang, J., Wang, Y., Xie, Z., Zhou, Y., Zhang, Y., & Wan, X. (2014). The anti-obesity effects of green tea in human intervention and basic molecular studies. *European Journal of Clinical Nutrition, 68*(10), 1075–1087. https://doi.org/10.1038/ejcn.2014.143

ISO publishes standard on classification of tea types. (2023). *China Standardization* (03), 13.

Ji, H. G., Lee, Y. R., Lee, M. S., Hwang, K. H., Park, C. Y., Kim, E. H., Park, J. S., & Hong, Y. S. (2018). Diverse metabolite variations in tea (*Camellia sinensis L.*) leaves grown under various shade conditions revisited: A metabolomics study. *Journal of Agricultural and Food Chemistry, 66*(8), 1889–1897. https://doi.org/10.1021/acs.jafc.7b04768

Joshi, R., & Gulati, A. (2015). Fractionation and identification of minor and aroma-active constituents in Kangra orthodox black tea. *Food Chemistry, 167*, 290–298. https://doi.org/10.1016/j.foodchem.2014.06.112

Jumtee, K., Komura, H., Bamba, T., & Fukusaki, E. (2011). Predication of Japanese green tea (*Sencha*) ranking by volatile profiling using gas chromatography mass spectrometry and multivariate analysis. *Journal of Bioscience and Bioengineering, 112*(3), 252–255. https://doi.org/10.1016/j.jbiosc.2011.05.008

Kaneko, S., Kumazawa, K., Masuda, H., Henze, A., & Hofmann, T. (2006). Molecular and sensory studies on the umami taste of Japanese green tea. *Journal of Agricultural and Food Chemistry, 54*(7), 2688–2694. https://doi.org/10.1021/jf0525232

Kang, S., Yan, H., Zhu, Y., Liu, X., Lv, H. P., Zhang, Y., Dai, W. D., Guo, L., Tan, J. F., Peng, Q. H., & Lin, Z. (2019). Identification and quantification of key odorants in the world's four most famous black teas. *Food Research International, 121*, 73–83. https://doi.org/10.1016/j.foodres.2019.03.009

Kapoor, M. P., Okubo, T., Rao, T. P., & Juneja, L. R. (2013). Green tea polyphenols in food and nonfood applications. *Green tea polyphenols: Nutraceuticals of modern life* (pp. 315–332). CRC Press. https://doi.org/10.1201/b14878

Katoh, Y., Katoh, M., Takeda, Y., & Omori, M. (2003). Genetic diversity within cultivated teas based on nucleotide sequence comparison of ribosomal RNA maturase in chloroplast DNA. *Euphytica, 134*(3), 287–295. https://doi.org/10.1023/B:EUPH.0000004970.39267.02

Kim, H. J., & Lim, Y. S. (2019). Korean tea culture. *International Journal of Humanities and Social Science Research.*

Kongpichitchoke, T., Chiu, M. T., Huang, T. C., & Hsu, J. L. (2016). Gallic acid content in Taiwanese teas at different degrees of fermentation and its antioxidant activity by inhibiting PKCdelta activation: In vitro and in silico studies. *Molecules, 21*(10). https://doi.org/10.3390/molecules21101346

Lawless, H. T., & Heymann, H. (2010). *Sensory evaluation of food: Principles and practices.* Springer.

Li, X., Feng, Y., Liu, J., Feng, X., Zhou, K., & Tang, X. (2013). Epigallocatechin-3-gallate inhibits IGF-I-stimulated lung cancer angiogenesis through downregulation of HIF-1alpha and VEGF expression. *Journal of Nutrigenetics and Nutrigenomics, 6*(3), 169–178. https://doi.org/10.1159/000354402

Li, C., & Qin, Y. (2015). Investigation on the De'ang nationality acid tea from the perspective of the anthropology of science and technology. *Journal of Yunnan Agricultural University (Social Science).*

Li, M., Kasun, M. M., Yan, L., Liu, J., & Gao, L. (2015). Genetic involvement of Camellia taliensis in the domestication of Camellia sinensis var. assamica (Assamica Tea) revealed by nuclear Microsatellite markers. *Plant Diversity and Resources,* 37(1): 29–37.

Li, Y., Zhang, L., Yang, X., LI, D., Zhang, K., & Wu, T. (2017). Study on spatial distribution and population dynamics of wild Tea in Hainan Island. *Forest and Grassland Resources Research* (2), 81.

Liu, Y., Yang, S., & Gao, L. (2010). Comparative study on the chloroplast RPL32-TRNL nucleotide variation within and genetic differentiation among ancient tea plantations of Camellia sinensis var. assamica and C. taliensis (Theaceae) from Yunnan, China. *Acta Botanica Yunnanica, 32*(05), 427–434.

Lv, H. P., Zhang, Y. J., Lin, Z., & Liang, Y. R. (2013). Processing and chemical constituents of Pu-erh tea: A review. *Food Research International, 53*(2), 608–618. https://doi.org/10.1016/j.foodres.2013.02.043

Manivel, L. (2022a). Chapter four—Mature tea (soil, water and shade) management. In *Scientific perspectives of tea plant horticulture and productivity.*

Manivel, L. (2022b). *Scientific perspectives of tea plant horticulture and productivity.* Academic Press.

Matsumoto, S., Kiriiwa, Y., & Takeda, Y. (2002). Differentiation of Japanese green tea cultivars as revealed by RFLP analysis of phenylalanine ammonia-lyase DNA. *Theoretical and Applied Genetics, 104*(6–7), 998–1002.

Meegahakumbura, M. K., Wambulwa, M. C., Thapa, K. K., Li, M. M., Möller, M., Xu, J. C., Yang, J. B., Liu, B. Y., Ranjitkar, S., Liu, J., Li, D. Z., Gao, L. M. (2016). Indications for three independent domestication events for the tea plant (Camellia sinensis (L.) O. Kuntze) and new insights into the origin of tea germplasm in China and India revealed by nuclear microsatellites. *Plos one, 11*(5). https://doi.org/10.1371/journal.pone.0155369

Meng, X. H., Li, N., Zhu, H. T., Wang, D., Yang, C. R., & Zhang, Y. J. (2019). Plant resources, chemical constituents and bioactivities of tea plants from the genus camellia section tea. *Journal of Agricultural and Food Chemistry, 67*(19), 5318–5349. https://doi.org/10.1021/acs.jafc.8b05037

Min, T. (1992). A revision of camellia sect. thea. *Acta Botanica Yunnanica, 14*, 115–132.

Min, T. (2000). *Monograph of the genus camellia.* Yunnan Science and Technology Press, Kunming.

Mo, H., Chen, Y., Huang, L., Zhang, H., Li, J., & Zhou, W. (2013). Neuroprotective effect of tea polyphenols on oxyhemoglobin induced subarachnoid hemorrhage in mice. *Oxidative Medicine and Cellular Longevity, 2013*, Article 743938. https://doi.org/10.1155/2013/743938

Nakagawa, M. (1975). Contribution of green tea constituents to the intensity of taste element of brew (*In Japanese*). *Nippon Shokuhin Kogyo Gakkaishi, 22*(2), 59–64. https://doi.org/10.3136/nskkk1962.22.59

Nowogrodzki, A. (2019). How climate change might affect tea. *Nature, 566*(7742), S10–S11. https://doi.org/10.1038/d41586-019-00399-0

Pang, X., Yu, W., Cao, C., Yuan, X., Qiu, J., Kong, F., & Wu, J. (2019). Comparison of potent odorants in raw and ripened Pu-Erh tea infusions based on odor activity value calculation and multivariate analysis: Understanding the role of pile fermentation. *Journal of Agricultural and Food Chemistry, 67*(47), 13139–13149. https://doi.org/10.1021/acs.jafc.9b05321

Pellissery, S., & Madaan, J. (2016). *Exploring the Indian tea culture: An overview.* Annals of the University of Oradea.

Raina, S. N., Ahuja, P. S., Sharma, R. K., Das, S. C., Bhardwaj, P., Negi, R., Sharma, V., Singh, S. S., Sud, R. K., Kalia, R. K., Pandey, V., Banik, J., Razdan, V., Sehgal, D., Dar, T. H., Kumar, A., Bali, S., Bhat, V., Sharma, S., & Mandi, S. S. (2012). Genetic structure and diversity of India hybrid tea. *Genetic Resources and Crop Evolution, 59*(7), 1527–1541. https://doi.org/10.1007/s10722-011-9782-6

Ritchie, C. (2018). The tea renaissance: Tea culture, globalization, and health. *Asian Social Science, 8*(7), 408–425.

Roberts, E., Cartwright, R., & Oldschool, M. (1957). The phenolic substances of manufactured tea. I.—Fractionation and paper chromatography of water-soluble substances. *Journal of the Science of Food and Agriculture, 8*(2), 72–80.

Scharbert, S., Holzmann, N., & Hofmann, T. (2004). Identification of the astringent taste compounds in black tea infusions by combining instrumental analysis and human bioresponse. *Journal of Agricultural and Food Chemistry, 52*(11), 3498–3508. https://doi.org/10.1021/jf049802u

Scharbert, S., & Hofmann, T. (2005). Molecular definition of black tea taste by means of quantitative studies, taste reconstitution, and omission experiments. *Journal of Agricultural and Food Chemistry, 53*(13), 5377–5384. https://doi.org/10.1021/jf050294d

Schuh, C., & Schieberle, P. (2006). Characterization of the key aroma compounds in the beverage prepared from Darjeeling black tea: Quantitative differences between tea leaves and infusion. *Journal of Agricultural and Food Chemistry, 54*(3), 916–924. https://doi.org/10.1021/jf052495n

Shitandi, & Anakalo., A. (2013). Tea processing and its impact on catechins, theaflavin and thearubigin formation (pp. 193–205). Elsevier Inc. https://doi.org/10.1016/B978-0-12-384937-3.00016-1

Shitandi, A. A., Muigai Ngure, F., & Mahungu, S. M. (2013). Tea processing and its impact on catechins, theaflavin and thearubigin formation. *Tea in Health and Disease Prevention, 193–205*. https://doi.org/10.1016/b978-0-12-384937-3.00016-1

Smith, A. (2016). Tea and the British empire: The contribution of tea to Britishculture and identity. *Colonial and Postcolonial Studies, 4*(2), 12–29.

Taniguchi, F., Kimura, K., Saba, T., Ogino, A., Yamaguchi, S., & Tanaka, J. (2014). Worldwide core collections of tea (Camellia sinensis) based on SSR markers. *Tree Genetics and Genomes, 10*(6), 1555–1565.

Teranishi, R., Wick, E. L., & Hornstein, I. (1999). *Flavor chemistry: Thirty years of progress.* Kluwer Academic/Plenum Publishers.

Tian, Y., Liang, Y., Yan, D., Luo, X., & Zhou, G. (2008). Geography distribution and ecotypes of wild tea resources in Guizhou. *Guizhou Agricultural Sciences, 2*, 160–161.

Uchiyama, S., Taniguchi, Y., Saka, A., Yoshida, A., & Yajima, H. (2011). Prevention of diet-induced obesity by dietary black tea polyphenols extract in vitro and in vivo. *Nutrition, 27*(3), 287–292. https://doi.org/10.1016/j.nut.2010.01.019

Wambulwa, M. C., Meegahakumbura, M. K., Chalo, R., Kamunya, S., Muchugi, A., Xu, J. C., Liu, D. Z., & Gao, L. M. (2016a). Nuclear microsatellites reveal the genetic architecture and breeding history of tea germplasm of East Africa. *Tree Genetics and Genomes, 12*(1), 11. https://doi.org/10.1007/s11295-015-0963-x.

Wambulwa, M. C., Meegahakumbura, M. K., Samson, K., Alice, M., Michael, M. L., Jie, L., Jian-Chu, X., Sailesh, R., De-Zhu, L., & Lian-Ming, G. (2016b). Insights into the genetic relationships and breeding patterns of the African tea germplasm based on nSSR markers and cpDNA sequences. *Frontiers in Plant Science, 7*, 1244–1255. https://doi.org/10.3389/fpls.2016.01244

Wang, P., & Yu, F. (2002). The geographic distribution, diversity and utilization of wild tea Camellias in China. *Journal of Tea Science, 22*(2), 105–108.

Wang, Y., Kan, Z., Thompson, H. J., Ling, T., Ho, C. T., Li, D., & Wan, X. (2019). Impact of six typical processing methods on the chemical composition of tea leaves using a single *Camellia sinensis* cultivar, Longjing 43. *Journal of Agricultural and Food Chemistry, 67*(19), 5423–5436.

Wang, M. Q., Ma, W. J., Shi, J., Zhu, Y., Lin, Z., & Lv, H. P. (2020). Characterization of the key aroma compounds in Longjing tea using stir bar sorptive extraction (SBSE) combined with gas chromatography-mass spectrometry (GC-MS), gas chromatography-olfactometry (GC-O), odor activity value (OAV), and aroma recombination. *Food Research International, 130*, Article 108908. https://doi.org/10.1016/j.foodres.2019.108908

Wei, C., Yang, H., Wang, S., Zhao, J., & Wan, X. (2018). Draft genome sequence of Camellia sinensis var. sinensis provides insights into the evolution of the tea genome and tea quality. *Proceedings of the National Academy of Sciences of the United States of America, 115*(18), 201719622. https://doi.org/10.1073/pnas.1719622115

Xia, E. H., Li, F. D., Tong, W., Li, P. H., Wu, Q., Zhao, H. J., Ge, R. H., Li, R. P., Li, Y. Y., & Zhang, Z. Z. (2019). Tea Plant Information Archive: a comprehensive genomics and bioinformatics platform for tea plant. *Plant Biotechnology Journal*. https://doi.org/10.1111/pbi.13111

Yang, Y. (2023). Adhere to problem orientation and promote high-quality development of china's tea industry—Current situation and suggestions of tea industry development in China. *China Tea, 45*(1), 1–5.

Yang, Z., Baldermann, S., & Watanabe, N. (2013). Recent studies of the volatile compounds in tea. *Food Research International, 53*(2), 585–599. https://doi.org/10.1016/j.foodres.2013.02.011

Yang, P., Song, H., Lin, Y., Guo, T., Wang, L., Granvogl, M., & Xu, Y. (2021). Differences of characteristic aroma compounds in *Rougui* tea leaves with different roasting temperatures analyzed by switchable GC-O-MS and GC x GC-O-MS and sensory evaluation. *Food and Function, 12*(11), 4797–4807. https://doi.org/10.1039/d1fo00165e

Yao, M. Z., Ma, C. L., Qiao, T. T., Jin, J. Q., & Chen, L. (2012). Diversity distribution and population structure of tea germplasms in China revealed by EST-SSR markers. *Tree Genetics and Genomes, 8*(1), 205–220. https://doi.org/10.1007/s11295-011-0433-z

Yin, J. (2021). Current situation and development trend of novel-tea beverage industry. *China Tea, 43*(08), 1–6.

Yu, F. (1989). Wild Arbor tea in China. *China Tea, 02*, 6–7.

Yu, F. L. (2016). *Ancient tea plants in China*. Yunnan Scientific Publishing Press.

Yu, H. (2023). Guizhou tea brewing method. *Journal of Guizhou Tea, 03*, 20–23.

Yu, P., Yeo, A. S., Low, M. Y., & Zhou, W. (2014). Identifying key non-volatile compounds in ready-to-drink green tea and their impact on taste profile. *Food Chemistry, 155*, 9–16. https://doi.org/10.1016/j.foodchem.2014.01.046

Zhai, X., Zhang, L., Granvogl, M., Ho, C. T., Wan, X. J. C. R. i. F. S., & Safety, F. (2022). Flavor of tea (Camellia sinensis): A review on odorants and analytical techniques. *Comprehensive Reviews in Food Science and Food Safety, 21*(5), 3867–3909. https://doi.org/10.1111/1541-4337.12999

Zhang, H. (1981). A taxonomy of the genus camellia. *Journal of National Sun Yat-sen University.*

Zhang, Y. N., Yin, J. F., Chen, J. X., Wang, F., Du, Q. Z., Jiang, Y. W., & Xu, Y. Q. (2016). Improving the sweet aftertaste of green tea infusion with tannase. *Food Chemistry, 192*, 470–476. https://doi.org/10.1016/j.foodchem.2015.07.046

Zhang, C., Suen, L. C., Yang, C., & Quek, S. Y. (2018a). Antioxidant capacity and major polyphenol composition of teas as affected by geographical location, plantation elevation and leaf grade. *Food Chemistry, 244*, 109. https://doi.org/10.1016/j.foodchem.2017.09.126

Zhang, W., Rong, J., Wei, C., Gao, L., & Chen, J. (2018b). Domestication origin and spread of cultivated tea plants. *Biodiversity Science, 26*, 357–372. https://doi.org/10.17520/BIODS.2018006

Zhang, L., Ho, C. T., Zhou, J., Santos, J. S., & Granato, D. (2019a). Chemistry and biological activities of processed camellia sinensis teas: A comprehensive review. *Comprehensive Reviews in Food Science and Food Safety, 18*(5), 1474–1495. https://doi.org/10.1111/1541-4337.12479.

Zhang, L., Zhang, H., Chen, Y., Dai, S., Li, X., Kenji, I., Liu, Z., & Li, M. (2019b). Real-time monitoring of optimum timing for harvesting fresh tea leaves based on machine vision. *International Journal of Agricultural and Biological Engineering* (1). https://doi.org/10.25165/ijabe.v12i1.3418.

Zhang, L., Cao, Q.-Q., Granato, D., Xu, Y.-Q., & Ho, C.-T. (2020a). Association between chemistry and taste of tea: A review. *Trends in Food Science and Technology, 101*, 139–149. https://doi.org/10.1016/j.tifs.2020.05.015

Zhang, M., Yang, Y., Yuan, H., Hua, J., Deng, Y., Jiang, Y., & Wang, J. (2020b). Contribution of addition theanine/sucrose on the formation of chestnut-like aroma of green tea. *Lwt-Food Science and Technology, 129*, Article 109512. https://doi.org/10.1016/j.lwt.2020.109512

Zhao, D., Yang, J., Yang, S., Kenji, K., & Luo, J. (2014). Genetic diversity and domestication origin of tea plant Camellia taliensis (Theaceae) as revealed by microsatellite markers. *BMC Plant Biology, 14*(1), 14–25. https://doi.org/10.1186/1471-2229-14-14

Zhu, J., Chen, F., Wang, L., Niu, Y., Yu, D., Shu, C., Chen, H., Wang, H., & Xiao, Z. (2015). Comparison of aroma-active volatiles in oolong tea infusions using GC-Olfactometry, GC-FPD, and GC-MS. *Journal of Agricultural and Food Chemistry, 63*(34), 7499–7510. https://doi.org/10.1021/acs.jafc.5b02358

Zhu, Y., Lv, H. P., Shao, C. Y., Kang, S., Zhang, Y., Guo, L., Dai, W. D., Tan, J. F., Peng, Q. H., & Lin, Z. (2018). Identification of key odorants responsible for chestnut-like aroma quality of green teas. *Food Research International, 108*, 74–82. https://doi.org/10.1016/j.foodres.2018.03.026

Zhu, J., Niu, Y., & Xiao, Z. (2021). Characterization of the key aroma compounds in Laoshan green teas by application of odour activity value (OAV), gas chromatography-mass spectrometry-olfactometry (GC-MS-O) and comprehensive two-dimensional gas chromatography mass spectrometry (GC × GC-qMS). *Food Chemistry, 339*, Article 128136. https://doi.org/10.1016/j.foodchem.2020.128136

Chapter 2
Color Characteristics in Six Types of Teas

Piaopiao Long and Zisheng Han

Abstract Tea color, a critical quality attribute, consists of the color characteristics of dry leaves, infusion, and bottom of leaves, determined by the dynamic transformation of natural pigments and the formation of new chromophores during processing. This chapter systematically reviews the classification, chemical structures, and chromogenic mechanisms of color compounds in tea, focusing on catechin oxidation products (e.g., theaflavins, thearubigins, theabrownins), flavonoids (flavones, flavonols, and their glycosides), amino acid-derived pigments (e.g., Maillard reaction products, N-ethyl-2-pyrrolidinone-substituted flavan-3-ols), and other colorants (e.g., chlorophylls, carotenoids, anthocyanins). The coloration mechanisms are elucidated through molecular orbital theory (HOMO-LUMO energy gaps), absorption spectroscopy, and complementary color principles. Key factors influencing tea color, including processing techniques (fermentation, roasting, drying), pH, temperature, and ionic concentration, are discussed for their roles in modulating pigment formation and stability. Despite advances, structural elucidation of complex polymers (e.g., thearubigins) and their quantitative color contributions remain challenges. Future research leveraging molecular sensory science and advanced analytics is urged to decode structure-color relationships and optimize production for enhanced quality and functionality.

Keywords Tea · Color · Catechin oxidation products · Theaflavins · Maillard reaction

P. Long · Z. Han (✉)
State Key Laboratory of Tea Plant Germplasm Innovation and Resource Utilization, School of Tea Science, Anhui Agricultural University, Hefei, China
e-mail: hanzisheng@outlook.com

International Joint Laboratory on Tea Chemistry and Health Effects of Ministry of Education, Anhui Agricultural University, Hefei, China

X. Zhai (ed.), *Tea Chemistry and Health Benefits*,
https://doi.org/10.1007/978-3-032-11828-8_2

2.1 Introduction

Tea quality is the cornerstone of the tea industry and is intrinsically linked to cultivar breeding, cultivation practices, primary processing, and deep processing. Enhancing tea quality is fundamental to promoting the sustainable and healthy development of the tea industry and improving its economic benefits. As a globally popular beverage, flavor is regarded as the core attribute of tea quality. The quality assessment of tea primarily focuses on four key flavor attributes: color, aroma, taste, and appearance, which serve as critical criteria for tea quality analysis and evaluation. The flavor quality of tea is closely associated with its internal chemical composition, particularly small-molecule phytochemicals.

The color of tea encompasses the appearance of dry tea leaves, the infusion, and the bottom of leaves. The color of dry leaves and bottom of leaves is associated with primary plant pigments such as chlorophylls, carotenoids, and xanthophylls, while the color of the tea infusion is primarily attributed to pigment compounds formed during processing. Tea processing involves various techniques, including fermentation (enzymatic oxidation), post-fermentation, and drying. Fresh tea leaves contain substantial amounts of polyphenolic compounds, which undergo enzymatic oxidation and microbial transformation during processing. The resulting products, which are generally colored, collectively determine the final hue of the tea infusion. Furthermore, high-temperature roasting during tea processing promotes caramelization and the Maillard reaction between reducing sugars and amino acids, generating colored products such as melanoidins that influence infusion color. Although the contents of free amino acids and reducing sugars in tea are relatively low compared to other food products, the pigments derived from carbonyl-amine reactions still contribute significantly to the color of roasted teas. In recent years, with the growing application of molecular sensory science in the field of tea research, studies on color compounds in tea have progressively increased. Researchers have isolated and identified a series of new pigment components from both tea and model systems, and have investigated the effects of processing techniques and other factors on the color and flavor of tea.

2.2 The Chromogenic Mechanism of Color Compounds in Tea

Due to variations in tea plant cultivars and processing techniques, the characteristic color hues of the tea infusion differ significantly among the six major categories of tea. Green tea infusion typically exhibits greenish-yellow or yellowish-green hues, black tea presents a distinct red color, yellow tea displays a characteristic yellow infusion, while Pu-erh tea infusion is characterized by a bright, reddish-brown appearance. These diverse color manifestations result from the combined effects of multiple chromogenic compounds and their associated influencing factors.

Chromophores refer to chemical substances capable of absorbing light within the visible region of the electromagnetic spectrum (approximately 360–780 nm). Visible light is polychromatic and can be dispersed into monochromatic components including red, orange, yellow, green, cyan, blue, and violet. When illuminated by visible light, a chromophore absorbs specific wavelengths and reflects the unabsorbed light, known as the complementary light. The photoreceptor cells in the human eye detect this complementary light, and the brain processes the signal to perceive color. Low-molecular-weight chromophores in tea exhibit maximum absorption within the 360–420 nm range of the visible spectrum, contributing primarily to yellow and red color. In contrast, natural pigments such as chlorophylls produce green and red hues. Although some color-active compounds in tea (e.g., flavonoids) do not have their absorption maxima within the visible region, their spectral tailing into the 360–780 nm range results in perceived pale yellow to yellow colors.

The color intensity of a compound in solution deepens with increasing concentration. For epigallocatechin gallate (EGCG), the absorption peak at 280 nm increases in intensity with higher concentration while the peak position remains unchanged. In contrast, the absorption peak at 210 nm exhibits a red-shift and alters the absorption spectrum as concentration rises. On one hand, the absorbance values of multiple chromophores are additive. In black tea infusion, as the number and concentration of color compounds increase, the liquor tends to appear orange-red. With further increase, the solution may even exhibit deep red or reddish-brown hues due to the absorption of most incident light by these pigments. The same phenomenon occurs even with a single chromophore: a compound that appears yellow at low concentration may present as orange-yellow or orange-red at higher concentrations, and often shows brown or reddish-brown in solid powdered form. On the other hand, solutions containing yellow and red pigments exhibit an increase in perceived color intensity as their concentration rises. The aqueous solutions of catechin oxidation products are predominantly yellow or red. These oxidation products typically exhibit two characteristic absorption bands in the visible region, whose complementary colors correspond to yellow and red, respectively. As the concentration increases, the overall color of the solution deepens. The absorption band associated with red gradually dominates the overall hue, leading to a color shift from yellow to red. Furthermore, interference or masking effects can occur among different chromophores in tea. For instance, when green chlorophylls degrade, the yellow color of carotenoids becomes dominant, resulting in an overall yellowish appearance.

The visible light absorption spectra and color characteristics of chromophores are fundamentally governed by their molecular structures and the extent of π-electron conjugation. In organic compounds, an increase in conjugation leads to a narrowing of the energy gap between the highest occupied molecular orbital (HOMO) and the lowest unoccupied molecular orbital (LUMO). This results in a bathochromic shift (red shift) in absorption wavelength, effectively moving the absorption band from the ultraviolet into the visible region of the spectrum, thereby imparting color to the substance.

The perceived color depends not only on the inherent chromophore structure but also critically on the type, number, and substitution pattern of auxochromes (e.g., –

OH, $-OR$, $-NH_2$, $-NR_2$) attached to the system. These electron-donating groups, possessing lone pair electrons, can induce a bathochromic shift in the absorption maximum. The human eye is remarkably sensitive to these subtle electronic transitions, capable of discerning color differences resulting from spectral shifts of just a few nanometers, contributing to the rich and complex perception of color in tea.

The color compounds in tea can be categorized based on their solubility into lipid-soluble and water-soluble groups. They can also be classified according to their chemical structures into porphyrin-based compounds featuring metal coordination and chromophores possessing conjugated systems (Wong, 2018). Chlorophylls are metalloporphyrin derivatives classified as lipid-soluble pigments. Their structure, featuring a central magnesium ion coordinated within a cyclic conjugated system, results in absorption peaks at longer wavelengths, conferring a green color. They serve as the primary chromophores in fresh tea leaves. In contrast, carotenoids and xanthophylls, while also lipid-soluble, possess distinct chemical structures based on conjugated polyene systems.

Water-soluble chromophores with conjugated chromophoric systems are the major contributors to tea infusion color. Oxidation products of catechins represent the principal colorants in fermented and semi-fermented teas. Depending on their oxidation degree and structural features, these compounds impart hues ranging from pale yellow, orange-yellow, and orange-red to deep red and brown in solution, with higher concentrations yielding deep red or reddish-brown shades. Flavones, flavonols, and their glycosides typically contribute pale to bright yellow tones in the infusion. Thermal reaction products of amino acids (e.g., theanine) and sugars exhibit pale yellow or faint pink colors, which intensify to orange-yellow or orange-red at higher concentrations. Maillard reactions between amino acids and sugars yield Amadori rearrangement products in the initial stages, which appear light yellow to yellow. Anthocyanins, exhibiting strong absorption between 475 and 560 nm, display red, blue, or purple coloration depending on the pH. For instance, the cultivar "Zijuan" from Yunnan contains high anthocyanin levels, resulting in purple leaves and a characteristic purple infusion (Shen et al., 2018). Tea polysaccharides, complex water-soluble carbohydrates extractable in hot water, produce extracts in shades of off-white, light yellow, or grayish-brown, also influencing infusion color.

Chromophores generally contain unsaturated carbon chains or rings within conjugated systems, often influenced by heteroatoms like oxygen or nitrogen. The relationship between molecular structure and color is complex, primarily determined by the chromophore's core structure coupled with the type, number, and substitution patterns of auxochromic groups. Collectively, these diverse chemical entities determine the final color attributes of dry tea leaves, the infusion, and the bottom of leaves.

In general, four types of color compounds were mainly discussed in the present chapter, including catechins and their oxidation products, flavones, flavonols, and their glycosides, amino acid contributing compounds, as well as other colorants (Table 2.1).

Table 2.1 Main color compounds in tea and their color characteristics

Classification	Compounds	Color properties	References
Catechin oxidation products	Theaflavins	Orange-yellow	
	Thearubigins	Reddish-brown	
	Theabrownins	Dark reddish-brown	
	Dehydrotheaflavin	Reddish-brown	Tanaka et al., (2002a, 2002b)
	Hydroxytheaflavin	Red amorphous powder	Matsuo et al. (2006)
	Isotheaflavin 3′-O-gallate	Red	Lewis et al. (1998)
	Neotheaflavin 3′-O-gallate	Red	Lewis et al. (1998)
	Theaflavate A	Yellow	Wan et al. (1997)
	Theaflavate B	Orange-red amorphous powder	Lewis et al. (1998)
	Theaflavate C	Red	Kusano et al. (2007)
	Hydroxytheaflavin	Red	Matsuo et al. (2006)
	Theaflavic acid	Red	Bailey et al. (1991)
	Epitheaflavic acid	Red	Berkowitz et al. (1971)
	Epitheaflavic acid gallate	Bright red	Berkowitz et al. (1971)
	Epitheaflagallin	Orange	Itoh et al. (2007)
	Epitheaflagallin 3-*O*-gallate	Deep-orange	Itoh et al. (2007)
	Proepitheaflagallin	Brown amorphous powder	Matsuo et al. (2006)
	Theasinensin A	A pale brown amorphous powder	Nonaka et al. (1983)
	Theasinensin B	A pale brown amorphous powder	Nonaka et al. (1983)
	Theasinensin D	White	Morimoto et al. (1988)
	Theasinensin E	White	Morimoto et al. (1988)
	Theasinensin F	White	Morimoto et al. (1988)
	Theasinensin G	White	Morimoto et al. (1988)

(continued)

Table 2.1 (continued)

Classification	Compounds	Color properties	References
	Theasinensin H	Brown	Shii et al. (2011)
	Dehydrotheasinensin AQ	Reddish-orange	Tanaka et al. (2005)
	Dehydrotheasinensin C	White amorphous powder	Matsuo et al. (2006)
	Dehydrotheasinensin E	Yellow	Tanaka et al. (2005)
	Dehydrotheasinensin H	Yellow	Shii et al. (2011)
	Theacitrin A	Yellow	Matsuo et al. (2011)
	Theacitrin B	Yellow	Matsuo et al. (2011)
	Theacitrin C	Yellow	Matsuo et al. (2011)
	Theacitrinin A	Pale yellow	
	Theacitrinin B	Reddish-brown	Matsuo et al. (2011)
	Theadibenzotropolone A	Orange/bright red	Sang et al. (2002)
	Theanaphthoquinone	Reddish orange/ yellow	Long et al. (2023)
	Oolongtheanins	Off-white	Morimoto et al. (1988)
Flavones, flavonols, and their glycosides	Flavones	Yellow	
	Flavonols	Yellow	
	Flavone glycosides	Pale to bright yellow	
	Flavonol glycosides	Pale to bright yellow	
Amino acid contributing compounds	Prolin–glucose-derived amadori compound	Yellow	Wang et al. (2021)
	N-ethyl-2-pyrrolidinone-substituted flavan-3-ols	Pink	
Other colorants	Anthocyanidins	Red/purple/Blue	Yang et al. (2025)
	Chlorophyll a	Green	Yu et al. (2019)
	Chlorophyll b	Green	Yu et al. (2019)
	Xanthophylls	Yellow	
	Carotenoids	Yellow to orange-red	

2.3 Catechins and Their Oxidation Products

During the processing of fermented teas, such as black tea, catechins undergo oxidation, polymerization, and condensation reactions catalyzed by polyphenol oxidase (PPO), peroxidase (POD), and thermo-humid conditions (Dai et al., 2017). This generates complex and diverse oxidation products, which constitute the crucial material basis for the bright, luminous reddish infusion characteristic of black tea. Among these, theaflavins (TFs) represent the signature chromogenic compounds. Structurally, benzotropolone moieties formed via benzotropolone ring formation from catechin precursors exhibit characteristic absorption maxima at 380 and 480 nm. These absorption wavelengths correspond to complementary colors in the yellow and orange-red spectra. Oxidation products containing the benzotropolone moiety, such as theaflavin (TF1), theaflavin-3-*O*-gallate (TF2a), theaflavin-3′-*O*-gallate (TF2b), and theaflavin-3,3′-*O*-digallate (TF3), typically yield bright yellow solutions. In contrast, compounds including hydroxytheaflavin, theaflavates, theaflavic acid, and theadibenzotropolone A exhibit orange to red hues (Degenhardt et al., 2000). Theadibenzotropolone A, characterized by a polybenzotropolone structure, displays deeper coloration, manifesting as orange-yellow and vibrant red. Hydroxytheaflavin, with absorption maxima at 378 and 428 nm, presents a distinct red color. Theacitrins are formed via a bicyclo[3.2.1]octane intermediate generated from epigallocatechin gallate (EGCG) or epigallocatechin (EGC) and their corresponding *o*-quinones, followed by structural rearrangement. Theacitrinins, formed upon heating of theacitrins, share similar chromophoric structures with their precursors, of which theacitrinin A displays a pale yellow color (Wang et al., 2022).

The enzymatic oxidation products of catechins exhibit considerable complexity. Researchers globally have isolated and purified a series of these oxidation products through model reaction systems. However, accurate and detailed characterization of their color attributes remains lacking for many of these compounds. Nevertheless, based on their chemical structures, the coloration of these oxidation products can be reasonably inferred to a certain extent. For instance, neotheaflavate B and theaflavate C, both containing the benzotropolone moiety, are expected to exhibit colors ranging from yellow to orange-red. Furthermore, several studies have reported the characteristic UV-Vis absorption spectra of catechin oxidation products, allowing color prediction based on their spectral features. Kusano et al. (2008) isolated and identified Theaflavate C, noting its maximum absorption peaks at 279 and 405 nm, consistent with its classification as a red pigment.

A substantial portion of catechin oxidation products comprises colorless compounds lacking distinctive chromophores. For instance, Theasinensins, dimers formed by the condensation of two catechin molecules via C–C bonds, typically appear as white powders and yield colorless solutions. Representative theasinensins isolated from tea, including Theasinensin D, Theasinensin E, and Theasinensin F, consistently present as off-white powders (Long et al., 2023; Tanaka et al., 2002a, 2002b). Tanaka et al. (2005) isolated an EGCG trimer via a model reaction system. This trimer, formed through C–C bond condensation involving the B-ring of one

EGCG molecule and the galloyl moiety (D-ring) of another, exhibits a maximum absorption peak at 275 nm and is characterized as a colorless substance. During enzymatic oxidation of EGCG catalyzed by Japanese pear homogenate, Theasinensin A and Theasinensin D are generated. A key intermediate in this pathway, Dehydrotheasinensin A, is colorless. However, its oxidation product, the quinone Dehydrotheasinensin AQ, has been identified as a pigment. This compound possesses a 1,2-dione structural motif with absorption maxima at 447 and 277 nm, resulting in an orange-red color (Tanaka et al., 2005).

The characteristic reddish color of black tea infusion is primarily formed during the rolling and fermentation stages, during which a substantial quantity of chromogenic substances is generated. As fermentation progresses, low-polymerization catechin oxidation products undergo further oxidation and polymerization to form thearubigins and theabrownins, resulting in a progressively bright and reddish infusion. With extended fermentation time, thearubin content initially increases then decreases, while theabrownin content continuously accumulates (Hua et al., 2021). Thearubigins constitute the primary coloring compounds in black tea infusion, accounting for 9–19% of its dry mass. In solid form, thearubigins appear reddish-brown, while their saturated aqueous solution exhibits orange to reddish-brown coloration (Kuhnert, 2010). With a maximum UV absorption at 390 nm, thearubigins play a critical role in the color formation of black tea infusion (Long et al., 2023). The degree of fermentation during black tea processing crucially determines the final infusion color. The ratio of theaflavins to thearubigins serves as a key indicator for evaluating fermentation degree and predicting infusion color (Zhu et al., 2022). Theabrownins are highly polymerized brown pigments formed through polymerization of theaflavins and thearubigins. Excessive fermentation generates high levels of theabrownins, leading to a dull infusion that lacks the desirable bright reddish quality (Obanda et al., 2002). Furthermore, theabrownins accumulate substantially during the post-fermentation process of dark tea, contributing to its characteristically deep, rich, and bright reddish-brown infusion (Cheng et al., 2021), which is typically darker and more intense than that of black tea.

2.4 Flavones, Flavonols, and Their Glycosides

Flavones, flavonols, and their glycosides constitute approximately 3–4% of the dry weight of tea leaves. Predominantly yellow, these compounds serve as key constituents contributing to the characteristic infusion color of both green tea and black tea (Cui et al., 2022). All flavones, flavonols, and their glycosides share the common structural feature of a 2-phenylchromone backbone, formed by the fusion of the colorless chromone nucleus to the phenyl ring, features a cross-conjugated system comprising a cinnamoyl moiety (B-ring) and a benzoyl moiety (A-ring). Absorption bands originating from this conjugated system partially extend into the visible region, imparting hues of grayish-yellow, yellow, and bright yellow to these compounds.

Kaempferol, quercetin, and myricetin represent the principal flavonol aglycones in tea. Due to the presence of three hydroxyl groups on its B-ring (pyrogallol-type substitution), myricetin exhibits a deeper yellow hue compared to quercetin (catechol-type) and kaempferol (phloroglucinol-type). Flavonol glycosides primarily feature sugar conjugation at the C3 position, which minimally affects chromophore intensity. Consequently, these glycosides typically appear yellow to bright yellow.

Beyond their intrinsic color properties, flavonoids engage in reactions with other tea constituents to form chromogenic complexes. Dai et al. (2017) demonstrated that flavonols react with epigallocatechin gallate (EGCG) during thermal processing, resulting in significant darkening of green tea infusion. Furthermore, Shi et al. (2021) reported substantial variation in flavonoid content between green- and purple-leaf tea cultivars. Flavonoids function as co-pigments, complexing with proanthocyanidins to generate the distinctive purple coloration observed in young leaves and buds of purple cultivars.

2.5 Amino Acid Contributing Compounds

Beyond polyphenolic compounds, carbohydrates and amino acids in tea also significantly influence infusion color. During high-temperature drying, mono- and oligosaccharides undergo Maillard reactions with amino acids. This thermally driven process generates brown melanoidins through complex reaction cascades, impacting both tea infusion color and aroma (Jiang et al., 2019). Amadori rearrangement products (ARPs), formed during the initial Maillard stage, serve as key precursors for aroma development and browning. Previous studies have reported that Amadori compounds derived from umami-enhancing amino acids not only potentiated umami taste but also affected color; notably, the prolin-glucose-derived Amadori compound exhibits a distinct yellow hue (Wang et al., 2021).

Theanine, a signature amino acid in tea constituting approximately 50% of total free amino acids, participates in thermal reactions. Under high-temperature drying conditions, theanine condenses with catechins to form N-ethyl-2-pyrrolidinone-substituted flavan-3-ols (EPSFs). The N-ethyl-2-pyrrolidinone moiety in EPSFs contains a nitrogen atom with a high-energy lone pair of electrons. This electronic configuration reduces the energy gap between the highest occupied molecular orbital (HOMO) and lowest unoccupied molecular orbital (LUMO), enabling absorption of longer-wavelength visible light and conferring color to aqueous solutions. Researchers observed that the C-8 N-ethyl-2-pyrrolidinone-substituted epigallocatechin gallate (EGCG) derivative appeared as a distinct pink hue in its crude solid form (Zhou et al., 2019). Collectively, these findings demonstrate that amino acids contribute to tea liquor coloration through multiple pathways, suggesting undiscovered ARPs and EPSFs as potential key chromophores.

2.6 Other Colorants

2.6.1 *Chlorophyll a, Chlorophyll b, Carotenoids, and Xanthophylls*

Chlorophyll a, chlorophyll b, carotenoids, and xanthophylls constitute the primary lipid-soluble pigments in fresh tea leaves. Chlorophylls, accounting for approximately 0.24–0.85% of tea dry weight, serve as the dominant chromophores in dry tea leaves and leaf bottoms. Chlorophyll a exhibits absorption maxima in the range of 420–663 nm, while chlorophyll b absorbs maximally within 460–645 nm, resulting in subtly differential green hues. Significant variations in the ratios of chlorophyll a to chlorophyll b occur across different tea leaves due to factors like environment, climate, and cultivar, resulting in leaf colors ranging from deep green to light green and yellowish-green (Yu et al., 2019). Chlorophylls exhibit minimal direct contribution to infusion color. On one hand, a minor fraction of free chlorophylls a and b released from ruptured cells may impart a slight greenish tint to the infusion. On the other hand, portions of chlorophyll a and b undergo acid degradation during processing to form pheophytin a and pheophytin b, or hydrolysis yielding water-soluble chlorophyll derivatives (e.g., chlorophyllins), both of which can influence infusion color. Notably, during green tea beverage production, the leaching of chlorophylls can significantly increase infusion cloudiness, thereby diminishing the clean, bright visual quality characteristic of premium green tea infusions (Roshanak et al., 2016). Similarly, during the yellowing stage of yellow tea processing, the progressive decline in chlorophyll a and b content contributes to the reduction of green intensity in the infusion. Carotenoids are present at substantially lower levels ($\approx$0.06% of dry weight) in most tea leaves. However, in distinct cultivars like "Huangjinya" (*Camellia sinensis cv. Huangjinya*), elevated carotenoid concentrations confer the distinctive yellow coloration observed in both fresh leaves and processed dry tea (Yangen et al., 2019).

2.6.2 *Anthocyanins*

Anthocyanins exhibit exceptional water solubility and are predominantly found in purple sweet potatoes, grapes, red cabbage, and eggplant skins. While typically present at approximately 0.01% in processed dry tea, their concentration in purple-leaf tea cultivars can reach 0.5–1.0% (Yang et al., 2025). This elevated level confers a distinct purple hue to both fresh leaves and the brewed infusion. Cultivars like "Zijuan" and purple-leaf tea have garnered significant attention due to their unique purple leaf color, primarily attributed to high anthocyanin content. Anthocyanins possess a highly conjugated molecular system rich in double bonds, augmented by auxochromic groups (e.g., hydroxyl, methoxy substituents). This extensive conjugation results in strong absorption within the visible light spectrum. Crucially, anthocyanidin chromophores exhibit pH-dependent coloration, displaying red, purple, or blue hues in response to solution pH. However, anthocyanins contribute bitterness to tea infusions; concentrations exceeding optimal levels negatively impact sensory quality due to increased astringency and bitterness (Yang et al., 2025). Shi et al. (2021) identified significantly elevated levels of anthocyanins and flavonoids in purple-leaf teas. Delphinidin 3-*O*-galactoside and Cyanidin 3-*O*-galactoside emerged as the predominant anthocyanin glycosides. Concurrently, higher concentrations of chlorophyll a and chlorophyll b compared to standard green-leaf cultivars were observed, collectively contributing to the characteristic purple-green coloration of fresh purple tea leaves. The novel cultivar "Ziyan"demonstrates even greater anthocyanin richness, characterized by elevated levels of delphinidin, cyanidin, and pelargonidin aglycones. Its total anthocyanin content exceeds that of "Zijuan", resulting in a noticeably deeper purple coloration (Lai et al., 2016).

2.7 Factors Influencing Tea Color

2.7.1 Effect of Processing Techniques on Color

During tea processing, the actions of enzymes, microorganisms, and heat induce substantial transformations in the chemical composition of fresh leaves, leading to the formation of various color-active compounds and distinct color characteristics. Notably, the contents of chlorophylls and carotenoids decrease significantly. Carotenoids are further converted into volatile compounds, contributing to the formation of tea aroma. In green tea processing, chlorophyll is hydrolyzed during the de-enzyming stage to form chlorophyllin and phytol. Chlorophyllin exhibits partial hydrophilicity, allowing a fraction of chlorophyll-derived pigments to dissolve into the tea infusion. Meanwhile, the rapid inactivation of polyphenol oxidase (PPO and peroxidase (POD) under high-temperature de-enzyming prevents the oxidation of polyphenols into small molecular chromophores. As a result, green tea dry leaves retain a green color, and the infusion appears yellow-green and bright, maintaining the characteristic "clear soup with green leaves" appearance (Wang et al., 2020). In contrast, the color transformation mechanism in black tea processing is fundamentally different. Chlorophyll is extensively degraded, while the enzymatic oxidation of polyphenols is promoted, yielding a large quantity of yellow and red pigments. Specifically, the contents of reddish-brown thearubigins and dark brown theabrownins increase significantly. These compounds, along with other colored oxidation products, collectively contribute to the bright and vibrant reddish color characteristic of black tea infusion.

Both white tea and yellow tea belong to the category of lightly fermented teas. During white tea processing, the degradation of chlorophyll is minimal, and the degree of oxidation is significantly lower than that of black tea. Consequently, the color of white tea infusion is composed of green, yellow, and red pigments, yet green tones remain dominant. In contrast, the infusion of yellow tea is characterized by a dominant yellow hue. The unique yellowing process induces substantial degradation of chlorophyll, while polyphenols undergo partial color formation under the combined effects of heat and moisture, resulting in an overall yellow infusion. Studies indicate that large-leaf yellow tea exhibits a darker color compared to other yellow teas. During the final high-temperature drying stage, the contents of chlorophyll a, chlorophyll b, and total chlorophyll decrease significantly, transforming the dry leaves from dark green to brownish-yellow and yielding a distinctly yellow infusion (Zhou et al., 2019).

Oolong tea is a semi-fermented type of tea. The plucking standard generally consists of one bud with three or four leaves. The leaf-aging process is the key stage for color development in oolong tea. During this process, chlorophyll is degraded, changing the leaf color from green to yellowish-green. Mechanical disruption from tumbling and turning causes bruising along the leaf edges, leading to reddening and forming the characteristic appearance of "green leaves with red edges". The ideal degree of redness is achieved when the reddened area accounts for approximately 30% of the total leaf surface area, with the leaf center appearing yellowish-green and the margins red.

The characteristic color of dark tea infusion is primarily formed during the pile-fermentation process. Following pile-fermentation, the leaf color changes from dark green to yellowish-brown, accompanied by a significant increase in theabrownin content. The processed dry leaves exhibit a blackish-brown and glossy appearance, while the infusion typically presents a brownish-yellow or brownish-red hue.

2.7.2 Influence of pH and Temperature on Color

During tea brewing, the color of the infusion is influenced by factors such as water quality, temperature, and brewing time. Extensive research indicates that pH is a critical factor affecting tea infusion color. Organic acids and phenolic acids can alter the pH of the infusion, thereby modifying its color. Changes in pH can lead to variations in the UV-Vis spectra of chromophores. Some compounds may undergo cis–transisomerization under different acid-base conditions, resulting in bathochromic (red) shifts or hypsochromic (blue) shifts, which ultimately influence the perceived color of the tea. The structure of flavonoids may also transition between a pyran ring form and a chalcone-type configuration (which appears orange or brown) under varying pH conditions.

As pH increases, the absorbance of crude theaflavin extracts gradually rises, accompanied by a darkening in color. Quinic acid and gallic acid are the primary organic acids in tea. An increase in their content lowers the infusion pH, thereby altering its color. Cao et al. (2021) demonstrated that different types of brewing water, varying in pH and ion concentration, affect the sensory properties, including color, of tea infusions. Their findings confirmed that using high-pH water resulted in significantly darker infusions compared to low-pH conditions. Among tea types, green tea was most susceptible to pH changes, followed by yellow tea, while dark tea was largely unaffected. Xu et al. (2017) reported that under high-pH conditions, catechins in tea infusions undergo oxidation to form colored compounds, reducing the brightness (L^* value) of the infusion. Their recent research further indicated that natural spring water with low pH and low total dissolved solids (TDS) is more suitable for brewing green tea and lightly fermented teas, significantly enhancing their unique flavor profiles. In contrast, infusions brewed with high pH, high-TDS mountain spring water showed that epicatechin gallate (ECG) and epicatechin (EC) may undergo oxidative polymerization, degradation, and epimerization at room temperature, promoting the formation of theabrownins. This resulted in darker infusions and lower sensory scores (Deng et al., 2023).

Temperature is another major factor influencing tea infusion color. In black tea, as temperature decreases, theaflavins, thearubigins, and caffeine form light-brown creamy complexes, leading to reduced clarity, turbidity, and darkening of the infusion, substantially altering its visual quality (Qin et al., 2022). Additionally, lower temperatures cause a decrease in the brightness (L^* value) of black tea infusion, along with an overall darkening manifested as increased a^* (redness) and b^* (yellowness) values, resulting in a less bright and vivid reddish appearance. The ionic concentration in water also affects infusion color. Due to the polyhydroxy structures of catechins and flavonoid glycosides, they readily form complexes with metal ions, yielding compounds of varying colors. Specifically, Ca^{2+} ions promote the formation of tea cream, further influencing the color properties of the infusion.

2.8 Conclusion and Perspectives

The color of tea is fundamentally determined by the degradation of natural pigments present in the leaves, coupled with the formation of new colored compounds during processing. Typical chromophores in tea infusions often contain characteristic structural motifs such as the benzotropolone core, the 2-phenylchromone backbone, and the 5,7,8a-trihydroxycycloalkanol[α]indene-1,6,8(3aH,3bH,8aH)-trione system. Nevertheless, research on tea colorants still faces several unresolved scientific challenges.

The precise structures of complex oxidation products, particularly thearubigins and theabrownins, remain largely unelucidated, and their specific chromogenic activity and color-contribution thresholds have not been fully characterized (Matsuo et al., 2006). Due to variations in tea cultivars, growing environments, plucking seasons, and processing techniques, the composition and abundance of colorants are extremely complex. New chromophores with novel chromophoric systems continue to be identified (Itoh et al., 2007). Although these compounds are often present at low concentrations, methods for evaluating their combined color contribution are still lacking. Furthermore, it remains unclear whether these intermediate colored products contribute to the overall chromogenic properties of thearubigins and theabrownins, warranting further investigation.

Chromogenic substances formed during tea processing exhibit certain physiological functions and flavor-enhancing effects. Key challenges in the tea industry include: How to regulate the content and proportion of intrinsic components in fresh tea leaves through cultivar breeding and cultivation management to establish a material foundation for the formation of characteristic colors, and how to precisely control the formation of chromogenic substances by adjusting processing parameters, thereby enhancing the color score, flavor quality, and health benefits of the final product. Addressing these issues is essential for guiding actual tea production practices.

In summary, research on tea color components involves a large number of known and unknown substances with complex structures. The formation pathways, combined color effects, interaction mechanisms, and practical application value of these chromogenic substances warrant further in-depth investigation. With the advancement of molecular sensory science, it is anticipated that these scientific questions will be elucidated at the molecular level.

References

Bailey, R. G., Nursten, H. E., & McDowell, I. (1991). Comparative study of the reversed-phase high-performance liquid chromatography of black tea liquors with special reference to the thearubigins. *Journal of Chromatography A.*

Berkowitz, J. E., Coggon, P., & Sanderson, G. W. (1971). Formation of epitheaflavic acid and its transformation to thearubigins during tea fermentation. *Phytochemistry, 10*(10), 2271–2278. https://doi.org/10.1016/S0031-9422(00)89866-0

Cao, Q. Q., Wang, F., Wang, J. Q., Chen, J. X., & Xu, Y. Q. (2021). Effects of brewing water on the sensory attributes and physicochemical properties of tea infusions. *Food Chemistry, 364*(2), Article 130235.

Cheng, L., Wang, Y., Zhang, J., Xu, L., & Wei, X. (2021). Integration of non-targeted metabolomics and E-tongue evaluation reveals the chemical variation and taste characteristics of five typical dark teas. *LWT-Food Science and Technology, 150*(19), Article 111875.

Cui, Y., Lai, G., Wen, M., Han, Z., & Zhang, L. (2022). Identification of low-molecular-weight color contributors of black tea infusion by metabolomics analysis based on UV-visible spectroscopy and mass spectrometry. *Food Chemistry, 386*, 132788. https://doi.org/10.1016/j.foodchem.2022.132788

Dai, Q., He, Y., Ho, C. T., Wang, J., Wang, S., Yang, Y., Gao, L., & Xia, T. (2017). Effect of interaction of epigallocatechin gallate and flavonols on color alteration of simulative green tea infusion after thermal treatment. *Journal of Food Science & Technology.*

Degenhardt, A., Engelhardt, U. H., Wendt, A. S., & Winterhalter, P. (2000). Isolation of black tea pigments using high-speed countercurrent chromatography and studies on properties of black tea polymers. *Journal of Agricultural & Food Chemistry, 48*(11), 5200–5205.

Deng, S., Cao, Q.-Q., Zhu, Y., Wang, F., Chen, J.-X., Zhang, H., Granato, D., Liu, X., Yin, J.-F., & Xu, Y.-Q. (2023). Effects of natural spring water on the sensory attributes and physicochemical properties of tea infusions. *Food Chemistry, 419*, Article 136079. https://doi.org/10.1016/j.foodchem.2023.136079

FAOSTAT statistics database. Food and Agriculture Organization of the United Nations; crops-world production quantity (tea). Retrieved August 2021, from https://www.fao.org/faostat/

GB/T 23776-2018. Methodology for sensory evaluation of tea. Chinese National Standard. Retrieved August 2021, from http://www.foodmate.net

Hua, J., Wang, H., Yuan, H., Yin, P., Wang, J., Guo, G., & Jiang, Y. (2021). New insights into the effect of fermentation temperature and duration on catechins conversion and formation of tea pigments and theasinensins in black tea. *Journal of the Science of Food and Agriculture*.

Itoh, N., Katsube, Y., Yamamoto, K., Nakajima, N., & Yoshida, K. (2007). Laccase-catalyzed conversion of green tea catechins in the presence of gallic acid to epitheaflagallin and epitheaflagallin 3-O-gallate. *Tetrahedron, 63*(38), 9488–9492. https://doi.org/10.1016/j.tet.2007.06.093

Jiang, H., Yu, F., Qin, L., Zhang, N., Cao, Q., Schwab, W., Li, D., & Song, C. (2019). Dynamic change in amino acids, catechins, alkaloids, and gallic acid in six types of tea processed from the same batch of fresh tea (Camellia sinensis L.) leaves. *Journal of Food Composition and Analysis, 77*, 28–38. https://doi.org/10.1016/j.jfca.2019.01.005

Kuhnert, N. (2010). Unraveling the structure of the black tea thearubigins. *Archives of Biochemistry & Biophysics, 501*(1), 37–51.

Kusano, R., Tanaka, T., Matsuo, Y., & Kouno, I. (2007). Structures of epicatechin gallate trimer and tetramer produced by enzymatic oxidation. *Chemical and Pharmaceutical Bulletin, 55*(12), 1768–1772. https://doi.org/10.1248/cpb.55.1768

Kusano, R., Tanaka, T., Matsuo, Y., & Kouno, I. (2008). Structures of epicatechin gallate trimer and tetramer produced by enzymatic oxidation. *Chemical & Pharmaceutical Bulletin, 55*(12), 1768–1772.

Lai, Y. S., Li, S., Tang, Q., Li, H. X., & Chen, S. X. (2016). The Dark-purple tea cultivar 'Ziyan' accumulates a large amount of delphinidin-related anthocyanins. *Journal of Agricultural and Food Chemistry, 64*(13), 2719–2726.

Lewis, J. R., Davis, A. L., Cai, Y., Davies, A. P., Wilkins, J. P. G., & Pennington, M. (1998). Theaflavate B, isotheaflavin-3′-O-gallate and neotheaflavin-3-O-gallate: Three polyphenolic pigments from black tea. *Phytochemistry (Oxford), 49*(8), 2511–2519. https://doi.org/10.1016/S0031-9422(98)00153-8

Long, P., Rakariyatham, K., Ho, C.-T., & Zhang, L. (2023). Thearubigins: Formation, structure, health benefit and sensory property. *Trends in Food Science & Technology, 133*, 37–48. https://doi.org/10.1016/j.tifs.2023.01.013

Matsuo, Y., Li, Y., Watarumi, S., Tanaka, T., & Kouno, I. (2011). Production and degradation mechanism of theacitrin C, a black tea pigment derived from epigallocatechin-3-O-gallate via a bicyclo[3.2.1]octane-type intermediate. *Tetrahedron, 67*(11), 2051–2059. https://doi.org/10.1016/j.tet.2011.01.058

Matsuo, Y., Tanaka, T., & Kouno, I. (2006). A new mechanism for oxidation of epigallocatechin and production of benzotropolone pigments. *Tetrahedron, 62*(20), 4774–4783.

Morimoto, S., Nonaka, G.-I., Chen, R.-F., & Nishioka, I. (1988). Tannins and related compounds. LXI.: Isolation and structures of novel Bi- and Triflavanoids from the leaves of Cassia fistula L. *Chemical & Pharmaceutical Bulletin, 36*(1), 39–47. https://doi.org/10.1248/cpb.36.39

Nonaka, G. I., Kawahara, O., & Nishioka, I. (1983). Tannins and related compounds. XV. A new class of dimeric flavan-3-ol gallates, theasinensins A and B, and proanthocyanidin gallates from green tea leaf. *Chemical and Pharmaceutical Bulletin, 31*(11), 3906–3914. https://doi.org/10.1248/cpb.31.3906

Obanda, M., Owuor, P. O., & Mang'oka, R. (2002). Changes in the chemical and sensory quality parameters of black tea due to variations of fermentation time and temperature. *Food Chemistry, 75*(4), 395–404.

Qin, C., Lian, L., Xu, W., Jiang, Z., Wen, M., Han, Z., & Zhang, L. (2022). Comparison of the chemical composition and antioxidant, anti-inflammatory, α-amylase and α-glycosidase inhibitory activities of the supernatant and cream from black tea infusion. *Food & Function, 13*(11), 6139–6151. https://doi.org/10.1039/d2fo00707j

Roshanak, S., Rahimmalek, M., & Goli, S. A. H. (2016). Evaluation of seven different drying treatments in respect to total flavonoid, phenolic, vitamin C content, chlorophyll, antioxidant activity and color of green tea (Camellia sinensis or C. assamica) leaves. *Journal of Food Science and Technology, 53*(1), 721–729. https://doi.org/10.1007/s13197-015-2030-x

Sang, S., Tian, S., Meng, X., Stark, R. E., Rosen, R. T., Yang, C. S., & Ho, C.-T. (2002). Theadibenzotropolone A, a new type pigment from enzymatic oxidation of (−)-epicatechin and (−)-epigallocatechin gallate and characterized from black tea using LC/MS/MS. *Tetrahedron Letters, 43*(40), 7129–7133. https://doi.org/10.1016/S0040-4039(02)01707-0

Shen, J., Zou, Z., Zhang, X., Zhou, L., Wang, Y., Fang, W., & Zhu, X. (2018). Metabolic analyses reveal different mechanisms of leaf color change in two purple-leaf tea plant (Camellia sinensis L.) cultivars. *Horticulture Research, 5*(1), 7.

Shi, J., Simal-Gandara, J., Mei, J., Ma, W., Peng, Q., Shi, Y., Xu, Q., Lin, Z., & Lv, H. (2021). Insight into the pigmented anthocyanins and the major potential co-pigmented flavonoids in purple-coloured leaf teas. *Food Chemistry, 363*, 130278. https://doi.org/10.1016/j.foodchem.2021.130278

Shii, T., Tanaka, T., Watarumi, S., Matsuo, Y., Miyata, Y., Tamaya, K., Tamaru, S., Tanaka, K., Matsui, T., & Kouno, I. (2011). Polyphenol composition of a functional fermented tea obtained by tea-rolling processing of green tea and loquat leaves. *Journal of Agricultural and Food Chemistry, 59*(13), 7253–7260. https://doi.org/10.1021/jf201499n

Tanaka, T., Mine, C., & Kouno, I. (2002a). Structures of two new oxidation products of green tea polyphenols generated by model tea fermentation. *Tetrahedron, 58*(43), 8851–8856. https://doi.org/10.1016/S0040-4020(02)01061-X

Tanaka, T., Mine, C., Watarumi, S., Fujioka, T., Mihashi, K., Zhang, Y. J., & Kouno, I. (2002b). Accumulation of epigallocatechin quinone dimers during tea fermentation and formation of theasinensins. *Journal of Natural Products, 65*(11), 1582–1587.

Tanaka, T., Matsuo, Y., & Kouno, I. (2005). A novel black tea pigment and two new oxidation products of epigallocatechin-3-O-gallate. *Journal of Agricultural and Food Chemistry, 53*(19), 7571–7578. https://doi.org/10.1021/jf0512656

Wan, X., Nursten, H. E., Cai, Y., Davis, A. L., Wilkins, J. P. G., & Davies, A. P. (1997). A new type of tea pigment—From the chemical oxidation of epicatechin gallate and isolated from tea. *Journal of the Science of Food and Agriculture, 74*(3), 401–408. https://doi.org/10.1002/(SICI)1097-0010(199707)74:3<401::AID-JSFA816>3.0.CO;2-J

Wang, H., Hua, J., Jiang, Y., Yang, Y., & Yuan, H. (2020). Influence of fixation methods on the chestnut-like aroma of green tea and dynamics of key aroma substances. *Food Research International, 136*(5), Article 109479.

Wang, Y., Cui, H., Zhang, Q., Hayat, K., Yu, J., Hussain, S., Usman Tahir, M., Zhang, X., & Ho, C.-T. (2021). Proline-glucose Amadori compounds: Aqueous preparation, characterization and saltiness enhancement. *Food Research International, 144*, 110319. https://doi.org/10.1016/j.foodres.2021.110319

Wang, J.-Q., Fu, Y.-Q., Granato, D., Yu, P., Yin, J.-F., Zeng, L., & Xu, Y.-Q. (2022). Study on the color effects of (-)-epigallocatechin-3-gallate under different pH and temperatures in a model beverage system. *Food Control, 139*, 109112. https://doi.org/10.1016/j.foodcont.2022.109112

Wong, D. W. S. (2018). *Mechanism and theory in food chemistry*.

Xu, Y. Q., Zou, C., Gao, Y., Chen, J. X., & Yin, J. F. (2017). Effect of the type of brewing water on the chemical composition, sensory quality and antioxidant capacity of Chinese teas. *Food Chemistry, 236*, 142.

Yang, G., Shi, J., Tan, L., Yang, Y., Guo, L., Wang, L., Zheng, X., Lin, Z., & Lv, H. (2025). A comprehensive review on the promising purple leaf tea. *Comprehensive Reviews in Food Science & Food Safety, 24*(2).

Yangen, F., Xiuxiu, Z., Hanyue, W., Yueyue, T., Qinzeng, X., & Lixia, Z. (2019). Effects of light intensity on metabolism of light-harvesting pigment and photosynthetic system in Camellia sinensis L. cultivar 'Huangjinya'. *Environmental and Experimental Botany, 166*, 103796. https://doi.org/10.1016/j.envexpbot.2019.06.009

Yu, X., Hu, S., He, C., Zhou, J., Qu, F., Ai, Z., Chen, Y., & Ni, D. (2019). Chlorophyll metabolism in postharvest tea (Camellia sinensis L.) leaves: Variations in color values, chlorophyll derivatives,

and gene expression levels under different withering treatments. *Journal of Agricultural and Food Chemistry, 67*(38), 10624–10636. https://doi.org/10.1021/acs.jafc.9b03477
Zhou, J., Wu, Y., Long, P., Ho, C.-T., Wang, Y., Kan, Z., Cao, L., Zhang, L., & Wan, X. (2019). LC-MS-based metabolomics reveals the chemical changes of polyphenols during high-temperature roasting of large-leaf yellow tea. *Journal of Agricultural and Food Chemistry, 67*(19), 5405–5412. https://doi.org/10.1021/acs.jafc.8b05062
Zhu, J., Wang, J., Yuan, H., Ouyang, W., Li, J., Hua, J., & Jiang, Y. (2022). Effects of fermentation temperature and time on the color attributes and tea pigments of Yunnan Congou black tea. *Foods, 11*(13), 1845. https://doi.org/10.3390/foods11131845

Chapter 3
The Effects of Processing on Tea Color

Yujie Wang

Abstract The combination of rich tea germplasm resources and exquisite processing technology contributes to the formation of the flavor of tea, a non-alcoholic beverage that is popular around the world. Color is an important attribute factor that constitutes the appearance quality of tea. Interestingly, tea products present different colors. Throughout history, names have been given to tea products based on their color and the processing techniques used, including green, black, dark, yellow, white, and cyan (oolong) teas. Tea processing technology plays a pivotal role in influencing color quality, especially in regulating the conversion of lipid-soluble and water-soluble pigments. Processing technologies have profound effects on tea color by altering enzymatic reactions, microbial activity, and thermal dynamics. This chapter briefly describes some of the known key factors regarding the effects of processing technologies on tea color.

Keywords Tea color · Lipid-soluble pigments · Water-soluble pigments · Processing technologies · Tea types

3.1 Introduction

China, renowned as the origin of tea cultivation, harbors vast germplasm resources of tea plants. Recent exploration of unique tea varieties—such as albino, yellow, and purple cultivars—has expanded the diversity of tea product coloration. Furthermore, processing methods profoundly influence tea color by altering enzymatic reactions, microbial activity, and thermal dynamics. These processes drive biochemical transformations in fresh leaves, generating color-specific compounds such as oxidized catechins (orange-yellow to reddish-brown), modified flavonoids and flavonols (yellow to pale-yellow hues), and degradation products of chlorophylls and carotenoids (Long et al., 2023). The color of tea is evaluated through three critical parameters: the appearance of dried leaves, the chromaticity of the infusion,

Y. Wang (✉)
College of Biosystems Engineering and Food Science, Zhejiang University, Hangzhou, China
e-mail: wangyjnet@163.com

X. Zhai (ed.), *Tea Chemistry and Health Benefits*,
https://doi.org/10.1007/978-3-032-11828-8_3

and the visual characteristics of infused leaves—all of which significantly influence perceived quality. Consequently, elucidating the relationship between processing methodologies and color development is imperative for optimizing tea production and ensuring superior product quality.

3.2 Pigments in Fresh Leaves

Freshly harvested tea leaves are processed into distinct categories, such as green, white, yellow, oolong, black, dark, and oolong tea (Fig. 3.1), which are distinguished by unique chromatic attributes. The pigments responsible for these variations are broadly categorized as lipid-soluble pigment and water-soluble pigment (e.g., anthocyanins, anthocyanidins, and catechin oxidation products).

(a) *Lipid-Soluble Pigments* Lipid-soluble pigments *include* chlorophyll a, chlorophyll b, lutein, *β*-carotene, etc.
(b) *Water-Soluble Pigments* water-soluble pigments include anthocyanins, anthocyanidins, and catechin oxidation products.

Chlorophyll a (dark green) and chlorophyll b (yellow-green) are inherently unstable under thermal, oxidative, and photolytic conditions, undergoing degradation to form pheophytins (brown pigments) (Anguelova & Warthesen, 2000).

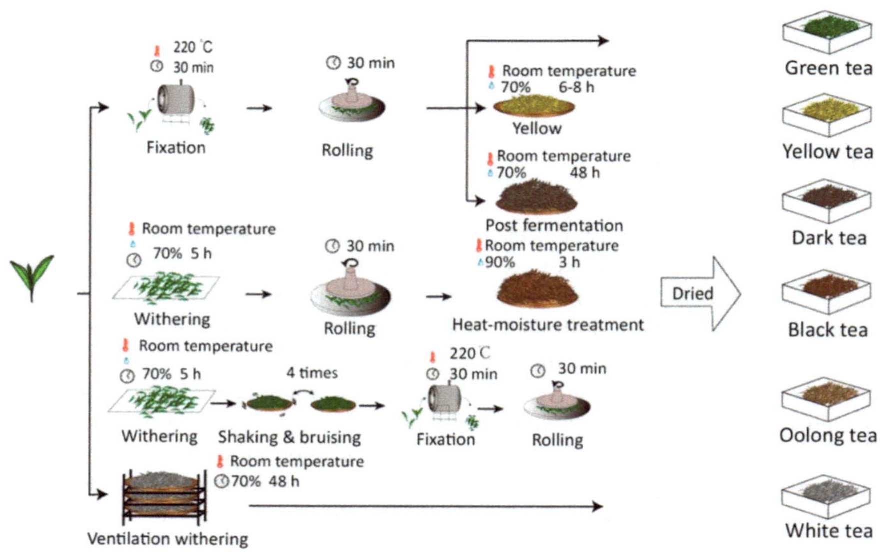

Fig. 3.1 Flow diagram depicting the manufacture processes to produce the six tea types (Wang et al., 2019)

3.3 Pigments Dynamics in Green Tea Production

In green tea production, rapid enzyme inactivation via microwave fixation or combined steam-hot air treatments is critical for preserving chlorophyll integrity, thereby maintaining the vibrant green coloration characteristic of high-quality green tea (Zhu et al., 2009). However, fixation duration must be meticulously controlled: insufficient heating diminishes leaf brightness, while excessive heating reduces green pigmentation (Zheng et al., 2020). Notably, fragmented chlorophyll particles released during the infusion of crushed green tea leaves contribute to increased turbidity and intensify greenish tones in the liquor (Wang et al., 2004).

3.4 Pigments Dynamics in Yellow Tea Production

Yellow tea undergoes a specialized "yellowing" process, wherein controlled heat and humidity accelerate chlorophyll degradation, transforming leaf color from green to yellow. Studies indicate that a two-step yellowing protocol is more effective than a single step in achieving optimal color transition (Wei et al., 2020; Xu et al., 2018). The moisture content during the second yellowing phase plays a pivotal role: excessively high moisture levels promote polyphenol oxidation and chlorophyll degradation, resulting in dull brown dry leaves and turbid, reddish infusions, whereas insufficient moisture impedes the desired color transformation. Additionally, high-temperature roasting—a hallmark of large-leaf yellow tea processing—reduces chlorophyll a, b, and total chlorophyll content by 74.6%, 87.5%, and 79.9%, respectively, yielding the distinctive "bronze" hue revered by local tea producers.

3.5 Enzymatic Browning

It is generally believed that water-soluble pigments are the main substances responsible for the color of tea infusion, including anthocyanins, anthocyanidins, and catechin oxidation products (Fu, 1997). Most changes of these colored substances are associated with enzymatic browning and non-enzymatic browning occurring during tea processing. Enzymatic reactions induced by physical damage in some fruits and vegetables lead to changes in color, texture, flavor, and nutrient content, which can be detrimental to the preservation of vegetable and fruit quality (Debelo et al., 2020). However, in tea, the enzymatic browning reaction is the basis for giving tea its unique color and flavor. During the withering process of tea, the synthase activities of catechins and flavonols are elevated, favoring the synthesis of these substances in a relatively short period of time. However, the increasing transcript levels of polyphenol oxidase (PPO), peroxidase (POD), and glutathione peroxidase (GPX) with increasing withering time promotes the oxidation of catechins to theaflavins

(TFs) and thearubigins (TRs) (Zhou et al, 2022), oxidized pigments that are believed to be positively correlated with the infusion color of black tea.

Different types of tea utilize the above conversion rules differently, with fermented teas (including black and dark teas) requiring more catechin oxidation products, while unfermented teas (green teas) inhibit these reactions for a green appearance. It has been shown that low-temperature withering and CO_2 intervention can inhibit the activity of PPO and the expression of chlorophyllase (CLH1), thus slowing down the oxidation of polyphenols and the degradation of chlorophyll, thus maintaining the green color of tea (Yu et al., 2020). The use of red-light withering facilitates the increase of PPO and POD activities and promotes the conversion of catechins into TFs, thus enhancing the color of black tea (Li et al., 2021). Both shaking and rolling can cause mechanical damage to tea leaves, but most of the cells of the shaken leaves remain intact, and only the edges of the leaves are subjected to mechanical force, resulting in a higher content of TFs and TRs at the edges than in the center of the leaves, which creates the unique characteristic "green leaves with red edges" of oolong tea.

The enzymatic browning reaction of black tea has been accompanied by the process before drying, and the activities of PPO and POD naturally increase during withering, peak at the rolling stage, and gradually decrease during subsequent fermentation (Hu et al., 2024), which ultimately results in the color appearance of black tea. In addition, re-rolling step allows the spatial position of enzymes and substrates in tea cells to be redistributed, allowing enzymes such as PPO and POD to promote the levels of TFs and TRs (Chen et al., 2024).

3.6 Microbe-Induced Transformation

Microbial community plays an important role for the formation of tea color. The yellowing process provides a favorable environment for the growth of microorganisms. These microorganisms secrete a series of extracellular enzymes, among which PPO secreted by *Aspergillus niger* contributes to polyphenol conversion, leading to the formation of TFs, which contributes to the enhancement of the yellowness of tea infusion (Xu et al., 2018). The pile-fermentation of dark tea is an ideal environment for microorganisms to multiply. The enzymes of microorganisms promote the decomposition of epigallocatechin gallate (EGCG) and epicatechin gallate (ECG) into epicatechin (EC) and epigallocatechin (EGC), which are converted into polymers such as TFs, TRs, and theabrownins (TBs), enhancing the brightness and redness of tea infusion (Hu et al., 2024).

3.7 Maillard Products in Thermal Processing

Tea is rich in amino acids and reducing sugars, which facilitates the occurrence of Maillard reactions during thermal processing. The previous study identified the Amadori rearrangement products (ARPs) of the Maillard reactions between amino acids and glucose in tea by liquid chromatography-mass spectrometry (LC-MS). It found that the formation of both derivatives of ARPs was most favored in a thermally reactive system at 100 °C (Han et al., 2022). In addition to this, it was found that pyroglutamic acid-glucose Amadori products were responsible for the main color of black tea (Long et al., 2024). In the drying process of tea leaves, ARP 1 peaked at 110–120 °C in the oven temperature and ARP 2 increased significantly with increasing temperature at 90–140 °C, and the increase of these ARPs was favorable to improve the yellowness and redness of tea broth (Su et al., 2024).

3.8 Conclusion

Tea coloration is a multifactorial phenomenon shaped by enzymatic oxidation, thermal degradation, microbial activity, and Maillard reactions. Strategic regulation of processing parameters—including fixation duration, yellowing conditions, and drying temperatures—enables precise modulation of chromatic characteristics. These insights establish a scientific foundation for optimizing tea quality, ensuring alignment with consumer preferences for specific visual and sensory attributes.

References

Anguelova, T., & Warthesen, J. (2000). Degradation of lycopene, α-carotene, and β-carotene during lipid peroxidation. *Journal of Food Science, 65*(1), 71–75. https://doi.org/10.1111/j.1365-2621.2000.tb15958.x

Chen, Q., Fu, Y., Heng, W., Yu, S., Xie, F., Dong, F., Lin, Z., Dai, W., & Fu, H. (2024). Re-rolling treatment in the fermentation process improves the taste and liquor color qualities of black tea. *Food Chemistry: X, 21*, Article 101143. https://doi.org/10.1016/j.fochx.2024.101143

Debelo, H., Li, M., & Ferruzzi, M. G. (2020). Processing influences on food polyphenol profiles and biological activity. *Current Opinion in Food Science, 32*, 90–102. https://doi.org/10.1016/j.cofs.2020.03.001

Fu, D. (1997). Water-soluble pigments in tea and their research progress. *Journal of Tea Communication, 01*, 5–10. (In Chinese).

Han, Z., Jiang, Z., Zhang, H., Qin, C., Rong, X., Lai, G., Wen, M., Zhang, L., Wan, X., & Ho, C.-T. (2022). Amadori reaction products of theanine and glucose: Formation, structure, and analysis in tea. *Journal of Agricultural and Food Chemistry, 70*(37), 11727–11737. https://doi.org/10.1021/acs.jafc.2c04560

Hu, J., Feng, X., Song, H., Hao, Z., Ma, S., Hu, H., Yang, Y., Zhou, S., Pan, Y., Fan, F., Gong, S., Chen, P., & Chu, Q. (2024). Enzymatic reactions throughout cultivation, processing, storage and post-processing: Progressive sculpture of tea quality. *Trends in Food Science & Technology, 143*, Article 104294. https://doi.org/10.1016/j.tifs.2023.104294

Li, Y., He, C., Yu, X., Zhou, J., Ran, W., Chen, Y., & Ni, D. (2021). Effects of red-light withering on the taste of black tea as revealed by non-targeted metabolomics and transcriptomics analysis. *LWT, 147*, Article 111620. https://doi.org/10.1016/j.lwt.2021.111620

Long, P., Su, S., & Zhang, L. (2023). Research progress on color components of tea. *Journal of Tea Science, 43*(05), 593–606. (In Chinese).

Long, P., Su, S., Wen, M., Liu, X., Han, Z., Ke, J.-P., Zhou, Y., Zhu, M., Cheng, Y., Shao, Y., Wan, X., & Zhang, L. (2024). An insight into trichomes-deficiency and trichomes-rich black teas by comparative metabolomics: The impact of oxidized trichomes on metabolic profiles and infusion color. *Food Research International, 190*, Article 114638. https://doi.org/10.1016/j.foodres.2024.114638

Su, S., Long, P., Zhang, Q., Wen, M., Han, Z., Zhou, F., Ke, J., Wan, X., Ho, C.-T., & Zhang, L. (2024). Chemical, sensory and biological variations of black tea under different drying temperatures. *Food Chemistry, 446*, Article 138827. https://doi.org/10.1016/j.foodchem.2024.138827

Wang, L. F., Park, S. C., Chung, J. O., Baik, J. H., & Park, S. K. (2004). The compounds contributing to the greenness of green tea. *Journal of Food Science, 69*(8), S301–S305. https://doi.org/10.1111/j.1365-2621.2004.tb09894.x

Wang, Y., Kan, Z., Thompson, H. J., Ling, T., Ho, C.-T., Li, D., & Wan, X. (2019). Impact of six typical processing methods on the chemical composition of tea leaves using a single Camellia sinensis cultivar, Longjing 43. *Journal of Agricultural and Food Chemistry, 67*(19), 5423–5436. https://doi.org/10.1021/acs.jafc.8b05140

Wei, Y., Fang, S., Jin, G., Ni, T., Hou, Z., Li, T., Deng, W.-W., & Ning, J. (2020). Effects of two yellowing process on colour, taste and nonvolatile compounds of bud yellow tea. *International Journal of Food Science & Technology, 55*(8), 2931–2941. https://doi.org/10.1111/ijfs.14554

Xu, J., Wang, M., Zhao, J., Wang, Y.-H., Tang, Q., & Khan, I. A. (2018). Yellow tea (Camellia sinensis L.), a promising Chinese tea: Processing, chemical constituents and health benefits. *Food Research International, 107*, 567–577. https://doi.org/10.1016/j.foodres.2018.01.063

Yu, X., Li, Y., He, C., Zhou, J., Chen, Y., Yu, Z., Wang, P., Ni, D. (2020). Nonvolatile metabolism in postharvest tea (Camellia sinensis L.) leaves: Effects of different withering treatments on nonvolatile metabolites, gene expression levels, and enzyme activity. *Food Chemistry, 327*, 126992. https://doi.org/10.1016/j.foodchem.2020.126992

Zheng, K., Du, X., Guo, Q., Tan, X., Ni, S., & Sun, J. (2020). Study on the effects of fixation time on the quality of matcha. *Chinese Wild Plant Resources, 39*(05), 14–18. (In Chinese).

Zhou, C.-z, Zhu, C., Li, X.-z, Chen, L., Xie, S.-y, Chen, G.-w, Zhang, H., Lai, Z.-x, Lin, Y.-l, & Guo, Y.-q. (2022). Transcriptome and phytochemical analyses reveal the roles of characteristic metabolites in the taste formation of white tea during the withering process. *Journal of Integrative Agriculture, 21*(3), 862–877. https://doi.org/10.1016/s2095-3119(21)63785-1

Zhu, D., Yue, P., & Di, Y. (2009). The influence of different pan-frying methods on the quality of green tea. *Transactions of the Chinese Society of Agricultural Engineering, 25*(8), 275–279. (In Chinese).

Chapter 4
Aroma Characteristics in Six Types of Tea

Dongzhu Huang and Xiaoting Zhai

Abstract This chapter summarizes the classic aroma characteristics of six types of tea (green tea, black tea, oolong tea, white tea, yellow tea, dark tea) and their key volatile contributors, and constructs a map correlating tea aroma perception with chemical composition. The formation and accumulation pathways of these odor-active compounds are also outlined, with influences of factors such as cultivars, processing parameters, and consumption practices on the formation and release of key odorants introduced. The development and application of technologies including metabolomics, sensomics, and molecular docking provide technical support for deepening the understanding of tea flavor and the targeted design of specific aroma characteristics.

Keywords *Camellia sinensis* · Tea types · Tea aroma · Aroma characteristics · Sensory analysis

This chapter summarizes the classic aroma characteristics of six types of tea (green tea, black tea, oolong tea, white tea, yellow tea, dark tea) and their key volatile contributors, and constructs a map correlating tea aroma perception with chemical composition in Table 4.1.

D. Huang · X. Zhai (✉)
State Key Laboratory of Tea Plant Germplasm Innovation and Resource Utilization, School of Tea Science, Anhui Agricultural University, Hefei, China
e-mail: xiaotingzhai@ahau.edu.cn

International Joint Laboratory on Tea Chemistry and Health Effects of Ministry of Education, Anhui Agricultural University, Hefei, China

X. Zhai (ed.), *Tea Chemistry and Health Benefits*,
https://doi.org/10.1007/978-3-032-11828-8_4

Table 4.1 Odor qualities and odor thresholds (OTs) in water of odorants in six types of tea

Group	Odorant	Odor quality[a]	OT (μg/L)[b]	Green tea	Black tea	Oolong tea	White tea	Yellow tea	Dark tea	References
Aldehydes	Acetaldehyde	Fatty, green	16			+				1
	Benzaldehyde	Bitter almond-like, marzipan-like	150	+		+		+	+	1, 2, 4, 5
	Butanal	Malty, sweaty	0.87			+				1
	Citral	Fruity	5		+					3
	β-Cyclocitral	Citrus-like	3						+	6
	(*E*,*E*)-2,4-Decadienal	Fatty, deep-fried	0.027	+	+					7, 21
	Cis-4,5-epoxy-(*E*)-2-decenal	Metallic	n.m.[c]							7, 8
	Trans-4,5-epoxy-(*E*)-2-decenal	Metallic	0.22	+	+					7, 8
	(*E*,*E*)-2,4-Heptadienal	Fatty, flowery	0.032	+		+			+	6, 9, 26
	Heptanal	Citrus-like, fatty	6.1	+		+				1, 13
	Hexanal	Green, grassy	2.4	+	+	+	+	+	+	1, 4, 6, 7, 10, 11
	(*E*)-2-Heptenal	Green, fatty	18		+	+				1, 12
	(*Z*)-4-Heptenal	Fishy, fish oil-like	0.0087	+		+			+	6, 7, 9
	(*E*,*E*)-2,4-Hexadienal	Green	94.8		+	+				1, 12
	(*E*)-2-Hexenal	Green apple-like, bitter almond-like	110		+	+				1, 11, 12
	2-Methyl butanal	Malty	1.5	+	+	+	+	+	+	4, 6, 7, 9, 10, 11

(continued)

Table 4.1 (continued)

Group	Odorant	Odor quality[a]	OT (μg/L)[b]	Green tea	Black tea	Oolong tea	White tea	Yellow tea	Dark tea	References
	3-Methyl butanal	Malty	0.5	+	+	+	+	+	+	4, 6, 7, 9, 10, 11
	(*E*)-2-Methyl butenal	Nutty, green	458.9							6
	2-Methyl propanal	Malty	0.49	+	+	+		+		4, 7, 8, 9
	(*E*,*E*)-2,4-Nonadienal	Fatty	0.06	+	+					7, 12
	(*E*,*Z*)-2,6-Nonadienal	Cucumber-like	0.0045	+	+	+			+	6, 12, 14, 26
	Nonanal	Citrus-like, soapy	2.8	+	+		+			9, 10, 11, 13
	(*E*,*E*,*Z*)-2,4,6-Nonatrienal	Oat flake-like	0.067	+	+					7, 8
	(*E*)-2-Nonenal	Fatty, green	0.4	+					+	6, 7
	Octanal	Citrus-like, soapy	3.4	+				+		1, 14, 24
	(*E*)-2-Octenal	Fatty, nutty	1.7		+	+				3, 13, 9
	Pentanal	Almond-like, malty	12			+				9, 10, 14
	(*E*)-2-Pentenal	Grassy, green	310			+				1
	Phenylacetaldehyde	Honey-like, beeswax-like	5.2	+	+	+	+	+	+	6, 7, 9, 10, 11, 24
	Safranal	Saffron-like	n.m.[c]							6, 11, 14
	Vanillin	Vanilla-like, sweet	53						+	4, 10, 26

(continued)

Table 4.1 (continued)

Group	Odorant	Odor quality[a]	OT (μg/L)[b]	Green tea	Black tea	Oolong tea	White tea	Yellow tea	Dark tea	References
Alcohols	Benzyl alcohol	Bitter almond-like, fruity	620							9, 10, 25, 26
	α-Cadinol	Woody, herb-like, flowery	n.m.[c]							14
	Cedrol	Woody	0.5							3, 26
	Geraniol	Rose-like, citrus-like	3.2	+	+	+	+	+	+	4, 9, 10, 11, 25, 26
	2-Heptanol	Coconut-like, dill-like	15			+				1
	1-Hexanol	Grassy, marzipan-like	590			+				1
	(*Z*)-3-Hexen-1-ol	Lettuce-like	3.9	+		+		+		9, 24, 26
	(*E*)-2-Hexen-1-ol	Green, leaf, walnut-like	3.9							9
	Hotrienol	Flowery	110		+	+				1, 20
	Linalool	Citrus-like, flowery	0.6	+	+	+	+	+	+	6, 9, 10, 11, 24, 26
	Nerol	Lemon-like, flowery	300							6, 12
	Nerolidol	Woody, flowery	10	+	+	+				1, 12, 14
	1-Octen-3-ol	Mushroom-like	45					+	+	24, 25

(continued)

Table 4.1 (continued)

Group	Odorant	Odor quality[a]	OT (μg/L)[b]	Green tea	Black tea	Oolong tea	White tea	Yellow tea	Dark tea	References
	1-Pentanol	Balsamic vinegar-like	5000			+				1
	1-Penten-3-ol	Butter-like, pungent	400	+	+	+				1, 12, 26
	(*Z*)-2-Penten-1-ol	Mushroom-like, green, grassy	720							23
	Phenethyl alcohol	Flowery, honey-like	140	+	+	+	+	+	+	1, 4, 10, 11, 25, 26
	α-Terpineol	Flowery	4.08			+			+	1, 6
Ketones	2,3-Butanedione	Butter-like	0.96	+		+				1, 7
	β-Damascenone	Baked apple-like, grape juice-like	0.006		+	+		+	+	1, 8, 24, 25
	2-Heptanone	Stale, cabbage-like	3000							9
	1-Hexen-3-one	Rubber-like, pungent	0.00069	+						7
	α-Ionone	Flowery, violet-like	0.4		+		+		+	6, 11, 12
	β-Ionone	Flowery, violet-like	0.021	+	+	+	+	+	+	9, 10, 11, 24, 25, 26
	(*Z*)-Jasmone	Flowery	0.26	+	+					10, 26
	Maltol	Caramel-like, fruity	5000	+						17

(continued)

Table 4.1 (continued)

Group	Odorant	Odor quality[a]	OT (µg/L)[b]	Green tea	Black tea	Oolong tea	White tea	Yellow tea	Dark tea	References
	4-Methyl-3-penten-2-one	Plastic, earthy	200						+	6
	6-Methyl-5-hepten-2-one	Nutty, fruity	0.16	+		+				9, 26
	2-Nonanone	Green	5		+					20
	(*E*,*E*)-3,5-Octadien-2-one	Fatty, musty	0.5	+		+				9, 26
	(*Z*)-1,5-Octadien-3-one	Geranium-like, metallic	0.00034	+						7
	1-Octen-3-one	Mushroom-like	0.016	+	+	+			+	9, 12, 25, 26
	4-Oxoisophorone	Woody, fruity, musty	25						+	6
	2,3-Pentanedione	Butter-like	3.9			+				1
	1-Penten-3-one	Pungent, train oil-like	0.94	+						7
	Geranyl acetone	Flowery	n.m.[c]						+	6
Esters	Butyrolactone	Caramel, sweet	n.m.[c]							19
	Coumarin	Woodruff-like	11	+	+			+		4, 10, 26
	Dihydroactinidiolide	Musty, pungent	500	+				+	+	6, 13, 24
	Ethyl 2-hydroxybutanoate	Fruity	0.15		+					3
	γ-Hexalactone	Coumarin, sweet	n.m.[c]							19
	(*Z*)-3-Hexenylbenzoate	Green, sweet, fruit	n.m.[c]							19
	Methyl anthranilate	Grape-like	n.m.[c]							26

(continued)

Table 4.1 (continued)

Group	Odorant	Odor quality[a]	OT (μg/L)[b]	Green tea	Black tea	Oolong tea	White tea	Yellow tea	Dark tea	References
	Methyl hexadecanoate	Roast, sweet	n.m.[c]							19
	(*E*)-Methyl jasmonate	Jasmine-like, flowery	70		+	+				1, 19
	(*Z*)-Methyl jasmonate	Jasmine-like, flowery	70		+	+				1, 19
	Methyl salicylate	Mint-like	40	+	+	+	+	+		1, 4, 10, 11, 17
	γ-Nonalactone	Coconut-like	9.7	+						7
	4-Nonanolide	Sweet	n.m.[c]							21
	γ-Octalactone	Coconut-like	6.5	+						7
	4-Octanolide	Sweet	n.m.[c]							21
	Phenethyl acetate	Rose, honey, tobacco	n.m.[c]							19
Pyrazines	2-Acetylpyrazine	Popcorn-like, roasty	2.2					+		4
	2-Sec-butyl-3-methoxypyrazine	Earthy, bell pepper-like, roasty	n.m.[c]							7
	2,5-Diethylpyrazine	Roasty, nutty	7	+						15
	2,3-Diethyl-5-methylpyrazine	Earthy, musty, roasty	0.031					+		4
	2,5-Dimethylpyrazine	Earthy, roasty, nutty	1700					+		4

(continued)

Table 4.1 (continued)

Group	Odorant	Odor quality[a]	OT (μg/L)[b]	Green tea	Black tea	Oolong tea	White tea	Yellow tea	Dark tea	References
	2,6-Dimethylpyrazine	Roasty, nutty	1700					+		4
	Ethylpyrazine	Roasty, potato-like	n.m.[c]							15
	2-Ethyl-3,5-dimethylpyrazine	Earthy, roasty	n.m.[c]							4, 23
	2-Ethyl-3,6-dimethylpyrazine	Earthy, roasty	0.28			+				4
	2-Ethyl-5-methylpyrazine	Roasty	57	+				+		4, 15
	2-Ethyl-6-methylpyrazine	Roasty	6.9	+				+		4, 15
	2-Isobutyl-3-methoxypyrazine	Bell pepper-like, earthy	0.0062	+						7
	2-Isopropyl-3-methoxypyrazine	Earthy, pea-like	0.0039	+						7
	Methylpyrazine	Burnt, roasty	110		+					20
	2,3,5-Trimethylpyrazine	Earthy	11	+	+					15, 20
Sulfur-containing compounds	Bis(2-methyl-3-furyl) disulfide	Sulfury	0.000022		+					22
	Diethyl disulfide	Onion-like	30	+						16
	Dimethyl disulfide	Cabbage-like, sulfury	1.7	+						16
	Dimethyl sulfide	Asparagus-like, putrid, cooked corn-like	0.3	+		+				1, 16
	Dimethyl sulfoxide	Garlic-like	n.m.[c]							16
	Dimethyl trisulfide	Cabbage-like, sulfury	0.0099	+		+				1, 16

(continued)

Table 4.1 (continued)

Group	Odorant	Odor quality[a]	OT (μg/L)[b]	Green tea	Black tea	Oolong tea	White tea	Yellow tea	Dark tea	References
	Ethanethiol	Rotten, onion-like	n.m.[c]							1
	Furfurylthiol	Coffee-like, roasty	n.m.[c]							1
	2-Furanmethanethiol	Sulfury, burnt	0.036	+						7
	3-Mercaptohexanol	Passionfruit, grapefruit	n.m.[c]							1
	4-Mercapto-4-methylpentan-2-one	Black currant-like, sulfury	0.0016	+						7
	Methanethiol	Sulfury, cabbage-like, gasoline-like, garlic-like	4	+						1, 16
	Methional	Cooked potato-like	0.43	+						1, 16
Phenols	Phenol	Phenolic	5000							4, 6, 10, 26
	2,6-Dimethoxyphenol	Smoky, sweet, clove-like	29						+	6
	3,4-Dimethoxyphenol	Smoky	5.57						+	6
	4-Ethyl-2-methoxyphenol	Smoky	16						+	6
	Eugenol	Clove-like	1.8		+					3
	Trans-isoeugenol	Clove-like	0.71						+	6
	2-Ethylphenol	Phenolic, smoky	4.8						+	6

(continued)

Table 4.1 (continued)

Group	Odorant	Odor quality[a]	OT (μg/L)[b]	Green tea	Black tea	Oolong tea	White tea	Yellow tea	Dark tea	References
	3-Ethylphenol	Phenolic, leather-like	0.85	+						7
	4-Ethylphenol	Phenolic	13	+						7
	2-Methoxyphenol	Smoky, gammon-like	0.84						+	6
	3-Methylphenol	Smoky, phenolic	15						+	6
	4-Methylphenol	Phenolic, horse stable-like	3.9						+	6
	2-Methoxy-4-vinylphenol	Smoky, clove-like	21							6
	4-Methyl-2-methoxyphenol	Vanilla-like, clove-like, smoky	21							6
Acids	Acetic acid	Vinegar-like	5600							6
	Butanoic acid	Sweaty	2400							6
	Hexanoic acid	Sweaty	4800							6, 26
	Octanoic acid	Carrot-like, musty	190			+				6
	Nonanoic acid	Moldy, pungent	26			+				6
	Decanoic acid	Soapy, musty	3.5							6
	2-Methylbutanoic acid	Sweaty	3100							7
	3-Methylbutanoic acid	Sweaty, fruity	490							7

(continued)

Table 4.1 (continued)

Group	Odorant	Odor quality[a]	OT (μg/L)[b]	Green tea	Black tea	Oolong tea	White tea	Yellow tea	Dark tea	References
	Pentanoic acid	Sweaty, fruity	11,000							7
	Phenylacetic acid	Beeswax-like, honey-like	68	+						7
	Propanoic acid	Rancid, cheese-like, fruity	16000							6
Furans and furanones	2-Acetylfuran	Balsamic vinegar-like	n.m.[c]							17
	2-Ethylfuran	Caramel-like, burnt	17			+				1
	Furfural	Bread-like, almond-like, sweet	282			+		+		4, 9
	2-Methylfuran	Caramel-like	11		+					19
	5-Methylfurfural	Caramel-like	n.m.[c]		+					19
	2-Pentylfuran	Fruity	4.8			+				9
	3-Hydroxy-4,5-dimethyl-2(*H*)-furanone	Seasoning-like, spicy	1.7	+						7
	4-Hydroxy-2,5-dimethyl-3(2*H*)-furanone	Caramel-like	87	+						7
Benzenes	1,2-Dimethoxybenzene	Stale	3.71						+	6
	1,2-Dimethoxy-4-methylbenzene	Stale, musty	2.71						+	6
	1,2,3-Trimethoxybenzene	Stale, musty	0.75						+	26
	1,2,3-Trimethoxy-5-methylbenzene	Stale, musty	4.45						+	6
	1,2,4-Trimethoxybenzene	Stale, musty	3.06						+	6

(continued)

Table 4.1 (continued)

Group	Odorant	Odor quality[a]	OT (μg/L)[b]	Green tea	Black tea	Oolong tea	White tea	Yellow tea	Dark tea	References
	1,2,3,4-Tetramethoxybenzene	Stale, musty	0.64						+	26
	4-Ethyl-1,2-dimethoxybenzene	Stale, musty	3.4						+	26
	Trans-methyl isoeugenol	Stale, woody	1600						+	6
Oxides	Linalool oxide I	Woody, flowery	60		+	+				1, 20
	Linalool oxide II	Woody, flowery	100		+	+			+	1, 20, 25
	Linalool oxide III	Woody, flowery	6		+	+				1, 20
	Linalool oxide IV	Woody, flowery	100		+	+				1, 20
	Cis-pyranoid linalool oxide	Woody, tea-like	5000						+	6
	Trans-pyranoid linalool oxide	Woody, tea-like	3000						+	6
Thiazoles, thiazoline, and thiophenone	2-Acetylthiazole	Roasty, popcorn-like	20	+						16
	2-Acetyl-2-thiazoline	Roasty, popcorn-like	0.079		+					3
	Benzothiazole	Gasoline, rubber	n.m.[c]							16
	2-Methyl-thiazole	Green vegetable-like	0.06	+						18
	4,5-Dihydro-3(2*H*)-thiophenone	Roasty	n.m.[c]							21
Pyrrole and pyrroline	2-Acetylpyrrole	Roasty, rice-like	n.m.[c]							4, 9
	2-Acetyl-1-pyrroline	Roasty, popcorn-like	0.053			+				9
Indoles	Indole	Fecal-like, mothball-like	11	+	+	+			+	6, 9, 10, 26
	3-Methyl-1*H*-indole	Fecal-like, mothball-like	0.13							7

(continued)

Table 4.1 (continued)

Group	Odorant	Odor quality[a]	OT (μg/L)[b]	Green tea	Black tea	Oolong tea	White tea	Yellow tea	Dark tea	References
Terpenes	Limonene	Citrus-like, carrot-like	13		+	+				9, 20
	Myrcene	Geranium-like, carrot-like, hop-like	1.2		+	+				9, 20

[a] Odor qualities were mentioned in the respective references

[b] Odor thresholds were obtained from following cited references

[c] n.m.: not mentioned in any of following cited references.+: the odorant was presented with an odor activity value (OAV) ≥1 in the tea leaves or infusion, indicating the compounds as key odorant of tea aroma

References: 1. Zhu et al. (2015), 2. Wang et al. (2020a, 2020b), 3. Chen et al. (2020a, 2020b), 4. Zhai et al. (2023), 5. Chen et al. (2025a, 2025b), 6. Pang et al. (2019), 7. Flaig et al. (2020a, 2020b), 8. Schuh and Schieberle (2006), 9. Yang et al. (2022a, 2022b, 2022c, 2022d), 10. Wang et al. (2024a, 2024b, 2024c, 2024d, 2024e). 11. Wu et al. (2022a, 2022b), 12. Chen et al. (2019), 13. Zhu et al. (2018), 14. Gan et al. (2024), 15. Zhu et al. (2021a, 2021b), 16. Yao et al. (2025), 17. Gong et al. (2017), 18. Mizukami et al. (2008), 19. Xiao et al. (2017), 20. Kang et al. (2019), 21. Kumazawa et al. (2006), 22. Wang et al. (2022a, 2022b, 2022c, 2022d), 23. Guo et al. (2022), 24. Wei et al. (2024), 25. Li et al. (2024b); 26. Yu et al. (2023)

4.1 Green Tea

4.1.1 Tender Aroma

Tender aroma is a delicate and unique odor impression of fresh green tea, making it a coveted premium aroma among consumers. The development of tender aroma is influenced by various factors, ranging from tea plant cultivars selection and harvesting criteria, to the plucking time, processing intricacies, and the ambient conditions during production. *Zhonghuang 1* (*Camellia sinensis* var. *sinensis* cv. *Zhonghuang 1*) is an exclusive albino tea cultivar in China, named for its yellow young shoots with lower content of chlorophyll. Green tea made from *Zhonghuang 1* is earning growing acclaim due to its vibrant freshness, naturally sweet flavor, and a distinctive cooked corn-like note. Dimethyl sulfide (DMS), an odorant exhibiting a typical sweet corn-like scent, has been identified as the principal contributor responsible for the tender aroma of high-grade green tea produced from the *Zhonghuang 1* cultivar (Liao et al., 2020). While DMS is commonly found in many teas, its stupendous presence in tender aroma green tea is attributed to the genetic traits of the specific tea plant variety, which result in the substantial synthesis of its precursor, *S*-methylmethionine (SMM). Furthermore, the high-temperature during tea thermal processing and brewing facilitates the degradation of SMM into DMS, and proper processing parameters can help shape the tender aroma of green tea (Zhai et al., 2022). Taking into account the threshold of volatile substances as a key factor characterizing their fundamental sensory properties, odorants at different concentrations exhibit varying aroma attributes. For DMS, it is recommended to keep its concentration below 165 μg/L in green tea infusion to maintain its popular sweet corn-like aroma. Exceeding 165 μg/L, the olfactory signature of DMS gradually transitions into undesirable notes akin to "cooked cabbage-like" and "asparagus-like". Furthermore, DMS of even higher content emits a pungent and rotten off-odor (Yao et al., 2025). *Anji* Bai Tea (AJBT) is a valuable and intriguing green tea in China. It is produced from a rare mutant tea cultivar *Baiye 1* (*Camellia sinensis* var. *sinensis* cv. *Baiye 1*). *Baiye 1* tea cultivar has earned increasing attention from both of industry and academic research due to its unique biological characteristics—the color of bud and leaf shifts from jade white to light green with the maturity of fresh leaf and the rise of environmental temperature, along with the disappear of its mild taste and tender aroma. But no worries, various processing techniques can reshape its aroma property. For example, roller-hot air fixation bestows a roasty characteristic on AJBT, whereas carding machine fixation retains the green odor, closely aligning with its natural quality (Xia et al., 2024).

Xinyang Maojian (XYMJ) green tea, ranked among the top ten famous teas in China, is renowned for its lasting and tender aroma. Its unique aroma is not bestowed by a particular tea cultivar, but rather achieved through rigorous criteria for raw material selection. Specifically, superior XYMJ is meticulously crafted from over 85% single bud fresh leaves, complemented by a select few one bud and one leaf combinations. The artful blend of key odorants such as nonanal, (*E*)-β-ionone, and linalool endows XYMJ with its tender and elegant aroma (Cui et al., 2022). In addition, the details of processing and the timing of raw materials collection greatly influence the contents of these odorants, thereby affecting the flavor of XYMJ. The thermal processing steps involving fixation and drying are pivotal in the unique flavor formation of XYMJ (Yin et al. 2022a, 2022b). Within an identical production framework, XYMJ manufactured from spring-harvested tea leaves performs superior aromatic qualities compared to that derived from autumn-harvested leaves (Yin et al., 2022a).

4.1.2 Faint Aroma

Faint aroma is one of the most common aroma types for green tea, which can be distinguished with a pure, fresh scent reminiscent of vibrant greenery. *Lushan Yunwu* (LSYW) green tea, originating from the prestigious *Lushan* Mountain in China, is highly cherished for its elegant and faint aroma. In LSYW, over fifty volatile compounds across eight chemical categories were identified, with *cis*-jasmone, linalool, and nonanal confirmed as key odor-active compounds that significantly contribute to the faint aroma. Moreover, linalool, along with certain specific terpene compounds and esters, can serve as metabolite markers to distinguish LSYW from different altitudes (Gao et al., 2024; Sun et al., 2024a, 2024b). As another representative of faint aroma green tea, Chinese *Jingshan* tea (JST) boasts a long history of processing and drinking, dating back to the Tang Dynasty over 1,000 years ago. However, twenty-seven key aroma compounds of JST including DMS, (*E*,*E*)-2,4-heptadienal, etc., were characterized very recently (Flaig et al., 2020a, 2020b). Besides, the research, meticulously tracked their alternation throughout the whole processing steps, and revealed the significant impact of thermal treatment on volatiles accumulation. The later research echoed the findings and highlighted that brief thermal processing is beneficial for shaping the faint aroma of green tea, whereas extended exposure to high temperature leads to irreversible heat damage to tea flavor (Yang et al., 2022a, 2022b, 2022c, 2022d).

Apart from the processing, storage conditions also have great effect on the faint aroma of green tea. Storing green tea at room temperature not only reduces its fresh and faint aroma but also promotes a stale flavor, thereby lowering the overall quality. Some volatile compounds, including hexanoic acid, *trans*-2-nonenal, 1-octen-3-ol, safranal, and (*E*,*E*)-3,5-octadien-2-one, have been implicated in the decline of flavor profile in different types of green tea. Notably, baking process can effectively accumulate the contents of 2-pentylfuran and limonene in green teas that suffer from poor aroma, subsequently ameliorating their stale flavor (Cao et al., 2024; Dai et al., 2020; Liu et al., 2023a, 2023b).

4.1.3 Seaweed-Like Aroma

Steamed green tea attracts Northeast Asians very much, especially Chinese and Japanese. Because steamed green tea evokes unique marine plant-like freshness and distinctive seaweed flavor, which is also called seaweed aroma. This aroma quality offers consumers a sensory journey that captures the essence of nature and evokes vivid memories of the ocean. Chinese *Enshi Yulu* (ESYL) green tea is famous for its distinctive steaming fixation and a milder de-enzyme environment. During the conventional industrial processing of ESYL, the critical stages for flavor development are steaming fixation and the final drying (Jiang et al., 2024). This unique fixation step not only differentiates it from other premium green teas but also ensures the preservation of natural integrity, while bestowing upon it an exceptional and characteristic seaweed aroma. During steaming fixation, non-enzymatic degradation of fatty acids is greatly enhanced, enriching key aroma compounds such as hexanal, nonanal, 1-octen-3-ol, geraniol, and linalool, which shape the unique flavor profile of ESYL. Recently, several non-origin tea cultivars have been utilized in the production of ESYL. The key aroma compounds of ESYL crafted from the cultivar *Longjing 43* was characterized as (*E*)-*β*-ionone and nonanal, which were starkly contrasts with the flavor styles derived from the cultivars of *Xiapu Chunbolv* and *Zhongcha 108* (Guo et al., 2023; Sui et al., 2025). The comparison of the aroma characteristics of steamed green tea and other Chinese green teas elucidated that the unique seaweed aroma imparted by steaming fixation has become a distinctive and classic signature of steamed green tea (Shi et al., 2022). Meanwhile, the retention of *cis*-3-hexenyl butanoate and *β*-cubebene provides the chemical foundation for the fresh and natural flavor.

Steamed green tea holds a major place in Japanese tea market, due to virtually all green teas undergoing the steaming fixation. Among them, *Sencha* is particularly prominent as the dominant subcategory, constituting over 80% of the total domestic tea output. A very early study has demonstrated the key odorants in *Sencha*, and they were indole and 4-mercapto-4-methyl-2-pentanone (Kumazawa & Masuda, 1999). Additionally, volatiles such as methional, *β*-damascone, and maltol were detected in green tea for the first time. Further, the dynamic changes of 4-mercapto-4-methyl-2-pentanone were monitored under different tea processing and crop timing in *Sencha* (Kumazawa et al., 2005). The results revealed that the best condition for the accumulation of 4-mercapto-4-methyl-2-pentanone is fresh leaves harvesting for the first time in spring and roasting at 112 °C. *Gyokuro* is another premium steamed green tea, which has a similar processing step with *Sencha*. While, the fresh leaves are preharvest shading treated for twenty days. This treatment boosts the contents of *α*-ionone, (*E*)-*β*-ionone, and indole, which promote its unique green and fresh flavor (Tan et al., 2019).

Matcha, an ancient beverage rooted in China and now cherished globally, is traditionally made from steamed green teas that are meticulously ground into a fine powder using a stone mill. Several odorants, including coumarin, *trans*-4,5-epoxy-(*E*)-2-decenal, *α*-ionone, and (*E*)-isoeugenol, were pivotal to *matcha*'s flavor (Baba et al., 2017). Furthermore, the variable concentrations of volatiles such as DMS, hexanal, and (*Z*)-4-heptenal in different grades of *matcha* account for the flavor nuances that distinguish high-grade from low-grade *matcha*. Very recently, a study compared and analyzed the differences in volatile and non-volatile compounds of *matcha* processed from four distinct tea cultivars. The effect of tea cultivars on the quality of *matcha* was emphasized, the advantage of the cultivar *Okumidori* in the total amount of product aroma was confirmed and the crucial roles of odorants like (*Z*)-4-heptenal, furaneol, hexanal, and 1-octen-3-ol were revealed (Huang et al., 2024a, 2024b, 2024c, 2024d).

4.1.4 Flowery Aroma

The flowery aroma is interpreted as a wonderful odor impression in green tea. It is the imagery of flowers such as orchids or jasmine. *Lu'an Guapian* (LAGP) green tea, steeped in a rich and illustrious history, hails from *Lu'an* City in Anhui Province, China. It is known for the exclusive materials picking criteria (single leaves without any buds and stems) and the distinctive "over-fired drying" step. (*E*)-*β*-ionone, geraniol, and (*E*,*E*)-2,4-heptadienal play major role in the flowery aroma of premium LAGP, and are crucial for discerning the superior quality of premium products from common LAGP (Yu et al., 2023). Besides, raspberry ketone, indole, and other substances significantly contribute to the overall flavor profile of LAGP (Xiao et al., 2022a, 2022b, 2022c). Additionally, with the industrialization and standardization of tea production, researches pay increasing attention on the optimal harvesting times and variety selection for LAGP's materials. Particularly, LAGP processed from fresh leaves collected before Grain Rain, one of the twenty-four solar terms in China, exhibits a strong flowery aroma (Wu et al., 2025a, 2025b). The tea varieties originating from the birthplace of LAGP are believed as the most suitable for crafting this tea, endowing it with exceptional qualities, including a rich flowery aroma (Feng et al., 2024a, 2024b, 2024c).

Taiping Houkui (TPHK), another representative green tea of Anhui Province, China, is famous for its noble orchid-like aroma and distinctive appearance. Its appearance is characterized by flat, smooth leaves, and an emerald hue. A seminal study juxtaposed the flavor profiles of premium TPHK with those of Huilan (*Cymbidium faberi* Rolfe), and determined that the characteristic orchid-like aroma of TPHK was owed to (*Z*)-methyl epijasmonate (epi-MeJA) (Feng et al., 2020). Furthermore, Zhou et al. (2022) found that indole, (*Z*)-jasmone, and methyl jasmonate were present in high concentrations in TPHK, potentially providing great contributions to its superior flavor. Lately, a new study compared the aroma differences between premium and common TPHK. The study highlighted (*Z*)-3-hexenyl hexanoate, DMS, and decanal. Despite not exhibiting typical floral scents, these odorants contributed positively to the overall flavor profile of TPHK, making the flowery aroma more complete and vivid (Jin et al., 2024). New green teas with flowery aroma were made by advancing tea cultivar selection and refining processing techniques. In these innovative teas, the key roles of volatiles such as linalool, geraniol, indole, α-farnesene, and phenylethyl alcohol are emphasized (Li et al., 2024a, 2024b, 2024c; Xie et al., 2023). The study not only corroborated from previous findings but also indicated a commonality in aromatic information among various flowery aroma green teas, underscoring the essential nature of certain key odorants.

4.1.5 Chestnut-Like Aroma

In the professional sensory evaluation, the chestnut-like aroma is described as a rich nutty and sweet note, accompanied by a slight baking odor, reminiscent of boiled or roasted chestnuts. As an important sensory indicator of high-quality green tea, chestnut-like aroma is composed of a series of volatile compounds in a specific proportion. By comparing with the authentic chestnut, seventeen key aroma compounds were identified in selected green teas with a typical chestnut-like aroma (Zhu et al., 2018). Among them, malty-smelling 3-methylbutanal and (*E*,*E*)-3,5-octadien-2-one remind people the flavor of chestnut. In contrast, phenylacetaldehyde and (*E*)-nerolidol exhibiting floral and citrus-like aroma also contribute to the chestnut-like aroma via aroma indirect and synergistic effect. For example, previous study indicated that the chestnut-like aroma is not the result of the direct action of a few odorants (Liu et al., 2023a, 2023b). Besides, geraniol, benzothiazole, 2-ethyl-3,6-dimethylpyrazine, coumarin, etc., were regarded as key odorants contributing to the chestnut-like aroma of green tea. The application of molecular docking technology further elucidated the interactions between these odorants and olfactory receptors, emphasizing the crucial role of sulfides and heterocyclic compounds in shaping the chestnut-like aroma (Zhu et al., 2024a, 2024b).

Thermal processing commonly leads to the loss of alcohols and aldehydesas as well as the accumulation of heterocyclic volatile compounds, thus transforming green tea's aroma from fresh to chestnut-like (Wang et al., 2024a, 2024b, 2024c, 2024d, 2024e). Meanwhile, during the thermal processing stage, i.e., fixation and drying, intense pyrolysis and Maillard reaction occur, resulting in the substantial synthesis of volatile compounds such as pyrroles, pyrazines, and sulfides, forming the chestnut-like aroma of green tea (Wang et al., 2020a, 2020b; Yang et al., 2022a, 2022b, 2022c, 2022d).

4.1.6 Tender Bean-Like Aroma

Over China's extensive history of tea production and consumption, tender bean-like aroma green tea is a distinctive category. The flavor is precisely described as the fragrance of freshly cooked or tender beans and is considered a highly desirable quality in green tea. As an outstanding representative of tender bean-like aroma green tea, Western Lake Longjing (WLLJ) is renowned for its exceptional quality and distinctive flavor, and can be considered a calling card of Chinese tea. Key aroma compounds in WLLJ tea infusion including DMS, (*E*,*E*)-2,4-heptadienal, methanethiol, and 1-hexen-3-one, were determined using sensomics approach (Flaig et al., 2020a, 2020b). Recently, the understanding of the aroma of WLLJ tea has deepened with the development of enantiomer analysis and innovative aroma collection methods, such as stir bar sorptive extraction (SBSE) and thin film-solid phase microextraction (TF-SPME). In addition to the well-documented roles of 2-methylbutanal, 3-methylbutanal, dimethyl sulfide, and (*E*,*E*)-2,4-heptadienal, volatile compounds with floral and fruity characteristics, such as geraniol and linalool, have also been confirmed as key odorants in WLLJ tea infusion. Furthermore, methyl salicylate, nonanal, and (*E*)-β-ionone have been identified as important contributors to the aroma of WLLJ, collectively enhancing its distinctive tender bean-like aroma (Zhu et al., 2021a, 2021b). These volatile compounds often have little intuitive connection with the bean-like flavor in the subconscious mind. However, the application of molecular docking and molecular dynamics simulation suggested that synergistic, additive, and masking effects occurs between these odorants. This provided initial insights into how a complex mixture system of volatile compounds leads to specific sensory experience, and offered a theoretical basis for understanding how the human sensory organs perceive the tender bean-like aroma of green tea (Xiao et al., 2024; Zhu et al., 2025a, 2025b). A latest investigation explored the formation mechanism of the tender bean-like aroma in Longjing green tea from the perspective of brewing techniques. It found that at higher steeping temperatures, the concentrations of key odorants such as 3,5-octadien-2-one and 2-ethyl-3,5-dimethylpyrazine significantly increased, effectively promoting the release and perception of the bean aroma (Zhang et al., 2024).

Laoshan green tea is produced in Shandong province of China. Despite the low-temperature environment and long growth cycle, it develops a premium and fresh bean-like flavor. The key aroma compounds in *Laoshan* green tea share significant similarities with those in Longjing, such as DMS, 2-methylbutanal, and 3-methylbutanal, while skatole is unique to the former (Zhu et al., 2021a, 2021b). Skatole is an interesting compound. It emits an unpleasant fecal odor at high concentrations but presents a pleasant floral or fruity aroma at low concentrations (Ma et al., 2021). In some industrialized foods involving fermentation processes, such as Baijiu (Dong et al., 2018) or white pepper (Steinhaus & Schieberle, 2005), high content of skatole was the primary source of off-odor. However, in tender bean-like aroma green tea, it commonly exists at low concentrations and enhances the delightful aroma.

4.2 Black Tea

4.2.1 *Flowery & Fruity Aroma*

In the early seventeenth century, the Dutch East India Company firstly introduced black tea to Europe, marking the initial encounter between the Western world and Chinese tea. Over the past few decades, international tea trade has expanded rapidly, and black tea has become one of the most important teas in this trade. Currently, over sixty countries and regions consume black tea, and its export volume accounts for nearly 99% of the global total tea export (Yong-Mei et al., 2022). According to the Food and Agriculture Organization of the United Nations (FAO, https://www.fao.org/home/en), the global consumption of black tea was approximately 6.4 million tons in 2021. Countries such as China, India, Kenya, and Sri Lanka are expected to play key role in the future development of black tea.

The flowery and fruity aroma is a common flavor characteristic of high-quality black tea. Among them, three most well-known high-aroma black teas are Keemun black tea, Darjeeling black tea, and Ceylon black tea. Keemun black tea, originated from *Keemun* County, Anhui Province, China, is distinguished by its unique "Keemun Aroma", characterized by flowery, fruity, and honey-like sensory attributes. Very recently, a total of 16 volatiles were demonstrated as key contributors for the "Keemun Aroma", among which geraniol, (*E*)-*β*-ionone, and linalool are the main sources of the flowery aroma (Wang et al., 2024a, 2024b, 2024c, 2024d, 2024e). In addition, (*Z*)-rose oxide was reported to play a significant role in the formation of the flowery and fruity attributes of Keemun black tea (Zheng et al., 2023). (*Z*)-rose oxide is a monoterpene compound and has a rose-like aroma with subtle hints of lychee. It works in synergy with other odor-active compounds to enhance the overall flowery profile and harmonious quality of black tea. However, green smelling (*Z*)-3-hexenyl isovalerate and fennel and ginger-like smelling hotrienol were negatively correlated with the "Keemun Aroma" (Su et al., 2022). Unlike the "Keemun Aroma", the flavor profile of Darjeeling black tea is described as "muscat", which is a sophisticated blend of flowery, fruity, and sweet aroma. In contrast, Ceylon black tea exhibits an intensely rich and pure flowery fragrance. The application of SBSE combining with GC-O/MS revealed that 2-methylbutanal, linalool, and *β*-cyclocitral were main contributors to the distinctive aroma of Darjeeling black tea. In Ceylon black tea, particularly the high-altitude Uva tea, salicylaldehyde present in high concentrations and played a crucial role in its aroma (Xiao et al., 2022a, 2022b, 2022c). The Assam tea cultivar (*Camellia sinensis* var. *assamica*), native to Assam, India, is widely used in the production of CTC black tea and is renowned for yielding teas with a rich malt and rose-like aroma. Kang et al. (2019) compared the flavor profile of Assam black tea with those of three other high-aroma black teas and identified (*E*)-2-octenal, which has a fatty and nutty odor, as the most definitive odorants in Assam black tea.

Some unique tea cultivars are naturally suited for crafting black tea with distinct fruity and flowery characteristics. *Camellia sinensis* var. *Yinghong No. 9* is a unique tea cultivar from *Yingde* City, Guangdong Province, China. Black tea processed from this tea cultivar, known as Yinghong No. 9 black tea (YJBT), exhibits typical flowery and fruity notes. Its typical aroma was from high contents of linalool, particularly (*R*)-linalool, geraniol, and methyl jasmonate (Liu et al., 2025). The optimal harvesting and manufacturing season for YJBT is autumn, which is different from other teas. In autumn, the mild climate facilitates the accumulation and transformation of flowery precursors such as carotenoids and glycosides, while the mature fresh leaves contribute to a richer overall flavor chemical profile (Liu et al., 2021). Due to high levels of δ-decalactone, jasmine lactone and (*Z*)-jasmone, the *Jinmudan* cultivar, a hybrid tea cultivar from *Tieguanyin* and *Huangdan*, produces black tea with rich flowery and honey peach-like aroma (Li et al., 2022a, 2022b, 2022c; Wu et al., 2023a, 2023b). This makes it a promising candidate for mainstream the production of black tea with flowery and fruity aroma.

4.2.2 Sweet Aroma

The sweet aroma is another readily detectable flavor characteristic of black tea, not a mere sugary sweetness but a composite experience of caramel, grains, and honey. Dianhong black tea, a representative sweet aroma black tea of China, is produced from the local Yunnan large-leaf tea cultivar in Yunnan Province. It is also one of the world's recognized birthplaces of tea. Differently from small- and medium-leaf tea cultivars, which have undergone long-term selection and domestication, Yunnan large-leaf tea cultivar preserves many traits of the ancestral tea species. These plants produce larger and more robust leaves, which are rich in organic compounds, offering Dianhong black tea a distinctive flavor profile. Previous research reported the key aroma compounds in Dianhong black tea, pinpointing phenylacetaldehyde, 3-ethyl-2,5-dimethylpyrazine, β-damascenone, and 1-ethyl-1H-pyrrole-2-carbaldehyde as key compounds responsible for the sweet and caramel flavors (Ma et al., 2022). Besides, furan, methyl acetate, and 2,3-pentanedione enhanced the comprehensive aroma quality of Dianhong black tea (Chen et al., 2022a, 2022b). Amidst the uniformity of environmental conditions for growth and the consistent manufacturing technologies, Dianhong black teas crafted from different local tea cultivars, universally displayed a pronounced sweet aroma, with a high degree of consistency in the types of volatile compounds. Yet, subtle distinctions emerge among the teas: a higher proportion of methyl salicylate and geraniol signals a stronger sweet note, whereas (*E*)-2-decenal and (*E*)-2-hexenyl 3-methylbutanoate exert an opposing influence (Wang et al., 2017). *Camellia nanchuanica*, another large-leaf tea cultivar from Southwest China, has recently emerged as a research focus (Wei et al. 2025) developed a flavor wheel for *Camellia nanchuanica* black tea, demonstrating that sweet aroma is its most prominent feature. They further identified (*E*)-β-ionone, linalool and geraniol as crucial contributors to the sweet aroma. In addition, these volatiles can improve the sweet taste of black tea (Yu et al., 2021).

As a traditional Chinese black tea, congou black tea is widely produced in many major tea-producing regions, and the differences in tea materials and processing parameters result in a wide range of aroma styles. In a comparative study of the aroma profiles of Chinese congou black teas from four regions, Tanyang congou black tea from Fujian Province showed the most pronounced sweet aroma, surpassing Keemun, Yixing, and Dianhong Congou black teas. This distinctive sweet aroma may be linked to the unique *Tanyang-caicha* tea variety, which contains elevated high amounts of sweet compounds including benzyl alcohol, phenylethyl alcohol, benzaldehyde, and 2-methylpyrazine. Therefore, they play significant role in the exceptional sweet aroma of Tanyang black tea (Xiao et al., 2017). In addition, Yang et al. underscored the pivotal role of heterocyclic volatiles in shaping the sweet aroma of black tea. Specifically, during the thermal processing, the Maillard reaction and Strecker degradation catalyze the substantial generation of pyrazine and furan compounds (Yang et al. 2020, 2022d). The honey-like aroma serves as an important label of Zunyi black tea and originates from the combined effects of multiple aroma compounds, such as phenylethyl alcohol, furaneol, cinnamaldehyde, and ethyl cinnamate. Molecular docking studies have revealed that the interactions between these aroma compounds and human olfactory receptors predominantly take place within the TM3-TM5 regions of the receptors. The binding of these compounds to the receptors is primarily driven by hydrogen bonds and hydrophobic interactions (Sun et al., 2024a, 2024b).

4.2.3 Smoky Aroma

Generally, improper baking may impart a smoky aroma that is considered an unpleasant attribute, negatively affecting the overall tea flavor. However, in Lapsang Souchong black tea, the smoky aroma is a niche yet distinctive attribute, reminiscent of burning pine branches. Lapsang Souchong black tea originates from the Tongmu Pass area in Wuyi Mountain in China. Dating back to the late Ming and early Qing dynasties, Lapsang Souchong black tea is the world's earliest black tea. Its creation was accidental: in an attempt to mitigate losses, tea producers subjected fresh tea leaves to an innovative secondary processing method, inadvertently developing this new tea category and thereby initiating the history of black tea. Based on its original processing techniques, the emphasis on the smoking step has been further enhanced, resulting in smoked Lapsang Souchong black tea, which features a stronger and more enduring smoky aroma. In order to unravel the unique smoky flavor, (Yao et al., 2005) analyzed the volatile chemical profiles of Lapsang Souchong and Smoked Lapsang Souchong black teas using GC–MS, discovering that volatiles from pine branches and their pyrolysis products significantly contribute to the smoky aroma, with longifolene and α-terpineol being the predominant odorants. Further researches (Li et al., 2024a, 2024b, 2024c) through sensomics approach has demonstrated the important role of phenolic compounds, such as 2-methoxy-4-methylphenol and*p*-cresol, which exhibit smoky and phenolic notes and are also crucial factors in forming the smoky aroma. Monitoring the entire production process of Lapsang Souchong black tea has determined that the pine smoking step is the primary source of its unique smoky flavor. During this period, phenolic compounds such as guaiacol and creosol, as well as volatile substances exclusive to pine, like α-terpineol and longifolene mentioned above, are adsorbed by the tea leaves. These compounds endow Lapsang Souchong black tea with its distinctive flavor (Su et al., 2025). To preserve the traditional flavor of Souchong black tea, mitigate the reliance on pines, protect the ecological environment, and achieve sustainable production, researchers have developed a novel fuel rod to serve as an effective substitute for traditional pine branches in the smoking process (Tian et al., 2025). Its application not only maintains the distinctive smoky flavor of Souchong black tea but also markedly accentuates its flowery and fruity nuances, thus offering extensive potential for further development.

4.3 Oolong Tea

4.3.1 Flowery & Fruity Aroma

Oolong tea, as a semi-fermented tea, has a fermentation degree between green tea (unfermented) and black tea (fully fermented). With its unique shaking process and more flexible fermentation control, oolong tea earns a rich and varied flavor profile, yielding a diverse range of aroma types. The most common flavor characteristics of oolong tea are elegant flowery and fruity aromas, reminiscent of orchids or jasmine flowers/several fresh fruits, rather than being intense and overpowering odor. *Tieguanyin* oolong tea, originating from the southern part of Fujian province, can be classified into three types according to its flavor profile. Among these, the fresh-scented Tieguanyin tea, which has not been vigorously roasted, is a typical representative of the faint and flowery aroma oolong tea category. In Zeng's studies, according to the Chinese national standard, high-grade fresh-scented *Tieguanyin* tea exhibits an elegant flowery aroma, which diminishes with lower grades while unpleasant traits become more pronounced. This flowery aroma mainly comes from key compounds like linalool, (*E*)-nerolidol, indole, n-hexyl benzoate, and phenethyl isovalerate (Zeng et al., 2022a, 2022b, 2023). Similar findings were also observed in the comparison between roasted and unroasted oolong teas (Wang et al., 2023a, 2023b, 2023c): compared with those that have been heavily roasted and show a strong caramel characteristic, the unroasted fresh-scented Tieguanyin tea, with higher levels of linalool and phenylacetaldehyde, presents a more refreshing and natural flowery fragrance. Like Tieguanyin, *Zhangping Shuixian* oolong tea stems from the southern region of Fujian and is characterized by a pronounced orchid fragrance. This aroma is primarily determined by volatile phenylpropanoids/benzenoids, including phenylethyl alcohol, coumarin, and indole, and is associated with mechanical damage and sun-withering during tea processing (Wu et al., 2022a, 2022b). Taiwan oolong tea is an important branch of Chinese oolong teas and is primarily divided into flowery and heavily roasted types based on varying degrees of fermentation and roasting (Su et al., 2021). As a quintessential example of flowery aroma Taiwan oolong teas, high-grade

Wenshan Bao-Chung Tea is distinguished by its high abundance of α-farnesene, (Z)-β-ocimene, and (Z)-jasmone, which collectively enhance its exquisite floral aroma. The development and stabilization of this aromatic profile are significantly influenced by the precise application of pan-firing techniques (Chen et al., 2023; Lin et al., 2022). Due to its deepest fermentation degree (60–80%) and the employment of fresh leaves infested by the tea green leafhopper (*Empoasca onukii* Matsuda) as raw materials, *Oriental Beauty* oolong tea is renowned for its unique processing techniques and rich fruity aroma, with geraniol, linalool and hotrienol playing key roles (Zeng et al., 2022b). The infestation of tea green leafhoppers on *Oriental Beauty* tea has been verified to have a decisive impact on the flavor shaping of the final product. The physical damage caused by tea green leafhoppers to fresh tea leaves, as well as the saliva they secrete, jointly influence the metabolic pathways of carotenoids and fatty acids, hence promoting the synthesis and accumulation of downstream fruity volatile compounds (Zeng et al. 2024). Nonetheless, the degree of pest infestation should be maintained within a controllable range. Excessive infestation can lead to huge water loss in fresh leaves and trigger an overactive defense response in the plant, consequently disrupting the normal metabolism of flavor compounds and ultimately reducing the overall quality (Li et al., 2023a, 2023b).

In oolong tea manufacturing process, shaking is a distinctive and pivotal process that involves the methodical agitation of tea leaves to induce mutual collision and friction. This controlled mechanical stress leads to partial damage to the leaf margins, thereby catalyzing partial enzymatic oxidation. During this process, the expression levels of biosynthetic genes for key contributors to the flowery aroma, such as indole and methyl jasmonate, are significantly upregulated. This upregulation promotes the substantial accumulation of these volatile compounds, which in turn provides the chemical basis for the formation of the faint and flowery aroma in oolong tea (Feng et al., 2021; Zeng et al., 2016). In addition, withering conditions significantly influence the development of the flowery aroma in oolong tea. In this phase, intensified respiration and increased cell membrane permeability spur the synthesis of flowery volatile precursors and provide energy for subsequent reactions. Concurrently, the accumulation of jasmonic acid contributes to the formation of flowery note (Wu et al., 2023a, 2023b). Prolonged solar withering under natural conditions tends to suppress the formation of the flowery aroma due to the heightened activity of heat shock proteins and heat shock factors. Conversely, shorter durations of this process are conducive to the development of the flowery aroma (Deng et al., 2020). Recently, researchers have started investigating alternative methods to simulate solar withering effects under unfavorable weather conditions. For instance, UV-B treatment can be employed as an effective alternative to traditional withering methods to enhance the aroma quality of oolong tea (Wang et al. 2024d).

Besides processing details, the selection of the tea cultivar for fresh leaf materials also determines the flavor style of oolong tea. In addition to the well-known *Tieguanyin*, the *Meizhan* and *Huangdan* cultivars are also suitable for making oolong tea with a flowery aroma (He et al., 2023). The *Meizhan* cultivar (*Camellia sinensis*, cv. Meizhan), a classic variety in Chinese oolong tea production, is favored for its unique flowery aroma. The substantial accumulation of linalool, benzaldehyde, and indole, induced by the shaking process, provides the prerequisite for the formation of its flowery feature (Ou et al., 2025). Compared with well-known green tea cultivars, "*Huangdan*" has distinct advantages in the number, expression levels, and structural variations of aroma-related genes, especially the terpene synthase gene family, which promotes the superior synthesis and accumulation of flowery compounds (Wang et al., 2022a, 2022b, 2022c, 2022d). The *Dancong* tea cultivar, specifically used for crafting oolong tea, is not a single independent variety but a collection of fine cultivars bred from the Fenghuang Shuixian varieties. Based on their aroma profiles, *Dancong* teas can be classified into several types, such as flowery, fruity, and honey-scented, each with distinct flavor characteristics. (Chen et al., 2022a, 2022b) identified indole, (*E*)-nerolidol and phenylacetonitrile as key contributors to the flowery note of *Dancong* tea, while Zheng's investigation defined the roles of hotrienol, linalool, and linalool oxides in this fragrance feature (Zheng et al., 2024). Additionally, after tea consumption, the empty cup aroma of premium *Fenghuang Dancong* tea is often described as performing typical flowery and sweet notes. Through the application of sensomics, (Kun et al., 2025) have demonstrated that α-ionone, geraniol, 2-ethylhexanol, and other compounds collectively contribute to the characteristic profile of this cup aroma. Differing from alcohols and aldehydes mainly contributing to the flowery aroma, lactones are considered key players in the fruity aroma of *Dancong* tea. Compared with non-typical fruity Dancong teas, Song et al. (2025) and Qin et al. (2023) identified γ-hexalactone, γ-octalactone, δ-decalactone, and jasmine lactone as the key odorants responsible for the distinct fruity notes. These compounds have flavors similar to coconut and peach and work together to impart a fresh and vibrant fruity flavor to the tea. In recent years, scientists have developed a series of tea cultivars with enhanced aroma and distinct flowery notes through hybrid breeding and single-plant selection. The oolong teas produced from these new cultivars surpass traditional cultivars in aroma and taste. For example, compared with *Shuixian*, the oolong teas made from *Huangmeigui* and *Zimudan* exhibit richer flavor and stronger flowery aroma in sensory evaluations, indicating significant potential for broader promotion (Guo et al., 2022).

4.3.2 Baked Aroma

Baked aroma is a unique olfactory attribute that is specifically developed in oolong tea through the high-temperature roasting process. This process imparts a rich, multi-dimensional fragrance profile, often distinguished by pronounced nutty and caramel notes. These baked aromatic compounds are generated as a result of the Maillard reaction and caramelization that occur during roasting, creating a complex and enduring sensory experience. Wuyi rock tea is a quintessential example of a baked aroma oolong tea. It is renowned for its distinctive "rock flavor (Yan Rhyme)", which is a direct result of its meticulous roasting technique. This flavor is characterized by its depth and longevity, with the nutty and caramel notes being particularly prominent Guo et al. (2021a, 2021b) conducted a comprehensive analysis of the flavor characteristics and volatile compounds of Wuyi rock tea and found that 2-ethyl-3,5-dimethylpyrazine (nutty and roasted odor) is the most critical odorant of Wuyi rock tea and the crucial factor for the "rock flavor". Subsequently, Wang et al. (2022a, 2022b, 2022c, 2022d) utilized the sensomics approach to identify key odorants in the tea infusion of *Dahongpao*, a premium Wuyi rock tea. The study revealed that the baked aroma of *Dahongpao* mainly stems from thermally-induced volatiles, such as 2-ethyl-3,5-dimethylpyrazine and 4-hydroxy-2,5-dimethylfuran-3(2*H*)-one. Moreover, eight novel key odorants were identified for the first time, including 2,6-diethylpyrazine and vanillin, enriching the understanding of the flavor of rock tea. Rougui tea is another distinguished category of Wuyi rock tea, which, while exhibiting a strong baked aroma, also possesses a unique cinnamon-like odor, thereby amassing a considerable following. During the roasting process, a significant array of Maillard reaction products is synthesized, including 2-methylbutanal, 3-methylbutanal, furfural, methional, 5-methyl-2-ethylpyrazine and 2-acetylpyrrole. These compounds play a crucial role in shaping the baked aroma of Rougui tea. Notably, 2-methylbutanal and 3-methylbutanal, which impart a rich, malty note, are generated through the thermal decomposition of leucine and isoleucine, directly enhancing the flavor of the refined Rougui tea (Yang et al., 2022a, 2022b, 2022c, 2022d). The development of the *Rougui* tea aroma wheel offers a more nuanced understanding of this esteemed tea's flavor profile. Beyond their role in creating the baked aroma, certain Maillard reaction products might also improve the spice or cinnamon-like features (Liang et al., 2024). To optimize the roasting process and enable targeted production of rock tea with desired attributes, researchers investigated the aroma differences between *Dahongpao* and *Rougui* tea processed at gradient roasting temperatures. With the rise in roasting temperature, the proportion of heterocyclic compounds in the tea's volatile profile increased, shifting the tea's aroma from flowery and fruity to a stronger roasted character (Wang et al., 2023a, 2023b, 2023c). Shuixian, a tea cultivar with a long-standing history, is widely used for crafting Wuyi rock tea. Compared to *Rougui* and *Dahongpao*, it has a more complex aroma profile. Through sensory evaluation and instrumental analysis, Wang et al. (2024a, 2024b, 2024c, 2024d, 2024e) discovered that compounds such as 2-ethyl-3,5-dimethylpyrazine, 2,3-diethyl-5-methylpyrazine, and 5-methylfurfural, which have roasted or meaty odors, are key factors contributing to the baked aroma of Shuixian rock tea. During the "Chinese tea ceremony" process, in the early stages of brewing, the baked aroma of Shuixian rock tea is rich and long-lasting. However, as the number of consumption increases, this characteristic gradually fades due to the decreasing concentration of odorants like 2-methylbutanal and 1-ethylpyrrole. Instead, a mild flowery fragrance emerges (Yang et al., 2025). In the latest research, simulated roasting systems incorporating Maillard reaction precursors have been developed to elucidate the molecular mechanisms underlying the formation of the baked aroma in rock tea. The pivotal roles of alanine and theanine, in conjunction with glucose, in generating key aroma compounds have been underscored. Meanwhile, catechins—the most representative non-volatile metabolites in tea—have been shown to exert an inhibitory influence on this aroma formation process (Chen et al., 2025a, 2025b; Li et al., 2022a, 2022b, 2022c).

Aside from rock tea, *Nongxiang Tieguanyin* stands as a typical baked aroma oolong tea, celebrated for its rich and elegant flavor profile that evokes the delightful notes of nuts and caramel—hallmarks of its signature "Yin Rhyme". According to Wang's study, high-grade *Nongxiang Tieguanyin* tea is found to contain elevated levels of heterocyclic compounds, such as 1-ethylpyrrole and 2-methylfuran, which impart nutty and roasted odors. The presence of these compounds aligns with what has been reported in rock tea (Wang et al., 2019). *Dong Ding* oolong tea is a typical baked aroma member of Taiwan oolong teas due to its customized multiple roasting process, in which pyrazines with a roasted odor (2-ethyl-5-methylpyrazine, 2-ethyl-3,5-dimethylpyrazine, and 2,3-diethyl-5-methylpyrazine) play a significant role. The absence of any of these compounds would have a significant impact on the authentic flavor (Wang et al., 2023a, 2023b, 2023c).

4.3.3 Other Aroma

The aroma profile of oolong tea is remarkably diverse. Beyond the prevalent flowery, fruity, and baked notes, a variety of niche and distinctive flavor features have developed. This is attributed to the rich germplasm resources of tea cultivars suitable for oolong tea production, the intricate and extended processing techniques, and the unique climatic conditions in which these teas are grown. Aged aroma *Tieguanyin*, a special type of *Tieguanyin* tea, originates from semi-processed rough tea and is meticulously crafted through an extended period of storage. During the storage, the aroma of the tea undergoes a complex alteration, resulting in a unique aged flavor. Yang et al. (2023) revealed that during the latter phase of storage for aged aroma *Tieguanyin* tea, there is a notable surge in the concentration of certain volatile compounds, which are believed to be pivotal in building this tea's distinctive aged aroma profile. To elaborate, 2-heptanone exudes a moldy and stale odor, 1-octen-3-ol gives off an earthy and mushroom-like odor, while safranal carries a unique saffron-like odor. The primary genesis of these volatile compounds lies in the oxidation of fatty acids and degradation of carotenoids. These metabolic pathways are intricately modulated not only by the storage conditions, such as temperature, humidity, and oxygen levels, but also by the influence of core microbial communities. Additionally, the hydrolysis of glycosides and esterification reactions will also continue to occur during this storage phase, leading to an increase in the content of compounds such as α-terpineol, myrtenol, and terpinyl acetate, which enhance the aged flavor. At the same time, the fresh and roasted odorants will gradually fade away (Zhang et al., 2023).

In rock tea, the nutty aroma, closely tied to roasting techniques and tea cultivar choice, is a common flavor note. Yue et al. (2023) processed sixteen common rock tea cultivars identically, discovering that *Qidan* and *Meizhan* show the strongest nutty aroma, with heterocyclic compounds and phenylalanine derivatives as key contributors. Milky aroma, a sensory experience found from a few high-end Oolong teas, arises from a blend of volatile compounds. In the case of Chinese *Jin Xuan* oolong tea, monitoring its processing reveals that the milky aroma primarily stems from fatty acid metabolism. This aroma is formed through the dynamic changes and accumulation of compounds during key processing steps, such as shaking and thermal processing (Wang et al., 2025).

4.4 White Tea

4.4.1 Tender Aroma

Apart from green tea, which completely halts enzymatic activity through the fixation to prevent enzymatic fermentation, white tea stands as the least fermented among the six major traditional Chinese tea classifications. This is largely because of its minimalist processing method, involving just three steps: plucking, withering, and drying, that effectively preserve the natural flavors of tea to the greatest extent. The flavors of white tea can be categorized into two main types: the tender and natural type, which is described as pekoe, herbaceous and flowery notes and is typically found in freshly manufactured tea; and the aged type, which develops a rich and mellow aroma with features of reed leaf, jujube, and herbs after long periods of storage and maturation.

Based on the maturity of the raw materials, white tea can be classified into *Baihaoyinzhen* (Silver Needle with White Hair, single bud), *Baimudan* (White Peony, one bud with one or two leaves), and the relatively rougher *Gongmei* and *Shoumei*. Chen et al. (2020a, 2020b) reported that the aroma composition of white tea was predominantly constituted by aldehydes and alcohols, with higher concentrations observed in *Baihaoyinzhen*, which is characterized by its superior tenderness and grade. Notably, phenylethyl alcohol, γ-nonalactone, and (E)-β-ionone, with pleasant sensory experiences, have been identified as the key causes to the distinct flavor profiles among different grades of white tea. Further, by applying quantitative descriptive analysis and molecular sensory science, Feng et al. (2022) found that dimethyl sulfide plays an absolutely crucial role in the aroma profiles of different grades of white tea and is involved in the construction of the aroma framework. Simultaneously, in comparison with *Gongmei* and *Shoumei*, *Baihaoyinzhen* exhibits a higher aroma character impact of linalool and geraniol, leading to a more obvious flowery and fruity notes. For *Baimudan* tea, seven key aroma compounds, including cedrol, dihydroactinidiolide, α-ionone, and others, were identified as significantly enhancing the pleasurable sensory experience by binding to specific neuroreceptors, activating neural signaling pathways, and regulating neurotransmitter release, thereby eliciting a sense of pleasure in consumers (Wu et al., 2025a, 2025b). Withering is a pivotal process in the production of white tea, playing a decisive role in shaping its quality. Different withering methods can lead to markedly distinct flavor profiles. Solar withering, a traditional method, involves placing fresh tea leaves under appropriate sunlight, allowing them to lose moisture naturally and undergo complex biochemical reactions. To meet the growing demand for increased production capacity and to mitigate the constraints imposed by uncontrollable weather conditions, innovative methods such as withering tank processing have been developed and applied in white tea manufacturing. Research findings indicate that solar withering upregulates the biosynthetic pathways of terpenoid compounds in white tea, thereby enhancing the accumulation of flowery and fruity substances. In contrast, the higher temperatures within withering tanks accelerate the oxidation of fatty acids, resulting in the substantial production of green volatiles (Huang et al., 2024a, 2024b, 2024c, 2024d; Wu et al., 2024; Zou et al., 2024). The latest attempt explored the application of UV-B radiation during the withering process of white tea, showing that the additional treatment improved the flavor quality by boosting the levels of volatile compounds associated with fruity and flowery odors while weakening fatty and burnt odors (Tan et al., 2025). In addition to light, ventilation during withering also significantly affects the flavor of white tea. Feng et al. (2025) introduced environmental airflow into the industrial indoor withering process, which also optimized the aroma of white tea, achieving comparable effects to UV-B radiation.

The tea cultivar is one of the key factors that determine the flavor profile of final white tea, holding equal significance to the modes and parameters employed in the processing steps. In *Zhenghe* County, a renowned area for Chinese white tea production, *Fuan Dabaicha* and *Zhenghe Dabaicha* are two major tea cultivars widely planted and used for white tea processing. White tea made from the former is celebrated for its complex and rich aroma, with distinct flowery and fruity features while the latter product stands out for its typical pekoe aroma (Bao et al., 2025). The understanding of tea germplasm resources suitable for white tea production remains in its infancy. Beyond the well-known common cultivars, certain primitive population varieties have also demonstrated their potential as high-quality raw materials for white tea. Recently, scholars discovered that white tea made from the "*Caicha*" population variety exhibits a unique fresh and sweet flavor, which are rarely found among existing popular cultivars (Lin et al., 2024).

4.4.2 Aged Aroma

During extended storage, the chemical components within white tea gradually transform over time. The content of certain green, fresh volatiles decrease, while the content of volatiles with sweet and herbal aromas increases, altering the overall flavor profile toward a richer and more complex style. In Qi's study, compared with fresh white tea, the flavor of the twelve-year-old sealed and stored white tea has evolved from a green and herbaceous aroma to a sweeter and more richly woody aged scent. This transformation is attributed to the reduction of aldehydes, exemplified by (*E*)-2-hexenal, and the increase of alcohols, represented by phenylethyl alcohol (Qi et al., 2018). As white tea ages, the levels of key enantiomers like S-γ-octanolactone and S-linalool, which are higher in fresh tea, significantly decrease, diminishing the tea's fresh flowery aroma (Zhu et al., 2020). In particular, the potential conversion between

S-linalool, which has a sweet and floral aroma, and its enantiomer R-linalool, which has a woody and lavender-like scent, during storage can alter the aroma of white tea (Zhu et al., 2017).

To facilitate transportation and storage, steaming and compressing loose white tea into a compact form by referring to the aging method of dark tea has become a common industrial practice for producing aged aroma white tea. Over a long storage, the initial fruity and flowery notes of compressed white tea gradually give way to woody aroma. After more than a decade of aging, the flavor evolves into a herbal aged scent. Throughout this transformative journey, γ-butyrolactone has emerged as a key compound that significantly influences the dynamic shift in flavor profile (Wang et al., 2022a, 2022b, 2022c, 2022d). Recently, to accelerate the maturation and reduce costs in white tea production, the fungus *Eurotium Cristatum* has been employed. This fungus promotes the development of a complex aroma of sweet and jujube character by degrading unsaturated fatty acids in tea, leading to the abundant synthesis of odorants like hexanal, 1-octen-3-ol, hexanol and heptanol, which contribute to the aged aroma of white tea (Zhu et al., 2024a, 2024b, 2025a, 2025b). Wang et al. (2024e) drew a flavor wheel for Aged Jinhua white tea (*Eurotium Cristatum* inoculated) and demonstrated that, except the volatile metabolites of fatty acids, ethyl guaiacol, 2-methoxy-4-ethylphenol, dihydrocarvone, and other compounds also play a role in the aged aroma.

4.4.3 Other Aroma

With specific processing techniques and unique raw materials, white tea can develop some distinctive aromas, matching consumers demand for personalized tea beverages and thus gaining favor in niche markets. Hao's research indicates that during the withering stage, higher temperature and humidity conditions can significantly promote the accumulation of dihydro-5-pentyl-2(3*H*)-furanone (coconut-like and creamy odor) and 2-pentyl-furan (butter-like aroma), contributing greatly to the milky aroma of white tea (Hao et al., 2023). Moreover, the maturity of fresh tea leaves, processing methods, and shaping techniques also have an important impact on this milky aroma: utilizing mature tea leaves for withering under high-temperature and high-humidity conditions is particularly favorable for the development of the milky aroma. Additionally, minor shaping methods, such as light compressing, simple bundling, or maintaining the loose state of the leaves, can mitigate the physical pressure on the tea leaves, which in turn is more conducive to the formation of the milky flavor (Feng et al., 2024a, 2024b, 2024c). In contrast to the milky aroma, the manufacturing and storage of white tea under relatively low humidity conditions, coupled with high-temperature drying, result in the accumulation of key compounds such as isovaleraldehyde, 2-methylbutyraldehyde, and phenylacetaldehyde. These compounds, which impart distinct cocoa and nutty notes, play a crucial role in enhancing the cocoa aroma in white tea (Zheng et al., 2025a, 2025b, 2025c).

4.5 Yellow Tea

4.5.1 Tender & Sweet Aroma

Yellow tea, boasting a history that dates back to the Tang Dynasty, is among the most ancient tea types in China. Crafted from the buds, leaves, and shoots, it undergoes a meticulous production process including fixation, rolling, yellowing, and drying. The pivotal step in this process is "yellowing", which tea leaves are gently oxidized under warm and humid conditions (Feng et al., 2024c). This unique step bestows the signature "three yellowing" traits of yellow tea: golden dry tea leaves, bright yellow infusion, and light-yellow leaves after brewing. Depending on the tenderness of the tea leaves and the processing techniques, the aroma of yellow tea evolves in two distinct types: tender flowery aroma and strong baked aroma. *Pingyang Huangtang* (PYHT) yellow tea, originating from Zhejiang province, China, was first developed during the Qing Dynasty. PYHT tea was highly prized for its distinctive flowery and sweet aroma, and was designated as a royal tribute. Its aroma characteristic mainly comes from geraniol, linalool, phenylacetaldehyde, and linalool oxides which are released from their glycoside precursors during the yellowing process (Dong et al., 2023). However, yellowing over ten hours increases the content of isoamyl alcohol, causing a watery dull odor. The flavor quality, especially for sweet aroma intensity and alcohols content, as well as volatiles contents of *Mogan* yellow tea increased with yellowing time, while significantly decreased after yellowing for ten hours (Fan et al., 2022). Besides, roasting step is equally important for flavor quality. Shorter roasting duration yields higher intensities of fresh and green notes, leading to the final product close to green tea. Conversely, longer roasting duration enhances the sweet aroma, matching the conventional expectations of yellow tea among consumers (Ye et al., 2022). Commonly, yellowing parameters should be optimized to enhance production efficiency and precision, and targetedly improve the aroma quality of yellow tea (Wei et al., 2024). The study indicated that reducing yellowing duration to 20–30% of traditional methods accelerated carotenoid thermal degradation, increasing (E)-β-ionone accumulation, and enhancing the sweet and flowery aroma. In addition, tea cultivars also affect the tender and sweet aroma of yellow tea. In a comparison of several well-known yellow teas from different regions in China, *Mengding Huangya* (MDHY) from Sichuan exhibited a pronounced sweet aroma, while *Weishan Huangcha* from Hunan and PYHT from Zhejiang demonstrated stronger intensity of flowery and fruity notes (Shi et al., 2021). Despite commonalities in detected volatile compounds, key differences emerged in compound concentrations: MDHY had higher content of 2-methylbutanal, whereas the amounts of phenylethyl alcohol, geraniol, α-ionone, and (E)-β-ionone were more abundant.

4.5.2 Baked Aroma

The baked aroma of yellow tea, reminiscent of caramel and rice crust, is a particular characteristic of large-leaf yellow tea and primarily stems from its high-temperature roasting process. Very recently, seventeen key odor-active compounds in large-leaf yellow tea (LYT) infusion were identified and replicated its rice crust-like baked aroma via aroma recombination by application of the sensomics approach (Zhai et al., 2023). Among them, the most significant contributors were 2-methylbutanal, (*E*,*E*)-2,4-heptadienal, 2-methylpropanal, and 2,3-diethyl-5-methylpyrazine. Most of them are synthesized through Strecker degradation and Maillard reaction during the roasting process. With increasing roasting intensity, the flavor of LYT changed from the grassy, fruity, and fatty notes to baked aroma (Guo et al., 2025). High-temperature and long-duration roasting boosts the Maillard reaction, raising the contents of heterocycles such as pyrazines and pyrroles. Therefore, LYT evokes caramel-like and rice crust-like notes, yielding a richer and mellower flavor (Guo et al., 2021a, 2021b; Li et al., 2023a, 2023b). In addition, the baked aroma is closely related to the roasting method. Traditional charcoal roasting, leveraging radiant and convective heat, rapidly elevates leaf temperature, thereby promoting the Maillard reaction and enhancing the rice crust-like aroma. Conversely, drum roasting with lower heat transfer efficiency, preserves more flowery odorants, thus resulting in a more pronounced flowery profile in LYT (Sheng et al., 2024). The incorporation of thermal simulation reactions has further deepened the understanding of the aroma of LYT: during the high-temperature roasting, *L*-theanine undergoes thermal degradation to produce volatile compounds such as *N*-ethylacetamide and *N*-ethylformamide. Concurrently, lysine, particularly in combination with soluble sugars like *D*-glucose, can generate a large number of pyrazines, providing the material basis for the baked aroma (Guo et al., 2019; Yin et al., 2023). Apart from the roasting step, the maturity of raw materials is important for the development of the baked aroma in LYT. Recent research indicated that the high levels of theanine and soluble monosaccharides in tea stems supply sufficient substrates for the Maillard reaction, synthesizing substantial amounts of baked-smelling pyrazines, such as 2-ethyl-3,5-dimethylpyrazine and 2-ethyl-3,6-dimethylpyrazine (Huang et al., 2024a, 2024b, 2024c, 2024d; Liu et al., 2024a, 2024b).

4.6 Dark Tea

4.6.1 Stale Aroma

As a post-fermented tea, dark tea has a unique stale aroma defined as a complex interplay of woody, jujube-like, and sweet notes. Stale aroma mainly stems from the intricate chemical transformations during microbial fermentation and long-term aging. Ripened Pu'er tea is a typical dark tea with a stale aroma and rich, mellow, and long-lasting taste. During the pile fermentation, the highly volatile terpenes decrease, while lowly volatile methoxybenzenes accumulate, thus shifting the flavor profile from fresh and flowery to stale impression (Pang et al., 2019). These methoxybenzenes, which impart stale, woody, and musty notes, are produced through microbial degradation of polyphenols and subsequent esterification. In specific, 1,2,3,4-tetramethoxybenzene, 1,2-dimethoxybenzene, and 1,2,3-trimethoxybenzene were identified as key contributors to the stale aroma (Wang et al., 2022a, 2022b, 2022c, 2022d). Very recently, 1-ethyl-4-methoxybenzene and 1,2,4-trimethoxybenzene, synthesized from the activities of *Blastobotrys*, *Rasamsonia* and *Thermomyces*, were pointed out as significant sources of stale aroma (Li et al. 2024b). Pile-fermentation and storage aging are pivotal processes in shaping the quality of ripened Pu'er tea. During pile-fermentation, the diversity, richness, and composition of microbial communities undergo strong fluctuations, affecting the accumulation of volatile compounds (Hu et al., 2021). Influenced by environmental factors such as temperature and humidity, long-term storage can trigger a series of oxidation, esterification, and degradation reactions, thereby altering the overall flavor profile (Xu et al., 2021). Qingzhuan tea, produced in Hubei province, China, has been an important part of China's tea exports for a long history and is distinguished by its characteristic stale aroma (Liu et al., 2024a, 2024b). Pile fermentation is the decisive stage for the aroma formation of Qingzhuan tea. The oxidative degradation of fatty acids and the methylation of gallic acid are highly activated, leading to an enrichment of unsaturated aldehydes, ketones, and methoxybenzenes. Among them, (*E*,*Z*)-2,6-nonadienal with fatty odor and 1-octen-3-one with earthy odor, are responsible for the stale aroma of Qingzhuan tea, due to their low odor thresholds and high concentrations (Liu et al., 2022; Fang et al., 2024). The thermophilic fungi *Thermomyces* and *Rasamsonia* dominate the high-temperature fermentation stage of Qingzhuan tea, thus significantly influence its quality. Meanwhile, *Aspergillus* is also vital for aroma formation by secreting its extracellular enzymes (Zhao et al., 2022). An tea, originating from Anhui Province, China, is a relatively niche yet quintessential dark tea with a distinctive stale aroma. Dew wetting is its distinctive processing step, which involves spreading the tea leaves on a platform to absorb dew on a clear night after the beginning of Autumn. The unique process enhances the woody aroma and balances the quality of An tea. However, the process is limited by weather which makes the aroma unstable. To ensure weather-independent processing of An tea, researchers introduced the innovative "steaming and heaping" method, which allows tea leaves ferment in a high-temperature and high-humidity environment, and promote the accumulation of desirable aroma compounds such as α-ionone and linalool (Chen et al., 2025a, 2025b). Increasing storage time can reduce the contents of flowery compounds while increase the contents of naphthalene and 1-octen-3-ol, leading to the final flavor shift toward woody, stale, and herbal note (Shen et al., 2023a, 2023b).

4.6.2 Fungal-Flowery Aroma

The fungal-flowery aroma refers to the unique flavor type formed in raw dark tea via a specific processing technique called "flowering". During this "flowering" step, heightened metabolic activity of fungus *Eurotium cristatum* decomposed tea leaf macromolecules into aroma precursors and volatiles, yielding the characteristic fungal-flowery aroma (Zheng et al., 2025a, 2025b, 2025c). Methyl salicylate, 4-methoxybenzaldehyde, coumarin, nonanal, decanoic acid, α-terpineol, and so on are the key aroma compounds of the fungal-flowery aroma in dark tea through application of the sensomics approach (Shen & Zhang et al., 2023). Among them, methyl salicylate, 4-methoxybenzaldehyde, and coumarin are especially crucial, endowing the complex fungal-flowery aroma with minty, woody, and flowery attributes. Another parallel study identified a large number of fatty acid degradation products (e.g., (*E*,*Z*)-2,6-nonadienal, (*E*)-2-octenal) and glycoside hydrolysis products (e.g., linalool, geraniol) as key contributors to the fungal-flowery aroma (Zhao et al., 2024). These compounds interact with each other through complex synergistic effects, for example, modulatory, masking, compensatory, and novelty effects. Twenty key aroma compounds were identified for the fungal-flowery aroma of Fu brick tea, demonstrating the complex synergistic effects among them (Zheng et al., 2025b). Innovatively, the study also found that sub-threshold compounds can influence the overall aroma profile in conjunction with other odorants. *Eurotium cristatum* shapes the fungal-flowery aroma of Fu brick tea via secreted hydrolases. Specifically, β-glucosidase hydrolyzes salicin in raw dark tea to yield salicylic acid, which is subsequently converted into methyl salicylate. Additionally, polyphenol oxidase and catalase in the fungus catalyze polyphenolic oxidation to quinones, which react with amino acids to generate complex aroma compounds (Xiao et al., 2022a, 2022b, 2022c).

4.6.3 Betelnut-Like Aroma

The betelnut-like aroma is a unique but rare aroma type in dark tea. It is a complex blend of betelnut-like, woody, ginseng-like, minty and other notes (Feng et al., 2023). By the application of instrumental and statistical analysis, seven aroma compounds such as pulegone, naphthalene, 1- and 2-methyl-naphthalene were identified as major aroma contributors to the betelnut-like aroma of *Liu-pao* tea. These compounds exhibit camphor-like, pungent, and smoky notes (Li & Hong et al., 2022). Additionally, the volatile fingerprint and aroma wheel of *Liu-pao* tea suggested that cedrol and methyl salicylate also play a role in betelnut-like aroma (Ma et al., 2023). Besides, storage contributes a lot in *Liu-pao* tea's flavor. A long-term study monitoring flavor dynamic change revealed the aroma evoluted from flowery and fruity to woody, along with a series of reactions such as glycoside hydrolysis, sesquiterpene synthesis, gallic acid methylation, and carotenoid degradation (Huang et al. 2024a, 2024b).

During storage, the dominant fungi in *Liu-Bpao* tea, including *Aspergillus*, *Penicillium*, *Wallemia* and *Blastobotrys*, promote the biosynthesis and transformation of aroma compounds via decomposition of glycosidic precursors and lipid metabolism (Xu et al., 2025).

References

Baba, R., Amano, Y., Wada, Y., & Kumazawa, K. (2017). Characterization of the potent odorants contributing to the characteristic aroma of matcha by gas chromatography-olfactometry techniques. *Journal of Agricultural and Food Chemistry, 65*(14), 2984–2989. https://doi.org/10.1021/acs.jafc.7b00421

Bao, S., An, Q., Yang, Y., Li, X., Chen, G., Chen, Y., et al. (2025). Tea plant varieties influence the aroma characteristics of Zhenghe white tea: Based on Zhenghe Dabaicha and Fuan Dabaicha. *Food Research International, 208*, 116278. https://doi.org/10.1016/j.foodres.2025.116278

Cao, Y., Huang, C., Guo, Y., Xu, Y., Gong, S., Chu, Q., & Chen, P. (2024). Unraveling the contributing factors of stale odor in Longjing tea through a sensomics approach. *Food Chemistry, 441*, Article 138301. https://doi.org/10.1016/j.foodchem.2023.138301

Chen, X., Chen, D., Jiang, H., Sun, H., Zhang, C., Zhao, H., et al. (2019). Aroma characterization of Hanzhong black tea (*Camellia sinensis*) using solid phase extraction coupled with gas chromatography–mass spectrometry and olfactometry and sensory analysis. *Food Chemistry, 274*, 130–136. https://doi.org/10.1016/j.foodchem.2018.08.124

Chen, Q., Zhu, Y., Yan, H., Chen, M., Xie, D., Wang, M., et al. (2020a). Identification of aroma composition and key odorants contributing to aroma characteristics of white teas. *Molecules, 25*(24), 6050. https://doi.org/10.3390/molecules25246050

Chen, X., Sun, H., Qu, D., Yan, F., Jin, W., Jiang, H., et al. (2020b). Identification and characterization of key aroma compounds in Chinese high altitude and northernmost black tea (*Camellia sinensis*) using distillation extraction and sensory analysis methods. *Flavour and Fragrance Journal, 35*(6), 666–673. https://doi.org/10.1002/ffj.3605

Chen, J., Yang, Y., Deng, Y., Liu, Z., Xie, J., Shen, S., et al. (2022a). Aroma quality evaluation of Dianhong black tea infusions by the combination of rapid gas phase electronic nose and multivariate statistical analysis. *LWT, 153*, 112496. https://doi.org/10.1016/j.lwt.2021.112496

Chen, W., Hu, D., Miao, A., Qiu, G., Qiao, X., Xia, H., & Ma, C. (2022b). Understanding the aroma diversity of Dancong tea (*Camellia sinensis*) from the floral and honey odors: relationship between volatile compounds and sensory characteristics by chemometrics. *Food Control, 140*, Article 109103. https://doi.org/10.1016/j.foodcont.2022.109103

Chen, P., Liu, C., & Chen, K. (2023). Determining the relationship between aroma and quality of Bao-Chung tea by solid-phase microextraction (SPME) and electronic nose analyses. *Horticulturae, 9*(8), 930. https://doi.org/10.3390/horticulturae9080930

Chen, W., Liu, W., Liu, Z., Wang, D., Lan, X., Zhan, S., et al. (2025a). Insight into the mechanism of roasting-induced characteristic aroma formation in Wuyi rock tea using an "in-leaf" model with isotopic labeling. *Food Chemistry, 474*, 143174. https://doi.org/10.1016/j.foodchem.2025.143174

Chen, Z., Jiang, Y., Ye, Z., Zhao, W., Zhang, J., Wu, Y., et al. (2025b). Improving Ancha tea aroma quality via innovative processing: Characterization of key odorants by GC-MS, GC-O, Sensomics, and Chemometrics. *Food Chemistry, 486*, 144681. https://doi.org/10.1016/j.foodchem.2025.144681

Cui, J., Zhai, X., Guo, D., Du, W., Gao, T., Zhou, J., et al. (2022). Characterization of key odorants in Xinyang Maojian green tea and their changes during the manufacturing process. *Journal of Agricultural and Food Chemistry, 70*(1), 279–288. https://doi.org/10.1021/acs.jafc.1c06473.

Dai, Q., Jin, H., Gao, J., Ning, J., Yang, X., & Xia, T. (2020). Investigating volatile compounds' contributions to the stale odour of green tea. *International Journal of Food Science & Technology, 55*(4), 1606–1616. https://doi.org/10.1111/ijfs.14387

Deng, H., Chen, S., Zhou, Z., Li, X., Chen, S., Hu, J., et al. (2020). Transcriptome analysis reveals the effect of short-term sunlight on aroma metabolism in postharvest leaves of oolong tea (*Camellia sinensis*). *Food Research International, 137*, 109347. https://doi.org/10.1016/j.foodres.2020.109347

Dong, W., Shi, K., Liu, M., Shen, C., Li, A., Sun, X., et al. (2018). Characterization of 3-methylindole as a source of a "mud"-like off-odor in strong-aroma types of Base Baijiu. *Journal of Agricultural and Food Chemistry, 66*(48), 12765–12772. https://doi.org/10.1021/acs.jafc.8b04734

Dong, R., Sheng, X., Xie, Q., Huang, X., Yan, F., & Liu, S. (2023). Aroma formation and transformation during sealed yellowing process of Pingyang yellow tea. *Food Research International, 165*, Article 112535. https://doi.org/10.1016/j.foodres.2023.112535

Fan, F., Zhou, S., Qian, H., Zong, B., Huang, C., Zhu, R., et al. (2022). Effect of yellowing duration on the chemical profile of yellow tea and the associations with sensory traits. *Molecules, 27*(3), 940. https://doi.org/10.3390/molecules27030940

Fang, X., Xu, W., Jiang, G., Sui, M., Xiao, J., Ning, Y., et al. (2024). Monitoring the dynamic changes in aroma during the whole processing of Qingzhuan tea at an industrial scale: From fresh leaves to finished tea. *Food Chemistry, 439*, 137810. https://doi.org/10.1016/j.foodchem.2023.137810

Feng, Z., Li, M., Li, Y., Wan, X., & Yang, X. (2020). Characterization of the orchid-like aroma contributors in selected premium tea leaves. *Food Research International, 129*, Article 108841. https://doi.org/10.1016/j.foodres.2019.108841

Feng, Y., Wang, J., Zhao, Y., Zhang, M., Zhou, Z., Li, Y., et al. (2021). Biosynthesis of orchid-like volatile methyl jasmonate in tea (*Camellia sinensis*) leaves in response to multiple stresses during the shaking process of oolong tea. *LWT, 143*, 111184. https://doi.org/10.1016/j.lwt.2021.111184

Feng, Z., Li, M., Li, Y., Yin, J., Wan, X., & Yang, X. (2022). Characterization of the key aroma compounds in infusions of four white teas by the sensomics approach. *European Food Research & Technology, 248*(5), 1299–1309. https://doi.org/10.1007/s00217-022-03967-3

Feng, X., Chen, M., Song, H., Ma, S., Ou, C., Li, Z., et al. (2023). A systemic review on Liubao tea: A time-honored dark tea with distinctive raw materials, process techniques, chemical profiles, and biological activities. *Comprehensive Reviews in Food Science and Food Safety, 22*(6), 5063–5085. https://doi.org/10.1111/1541-4337.13254

Feng, J., Zhuang, J., Chen, Q., Lin, H., Chu, Q., Chen, P., et al. (2024a). The effect of maturity of tea leaves and processing methods on the formation of milky flavor in white tea—A metabolomic study. *Food Chemistry, 447*, 139080. https://doi.org/10.1016/j.foodchem.2024.139080

Feng, W., Zhou, H., Xiong, Z., Sheng, C., Xia, D., Zhang, J., et al. (2024b). Exploring the effect of different tea varieties on the quality of Lu'an Guapian tea based on metabolomics and molecular sensory science. *Food Chemistry: X, 23*, 101534. https://doi.org/10.1016/j.fochx.2024.101534

Feng, X., Yang, S., Pan, Y., Zhou, S., Ma, S., Ou, C., et al. (2024c). Yellow tea: more than turning green leaves to yellow. *Critical Reviews in Food Science and Nutrition, 64*(22), 7836–7853. https://doi.org/10.1080/10408398.2023.2193271

Feng, J., Ye, S., Wang, J., Wu, J., Zhao, J., Tian, W., et al. (2025). From water migration to aroma development: Revealing the influence of environmental airflow on the aroma of white tea during withering. *Food Chemistry, 479*, 143797. https://doi.org/10.1016/j.foodchem.2025.143797

Flaig, M., Qi, S. C., Wei, G., Yang, X., & Schieberle, P. (2020a). Characterisation of the key aroma compounds in a Longjing green tea infusion (*Camellia sinensis*) by the sensomics approach and their quantitative changes during processing of the tea leaves. *European Food Research and Technology, 246*(12), 2411–2425. https://doi.org/10.1007/s00217-020-03584-y

Flaig, M., Qi, S., Wei, G., Yang, X., & Schieberle, P. (2020b). Characterization of the Key Odorants in a high-grade Chinese green tea beverage (*Camellia sinensis; Jingshan cha*) by Means of the Sensomics approach and elucidation of Odorant changes in tea leaves caused by the tea manufacturing process. *Journal of Agricultural and Food Chemistry, 68*(18), 5168–5179. https://doi.org/10.1021/acs.jafc.0c01300

Gan, S., Chen, Y., Zhao, L., Zhao, X., Qiu, T., Zhai, X., & Dai, Q. (2024). Characterization of the aroma-active compounds in Xiaokeng green tea by three pretreatment methods combined with gas chromatography−olfactometry (GC-O). *Food Research International, 187*, Article 114359. https://doi.org/10.1016/j.foodres.2024.114359

Gao, Y., Lei, Z., Huang, J., Sun, Y., Liu, S., Yao, L., et al. (2024). Characterization of Key Odorants in Lushan Yunwu Tea in response to intercropping with flowering cherry. *Foods, 13*(8), 1252. https://doi.org/10.3390/foods13081252

Gong, X., Han, Y., Zhu, J., Hong, L., Zhu, D., Liu, J., et al. (2017). Identification of the aroma-active compounds in Longjing tea characterized by Odor activity value, gas chromatography-olfactometry, and aroma recombination. *International Journal of Food Properties, 20*(sup1), S1107–S1121. https://doi.org/10.1080/10942912.2017.1336719

Guo, X., Ho, C., Schwab, W., Song, C., & Wan, X. (2019). Aroma compositions of large-leaf yellow tea and potential effect of theanine on volatile formation in tea. *Food Chemistry, 280*, 73–82. https://doi.org/10.1016/j.foodchem.2018.12.066

Guo, X., Ho, C., Schwab, W., & Wan, X. (2021a). Effect of the roasting degree on flavor quality of large-leaf yellow tea. *Food Chemistry, 347*, Article 129016. https://doi.org/10.1016/j.foodchem.2021.129016

Guo, X., Ho, C., Wan, X., Zhu, H., Liu, Q., & Wen, Z. (2021b). Changes of volatile compounds and Odor profiles in Wuyi rock tea during processing. *Food Chemistry, 341*, Article 128230. https://doi.org/10.1016/j.foodchem.2020.128230

Guo, X., Schwab, W., Ho, C., Song, C., & Wan, X. (2022). Characterization of the aroma profiles of oolong tea made from three tea cultivars by both GC-MS and GC-IMS. *Food Chemistry, 376*, Article 131933. https://doi.org/10.1016/j.foodchem.2021.131933

Guo, Y., Shen, Y., Hu, B., Ye, H., Guo, H., Chu, Q., & Chen, P. (2023). Decoding the chemical signatures and sensory profiles of Enshi Yulu: Insights from diverse tea cultivars. *Plants, 12*(21), 3707. https://doi.org/10.3390/plants12213707

Guo, X., Schwab, W., Ho, C., Song, C., & Wan, X. (2025). Characterization of the changes of aroma profiles in large-leaf yellow tea during processing using GC-MS and electronic nose analysis. *Food Chemistry: X, 27*, Article 102507. https://doi.org/10.1016/j.fochx.2025.102507

Hao, Z., Feng, J., Chen, Q., Lin, H., Zhou, X., Zhuang, J., et al. (2023). Comparative volatiles profiling in milk-flavored white tea and traditional white tea Shoumei via HS-SPME-GC-TOFMS and OAV analyses. *Food Chemistry: X, 18*, 100710. https://doi.org/10.1016/j.fochx.2023.100710

He, C., Zhou, J., Li, Y., Zhang, D., Ntezimana, B., Zhu, J., et al. (2023). The aroma characteristics of oolong tea are jointly determined by processing mode and tea cultivars. *Food Chemistry: X, 18*, 100730. https://doi.org/10.1016/j.fochx.2023.100730

Hu, S., He, C., Li, Y., Yu, Z., Chen, Y., Wang, Y., & Ni, D. (2021). The formation of aroma quality of dark tea during pile-fermentation based on multi-omics. *LWT, 147*, Article 111491. https://doi.org/10.1016/j.lwt.2021.111491

Huang, D., Chen, X., Tan, R., Wang, H., Jiao, L., Tang, H., et al. (2024a). A comprehensive metabolomics analysis of volatile and non-volatile compounds in matcha processed from different tea varieties. *Food Chemistry: X, 21*, 101234. https://doi.org/10.1016/j.fochx.2024.101234

Huang, J., Zhang, J., Chen, Z., Xiong, Z., Feng, W., Wei, Y., et al. (2024b). Sensory-directed flavor analysis of Jinggu white tea: Exploring the formation mechanisms of sweet and fruity aromas. *Food Chemistry: X, 24*, 102026. https://doi.org/10.1016/j.fochx.2024.102026

Huang, W., Liu, Q., & Ning, J. (2024c). Effect of tea stems on the quality formation of large-leaf yellow tea: Sensomics and flavor omics approaches. *Food Chemistry: X, 24*, Article 101794. https://doi.org/10.1016/j.fochx.2024.101794

Huang, Y., Liu, H., Wang, J., Zhang, R., Zhang, Y., Liu, Z., et al. (2024d). Impact of the long-term storage on flavor quality of Liupao tea using sensory evaluation combined with metabolomics analysis. *Food Research International, 198*, 115386. https://doi.org/10.1016/j.foodres.2024.115386

Jiang, G., Xue, R., Xiang, J., Wang, Y., Liu, B., Yuan, Y., et al. (2024). Dynamic changes in the aroma profiles and volatiles of Enshi Yulu tea throughout its industrial processing. *Food Chemistry, 458*, 140145. https://doi.org/10.1016/j.foodchem.2024.140145

Jin, G., Li, T., Yang, C., Du, X., Dai, Q., Li, A., & Hou, R. (2024). Deciphering the aromatic signature of Taiping Houkui premium green tea: The role of key floral compounds and process control on volatile compounds. *LWT, 212*, Article 116970. https://doi.org/10.1016/j.lwt.2024.116970

Kang, S., Yan, H., Zhu, Y., Liu, X., Lv, H., Zhang, Y., et al. (2019). Identification and quantification of key odorants in the world's four most famous black teas. *Food Research International, 121*, 73–83. https://doi.org/10.1016/j.foodres.2019.03.009

Kumazawa, K., & Masuda, H. (1999). Identification of potent Odorants in Japanese Green Tea (Sen-cha). *Journal of Agricultural and Food Chemistry, 47*(12), 5169–5172. https://doi.org/10.1021/jf9906782

Kumazawa, K., Kubota, K., & Masuda, H. (2005). Influence of manufacturing conditions and crop season on the formation of 4-Mercapto-4-methyl-2-pentanone in Japanese Green Tea (Sen-cha). *Journal of Agricultural and Food Chemistry, 53*(13), 5390–5396. https://doi.org/10.1021/jf050392z

Kumazawa, K., Wada, Y., & Masuda, H. (2006). Characterization of epoxydecenal isomers as potent odorants in black tea (Dimbula) infusion. *Journal of Agricultural and Food Chemistry, 54*(13), 4795–4801. https://doi.org/10.1021/jf0603127

Kun, J., Luo, Z., Gong, M., Meng, Q., Chen, Y., Yuan, L., et al. (2025). Characterization of the key aroma compounds in the floral honey-like cup aroma of Fenghuang Dancong oolong tea by application of the sensomics approach. *Food Chemistry: X, 28*, 102538. https://doi.org/10.1016/j.fochx.2025.102538

Li, J., Hao, C., Jia, H., Zhang, J., Wu, H., Ning, J., et al. (2022a). Aroma characterization and their changes during the processing of black teas from the cultivar, *Camellia sinensis* (L.) O. Kuntze cv. Jinmudan. *Journal of Food Composition and Analysis, 108*, 104449. https://doi.org/10.1016/j.jfca.2022.104449

Li, M., Ho, C., Wang, J., Hu, Y., Zhai, X., Zhang, L., et al. (2022b). Formation of volatile heterocyclic compounds and open-chain amides of theanine in model systems with glucose, tea leaves, and tea extract under tea-roasting conditions. *Journal of Agricultural and Food Chemistry, 70*(22), 6737–6746. https://doi.org/10.1021/acs.jafc.2c02039

Li, Q., Hong, X., Zheng, X., Xu, Y., Lai, X., Teng, C., et al. (2022c). Characterization of key aroma compounds and core functional microorganisms in different aroma types of Liupao tea. *Food Research International, 152*, 110925. https://doi.org/10.1016/j.foodres.2021.110925

Li, M., Zhang, Y., Yan, J., Ding, F., Chen, C., Zhong, S., et al. (2023a). Comparative metabolomic analysis reveals the differences in nonvolatile and volatile metabolites and their quality characteristics in beauty tea with different extents of punctured leaves by tea green leafhopper. *Journal of Agricultural and Food Chemistry, 71*(43), 16233–16247. https://doi.org/10.1021/acs.jafc.3c01380

Li, Y., Zhang, J., Jia, H., Pan, Y., Xu, Y., Wang, Y., & Deng, W. (2023b). Metabolite analysis and sensory evaluation reveal the effect of roasting on the characteristic flavor of large-leaf yellow tea. *Food Chemistry, 427*, Article 136711. https://doi.org/10.1016/j.foodchem.2023.136711

Li, M., Wang, H., Wang, J., Zhang, X., Yao, X., Yu, J., et al. (2024a). Molecular insights into the contribution of specialty smoking process to the aroma of Lapsang Souchong black tea (*Camellia sinensis*). *Industrial Crops and Products, 216*, 118739. https://doi.org/10.1016/j.indcrop.2024.118739

Li, T., Wei, Y., Lu, M., Wu, Y., Jiang, Y., Ke, H., et al. (2024b). Exploring microbial and moist-heat effects on Pu-erh tea volatiles and understanding the methoxybenzene formation mechanism using molecular sensory science. *Food Chemistry: X, 23*, 101553. https://doi.org/10.1016/j.fochx.2024.101553

Li, X., Lei, W., You, X., Kong, X., Chen, Z., Shan, R., et al. (2024c). The tea cultivar 'Chungui' with jasmine-like aroma: From genome and epigenome to quality. *International Journal of Biological Macromolecules, 281*, 136352. https://doi.org/10.1016/j.ijbiomac.2024.136352

Liang, Y., Wang, Z., Zhang, L., Dai, H., Wu, W., Zheng, Z., et al. (2024). Characterization of volatile compounds and identification of key aroma compounds in different aroma types of Rougui Wuyi rock tea. *Food Chemistry, 455*, 139931. https://doi.org/10.1016/j.foodchem.2024.139931

Liao, X., Yan, J., Wang, B., Meng, Q., Zhang, L., & Tong, H. (2020). Identification of key odorants responsible for cooked corn-like aroma of green teas made by tea cultivar '*Zhonghuang 1*'. *Food Research International, 136*, Article 109355. https://doi.org/10.1016/j.foodres.2020.109355

Lin, S., Hsiao, Y., & Chen, P. (2022). Revealing the profound meaning of pan-firing of oolong tea—A decisive point in odor fate. *Food Chemistry, 375*, Article 131649. https://doi.org/10.1016/j.foodchem.2021.131649

Lin, Y., Huang, Y., Zhou, S., Li, X., Tao, Y., Pan, Y., et al. (2024). A newly-discovered tea population variety processed Bai Mu Dan white tea: Flavor characteristics and chemical basis. *Food Chemistry, 446*, 138851. https://doi.org/10.1016/j.foodchem.2024.138851

Liu, H., Xu, Y., Wu, J., Wen, J., Yu, Y., An, K., & Zou, B. (2021). GC-IMS and olfactometry analysis on the tea aroma of Yingde black teas harvested in different seasons. *Food Research International, 150*, Article 110784. https://doi.org/10.1016/j.foodres.2021.110784

Liu, P., Zheng, P., Feng, L., Gong, Z., Zheng, L., Gao, S., et al. (2022). Dynamic changes in the aroma profile of Qingzhuan tea during its manufacture. *Food Chemistry, 375*, 131847. https://doi.org/10.1016/j.foodchem.2021.131847

Liu, N., Shen, S., Huang, L., Deng, G., Wei, Y., Ning, J., & Wang, Y. (2023a). Revelation of volatile contributions in green teas with different aroma types by GC–MS and GC–IMS. *Food Research International, 169*, Article 112845. https://doi.org/10.1016/j.foodres.2023.112845

Liu, P., Feng, L., Xu, Y., Zheng, L., Yin, P., Ye, F., et al. (2023b). Characterization of stale odor in green tea formed during storage: Unraveling improvements arising from reprocessing by baking. *LWT, 174*, 114458. https://doi.org/10.1016/j.lwt.2023.114458

Liu, P., Feng, L., Chen, J., Wang, S., Wang, X., Han, Y., et al. (2024a). Unlocking the secrets of Qingzhuan tea: A comprehensive overview of processing, flavor characteristics, and health benefits. *Trends in Food Science & Technology, 147*, 104450. https://doi.org/10.1016/j.tifs.2024.104450

Liu, Q., Huang, W., Sheng, C., Wu, Y., Lu, M., Li, T., et al. (2024b). Contribution of tea stems to large-leaf yellow tea aroma. *Food Chemistry, 460*, 140472. https://doi.org/10.1016/j.foodchem.2024.140472

Liu, C., Li, J., Wang, M., Jian, G., Zhu, C., Li, H., et al. (2025). (R)-Linalool is a key indicator of aroma quality levels of a distinctive black tea (*Camellia sinensis* var. Yinghong No. 9). *Industrial Crops and Products, 225*, 120506. https://doi.org/10.1016/j.indcrop.2025.120506

Ma, Q., Meng, N., Li, Y., & Wang, J. (2021). Occurrence, impacts, and microbial transformation of 3-methylindole (skatole): A critical review. *Journal of Hazardous Materials, 416*, Article 126181. https://doi.org/10.1016/j.jhazmat.2021.126181

Ma, L., Gao, M., Zhang, L., Qiao, Y., Li, J., Du, L., et al. (2022). Characterization of the key aroma-active compounds in high-grade Dianhong tea using GC-MS and GC-O combined with sensory-directed flavor analysis. *Food Chemistry, 378*, 132058. https://doi.org/10.1016/j.foodchem.2022.132058

Ma, W., Zhu, Y., Ma, S., Shi, J., Yan, H., Lin, Z., & Lv, H. (2023). Aroma characterisation of Liu-pao tea based on volatile fingerprint and aroma wheel using SBSE-GC-MS. *Food Chemistry, 414*, Article 135739. https://doi.org/10.1016/j.foodchem.2023.135739

Mizukami, Y., Sawai, Y., & Yamaguchi, Y. (2008). Changes in the concentrations of acrylamide, selected odorants, and catechins caused by roasting of green tea. *Journal of Agricultural and Food Chemistry, 56*(6), 2154–2159. https://doi.org/10.1021/jf0731806

Ou, X., Li, Q., Yu, X., Hao, Y., He, J., Liang, J., et al. (2025). The regulatory network on characteristic aroma formation during the processing of Meizhan oolong tea. *Food Research International, 211*, 116373. https://doi.org/10.1016/j.foodres.2025.116373

Pang, X., Yu, W., Cao, C., Yuan, X., Qiu, J., Kong, F., & Wu, J. (2019). Comparison of potent odorants in raw and ripened Pu-Erh tea infusions based on odor activity value calculation and multivariate analysis: Understanding the role of pile fermentation. *Journal of Agricultural and Food Chemistry, 67*(47), 13139–13149. https://doi.org/10.1021/acs.jafc.9b05321

Qi, D., Miao, A., Cao, J., Wang, W., Chen, W., Pang, S., et al. (2018). Study on the effects of rapid aging technology on the aroma quality of white tea using GC–MS combined with chemometrics: In comparison with natural aged and fresh white tea. *Food Chemistry, 265*, 189–199. https://doi.org/10.1016/j.foodchem.2018.05.080

Qin, D., Wang, Q., Jiang, X., Ni, E., Fang, K., Li, H., et al. (2023). Identification of key volatile and odor-active compounds in 10 main fragrance types of Fenghuang Dancong tea using HS-SPME/GC-MS combined with multivariate analysis. *Food Research International, 173*, 113356. https://doi.org/10.1016/j.foodres.2023.113356

Schuh, C., & Schieberle, P. (2006). Characterization of the key aroma compounds in the beverage prepared from darjeeling black tea: quantitative differences between tea leaves and infusion. *Journal of Agricultural and Food Chemistry, 54*(3), 916–924. https://doi.org/10.1021/jf052495n

Shen, S., Wu, H., Li, T., Sun, H., Wang, Y., & Ning, J. (2023a). Formation of aroma characteristics driven by volatile components during long-term storage of An tea. *Food Chemistry, 411*, Article 135487. https://doi.org/10.1016/j.foodchem.2023.135487

Shen, S., Zhang, J., Sun, H., Zu, Z., Fu, J., Fan, R., et al. (2023b). Sensomics-assisted characterization of fungal-flowery aroma components in fermented tea using *Eurotium cristatum*. *Journal of Agricultural and Food Chemistry, 71*(48), 18963–18972. https://doi.org/10.1021/acs.jafc.3c05273

Sheng, C., Lu, M., Liu, Q., Zhou, H., Xiong, Z., Li, T., et al. (2024). Differences in the aroma quality of large-leaf yellow tea subjected to different roasting methods. *LWT, 204*, 116475. https://doi.org/10.1016/j.lwt.2024.116475.

Shi, Y., Wang, M., Dong, Z., Zhu, Y., Shi, J., Ma, W., et al. (2021). Volatile components and key odorants of Chinese yellow tea (*Camellia sinensis*). *LWT, 146*, 111512. https://doi.org/10.1016/j.lwt.2021.111512

Shi, Y., Zhu, Y., Ma, W., Lin, Z., & Lv, H. (2022). Characterisation of the volatile compounds profile of Chinese pan-fried green tea in comparison with baked green tea, steamed green tea, and sun-dried green tea using approaches of molecular sensory science. *Current Research in Food Science, 5*, 1098–1107. https://doi.org/10.1016/j.crfs.2022.06.012

Song, Y., Li, J., Ding, Q., Yang, X., Yu, J., Zhang, Y., et al. (2025). Exploring key aroma differences of three Fenghuang Dancong oolong teas and their perception interactions with caffeine via sensomics approach. *Food Research International, 202*, 115665. https://doi.org/10.1016/j.foodres.2025.115665

Steinhaus, M., & Schieberle, P. (2005). Role of the fermentation process in off-odorant formation in white pepper: On-site trial in Thailand. *Journal of Agricultural and Food Chemistry, 53*(15), 6056–6060. https://doi.org/10.1021/jf050604s

Su, T., Yang, M., Huang, H., Kuo, C., & Chen, L. (2021). Using sensory wheels to characterize consumers' perception for authentication of Taiwan specialty teas. *Foods, 10*(4), 836. https://doi.org/10.3390/foods10040836

Su, D., He, J., Zhou, Y., Li, Y., & Zhou, H. (2022). Aroma effects of key volatile compounds in Keemun black tea at different grades: HS-SPME-GC-MS, sensory evaluation, and chemometrics. *Food Chemistry, 373*, Article 131587. https://doi.org/10.1016/j.foodchem.2021.131587

Sun, Q., Wu, F., Wu, W., Yu, W., Zhang, G., Huang, X., et al. (2024a). Identification and quality evaluation of Lushan Yunwu tea from different geographical origins based on metabolomics. *Food Research International, 186*, 114379. https://doi.org/10.1016/j.foodres.2024.114379

Sun, Z., Lin, Y., Yang, H., Zhao, R., Zhu, J., & Wang, F. (2024b). Characterization of honey-like characteristic aroma compounds in Zunyi black tea and their molecular mechanisms of interaction with olfactory receptors using molecular docking. *LWT, 191*, Article 115640. https://doi.org/10.1016/j.lwt.2023.115640

Su, W., Liu, Z., Huang, S., Wang, D., Feng, X., Liu, Y., & Ni, L. (2025). Insight into aroma dynamic changes during the whole manufacturing process of smoked Lapsang Souchong tea. *Food Chemistry, 463*, Article 141498. https://doi.org/10.1016/j.foodchem.2024.141498

Sui, M., Wang, L., Xue, R., Xiang, J., Wang, Y., Yuan, Y., et al. (2025). The aroma formation from fresh tea leaves of Longjing 43 to finished Enshi Yulu tea at an industrial scale. *Journal of the Science of Food and Agriculture, 105*(1), 244–254. https://doi.org/10.1002/jsfa.13823

Tan, H. R., Lau, H., Liu, S. Q., Tan, L. P., Sakumoto, S., Lassabliere, B., et al. (2019). Characterisation of key odourants in Japanese green tea using gas chromatography-olfactometry and gas chromatography-mass spectrometry. *LWT, 108*, 221–232. https://doi.org/10.1016/j.lwt.2019.03.054

Tan, Y., Cao, J., Ma, Z., Sun, Y., Wan, X., & Liu, L. (2025). Post-harvest UV-B radiation enhances flavor quality of white and green teas. *Scientia Horticulturae, 345*, Article 114164. https://doi.org/10.1016/j.scienta.2025.114164

Tian, W., Feng, J., Wang, J., Lin, H., Chen, Q., Zhuang, J., et al. (2025). Different smoking processes with the special fuel rods: Impart a smoky aroma to Souchong black tea. *Food Chemistry: X, 25*, 102142. https://doi.org/10.1016/j.fochx.2024.102142

Wang, C., Zhang, C., Kong, Y., Peng, X., Li, C., Liu, S., et al. (2017). A comparative study of volatile components in Dianhong teas from fresh leaves of four tea cultivars by using chromatography-mass spectrometry, multivariate data analysis, and descriptive sensory analysis. *Food Research International, 100*, 267–275. https://doi.org/10.1016/j.foodres.2017.07.013

Wang, W., Jin, S., & Guo, Y. (2019). Exploration of a method of distinguishing different nongxiang tieguanyin tea grades based on aroma determined by GC-MS combined with chemometrics. *Molecules, 24*(9), 1707. https://doi.org/10.3390/molecules24091707

Wang, H., Hua, J., Jiang, Y., Yang, Y., Wang, J., & Yuan, H. (2020a). Influence of fixation methods on the chestnut-like aroma of green tea and dynamics of key aroma substances. *Food Research International, 136*, Article 109479. https://doi.org/10.1016/j.foodres.2020.109479

Wang, M., Ma, W., Shi, J., Zhu, Y., Lin, Z., & Lv, H. (2020b). Characterization of the key aroma compounds in Longjing tea using stir bar sorptive extraction (SBSE) combined with gas chromatography-mass spectrometry (GC-MS), gas chromatography-olfactometry (GC-O), odor activity value (OAV), and aroma recombination. *Food Research International, 130*, Article 108908. https://doi.org/10.1016/j.foodres.2019.108908

Wang, B., Meng, Q., Xiao, L., Li, R., Peng, C., Liao, X., et al. (2022a). Characterization of aroma compounds of Pu-erh ripen tea using solvent assisted flavor evaporation coupled with gas chromatography-mass spectrometry and gas chromatography-olfactometry. *Food Science and Human Wellness, 11*(3), 9.

Wang, J., Li, M., Wang, H., Huang, W., Li, F., Wang, L., et al. (2022b). Decoding the specific Roasty Aroma Wuyi Rock Tea (*Camellia sinensis*: *Dahongpao*) by the Sensomics approach. *Journal of Agricultural and Food Chemistry, 70*(34), 10571–10583. https://doi.org/10.1021/acs.jafc.2c02249

Wang, P., Yu, J., Jin, S., Chen, S., Yue, C., Wang, W., et al. (2022c). Genetic basis of high aroma and stress tolerance in the oolong tea cultivar genome. *Horticulture Research, 8*, 107. https://doi.org/10.1038/s41438-021-00542-x

Wang, Z., Wang, Z., Dai, H., Wu, S., Song, B., Lin, F., et al. (2022d). Identification of characteristic aroma and bacteria related to aroma evolution during long-term storage of compressed white tea. *Frontiers in Nutrition, 9*, 1092048. https://doi.org/10.3389/fnut.2022.1092048

Wang, B., Yu, M., Tang, Y., Wang, Y., Xia, T., & Song, H. (2023a). Characterization of odor-active compounds in Dahongpao Wuyi Rock Tea (*Camellia sinensis*) by sensory-directed flavor analysis. *Journal of Food Composition and Analysis, 123*, Article 105612. https://doi.org/10.1016/j.jfca.2023.105612

Wang, D., Liu, Z., Chen, W., Lan, X., Zhan, S., Sun, Y., et al. (2023b). Comparative study of the volatile fingerprints of roasted and unroasted oolong tea by sensory profiling and HS-SPME-GC-MS. *Current Research in Food Science, 6*, 100442. https://doi.org/10.1016/j.crfs.2023.100442

Wang, D., Wang, C., Su, W., Lin, C., Liu, W., Liu, Y., et al. (2023c). Characterization of the key aroma compounds in Dong Ding Oolong tea by application of the sensomics approach. *Foods, 12*(17), 3158. https://doi.org/10.3390/foods12173158

Wang, C. M., Tan, X. Q., Du, X., Hu, J. J., Li, X. Y., Yan, L. S., et al. (2024a). Dynamic changes and interaction between different aroma types during low-temperature roasting of bud green tea. *Food Science & Nutrition, 12*(9), 6367–6378. https://doi.org/10.1002/fsn3.4249

Wang, D., Liu, Z., Lan, X., Wang, C., Chen, W., Zhan, S., et al. (2024b). Unveiling the aromatic intricacies of Wuyi Rock Tea: A comparative study on sensory attributes and odor-active compounds of Rougui and Shuixian varieties. *Food Chemistry, 435*, 137470. https://doi.org/10.1016/j.foodchem.2023.137470

Wang, H., Bi, H., Wang, J., Li, M., Yao, X., Zhang, X., et al. (2024c). Key odorants of 'Keemun Aroma' and their changes during the preliminary processing of Keemun BLack Tea (*Camellia sinensis*). *Journal of Food Composition and Analysis, 132*, 106254. https://doi.org/10.1016/j.jfca.2024.106254

Wang, X., Cao, J., Cheng, X., Liu, X., Zhu, W., Li, Y., et al. (2024d). UV-B application during the aeration process improves the aroma characteristics of oolong tea. *Food Chemistry, 435*, 137585. https://doi.org/10.1016/j.foodchem.2023.137585

Wang, Z., Liang, Y., Gao, C., Wu, W., Kong, J., Zhou, Z., et al. (2024e). The flavor characteristics and antioxidant capability of aged Jinhua white tea and the mechanisms of its dynamic evolution during long-term aging. *Food Chemistry, 436*, 137705. https://doi.org/10.1016/j.foodchem.2023.137705

Wang, B., Huang, P., Lo, C., & Chang, W. (2025). Metabolomic analysis elucidates the dynamic changes in aroma compounds and the milk aroma mechanism across various portions of tea leaves during different stages of Oolong tea processing. *Food Research International, 209*, Article 116203. https://doi.org/10.1016/j.foodres.2025.116203

Wei, Y., Zhang, J., Li, T., Zhao, M., Song, Z., Wang, Y., & Ning, J. (2024). GC–MS, GC–O, and sensomics analysis reveals the key odorants underlying the improvement of yellow tea aroma after optimized yellowing. *Food Chemistry, 431*, Article 137139. https://doi.org/10.1016/j.foodchem.2023.137139

Wei, F., Luo, L., Sun, F., Zhang, B., Yang, J., Liu, Y., & Zeng, L. (2025). Flavor characteristic and characterization of key sweet aroma compounds in *Camellia nanchuanica* black tea. *Food Research International, 209*, Article 116179. https://doi.org/10.1016/j.foodres.2025.116179

Wu, H., Chen, Y., Feng, W., Shen, S., Wei, Y., Jia, H., et al. (2022a). Effects of three different withering treatments on the aroma of white tea. *Foods, 11*(16), 2502. https://doi.org/10.3390/foods11162502

Wu, L., Wang, Y., Liu, S., Sun, Y., Li, C., Lin, J., & Wei, S. (2022b). The stress-induced metabolites changes in the flavor formation of oolong tea during enzymatic-catalyzed process: A case study of Zhangping Shuixian tea. *Food Chemistry, 391*, Article 133192. https://doi.org/10.1016/j.foodchem.2022.133192

Wu, Q., Zhou, Z., Zhang, Y., Huang, H., Ou, X., & Sun, Y. (2023a). Identification of key components responsible for the aromatic quality of Jinmudan Black tea by means of molecular sensory science. *Foods, 12*(9), 1794. https://doi.org/10.3390/foods12091794

Wu, S., Gu, D., Chen, Y., Wang, F., Qian, J., Zeng, L., et al. (2023b). Variations in oolong tea key characteristic floral aroma compound contents among tea (*Camellia sinensis*) germplasms exposed to postharvest stress. *Postharvest Biology and Technology, 197*, 112201. https://doi.org/10.1016/j.postharvbio.2022.112201

Wu, H., Sheng, C., Lu, M., Ke, H., Li, T., Wei, Y., et al. (2024c). Identification of the causes of aroma differences in white tea under different withering methods by targeted metabolomics. *Food Bioscience, 59*, 104020. https://doi.org/10.1016/j.fbio.2024.104020

Wu, W., Zheng, Z., Wang, Z., He, B., Du, S., Zeng, W., & Sun, W. (2025a). Identification of key aroma compounds contributing to the pleasurable sensory experience of white Peony tea using GC–MS, computational modeling, and sensory evaluation. *Food Research International, 208*, Article 116280. https://doi.org/10.1016/j.foodres.2025.116280

Wu, Y., Li, T., Huang, W., Liu, Q., Deng, G., Zhang, J., et al. (2025b). Investigation of the aroma profile and blending strategy of Lu'an Guapian teas during grain rain period by sensory evaluation combined with SBSE-GC-MS, GC-O and OAV. *Food Chemistry, 463*, 141167. https://doi.org/10.1016/j.foodchem.2024.141167

Xia, D., Zhang, J., Xiong, Z., Huang, W., Wei, Y., Feng, W., et al. (2024). Effects of different fixation methods on the aroma quality of Anjibai tea. *LWT, 204*, 116430. https://doi.org/10.1016/j.lwt.2024.116430

Xiao, Z., Wang, H., Niu, Y., Liu, Q., Zhu, J., Chen, H., & Ma, N. (2017). Characterization of aroma compositions in different Chinese congou black teas using GC-MS and GC-O combined with partial least squares regression. *Flavour and Fragrance Journal, 32*(4), 265–276. https://doi.org/10.1002/ffj.3378

Xiao, M., Zheng, F., Xiao, M., Qi, A., Wang, H., & Dai, Q. (2022a). Contribution of aroma-active compounds to the aroma of Lu'an Guapian tea. *Flavour and Fragrance Journal, 37*(2), 83–95. https://doi.org/10.1002/ffj.3688

Xiao, Y., Huang, Y., Chen, Y., Zhu, M., He, C., Li, Z., et al. (2022b). Characteristic fingerprints and change of volatile organic compounds of dark teas during solid-state fermentation with *Eurotium cristatum* by using HS-GC-IMS, HS-SPME-GC-MS, E-nose and sensory evaluation. *Food science & technology, 169*, 113925. https://doi.org/10.1016/j.lwt.2022.113925

Xiao, Z., Cao, X., Zhu, J., Chen, F., & Niu, Y. (2022c). Characterization of the key aroma compounds in three world-famous black teas. *European Food Research and Technology, 248*(9), 2237–2252. https://doi.org/10.1007/s00217-022-04039-2

Xiao, Z., Shen, T., Niu, Y., Ke, Q., Sun, Z., Yang, E., et al. (2024). Unraveling the characteristic aroma compounds in Longjing tea and their interaction mechanisms with S-curve and broad-spectrum olfactory receptors using molecular docking. *Food Bioscience, 60*, 104423. https://doi.org/10.1016/j.fbio.2024.104423

Xie, J., Wang, L., Deng, Y., Yuan, H., Zhu, J., Jiang, Y., & Yang, Y. (2023). Characterization of the key odorants in floral aroma green tea based on GC-E-Nose, GC-IMS, GC-MS and aroma recombination and investigation of the dynamic changes and aroma formation during processing. *Food Chemistry, 427*, Article 136641. https://doi.org/10.1016/j.foodchem.2023.136641

Xu, S., Zeng, X., Wu, H., Shen, S., Yang, X., Deng, W., & Ning, J. (2021). Characterizing volatile metabolites in raw Pu'er tea stored in wet-hot or dry-cold environments by performing metabolomic analysis and using the molecular sensory science approach. *Food Chemistry, 350*, Article 129186. https://doi.org/10.1016/j.foodchem.2021.129186

Xu, C., Zhang, J., Pan, Y., Feng, X., Yan, A., Wang, X., et al. (2025). Formation of aroma characteristics driven by microorganisms during long-term storage of Liubao tea. *Food Chemistry, 476*, 143400. https://doi.org/10.1016/j.foodchem.2025.143400

Yang, Y., Hua, J., Deng, Y., Jiang, Y., Qian, M. C., Wang, J., et al. (2020). Aroma dynamic characteristics during the process of variable-temperature final firing of Congou black tea by electronic nose and comprehensive two-dimensional gas chromatography coupled to time-of-flight mass spectrometry. *Food Research International, 137*, 109656. https://doi.org/10.1016/j.foodres.2020.109656

Yang, P., Yu, M., Song, H., Xu, Y., Lin, Y., & Granvogl, M. (2022a). Characterization of key aroma-active compounds in rough and moderate Fire Rougui Wuyi Rock Tea (*Camellia sinensis*) by sensory-directed flavor analysis and elucidation of the influences of roasting on aroma. *Journal of Agricultural and Food Chemistry, 70*(1), 267–278. https://doi.org/10.1021/acs.jafc.1c06066

Yang, Y., Chen, J., Jiang, Y., Qian, M. C., Deng, Y., Xie, J., et al. (2022b). Aroma dynamic characteristics during the drying process of green tea by gas phase electronic nose and gas chromatography-ion mobility spectrometry. *Food science & technology, 154*, 112691. https://doi.org/10.1016/j.lwt.2021.112691

Yang, Y., Qian, M. C., Deng, Y., Yuan, H., & Jiang, Y. (2022c). Insight into aroma dynamic changes during the whole manufacturing process of chestnut-like aroma green tea by combining GC-E-Nose, GC-IMS, and GC × GC-TOFMS. *Food Chemistry, 387*, Article 132813. https://doi.org/10.1016/j.foodchem.2022.132813

Yang, Y., Zhu, H., Chen, J., Xie, J., Shen, S., Deng, Y., et al. (2022d). Characterization of the key aroma compounds in black teas with different aroma types by using gas chromatography electronic nose, gas chromatography-ion mobility spectrometry, and odor activity value analysis. *LWT, 163*, 113492. https://doi.org/10.1016/j.lwt.2022.113492

Yang, X., Peng, Z., Zhu, Q., Chen, C., Liu, J., Fu, S., & Zhang, J. (2023). Analysis of Tieguanyin aged aroma compounds and their correlation with microbial communities. *LWT, 185*, Article 115205. https://doi.org/10.1016/j.lwt.2023.115205

Yang, N., Simon, J., Fang, W., Ayed, C., Zhang, W. E., Axell, M., et al. (2025). Development of analytical "aroma wheels" for Oolong tea infusions (Shuixian and Rougui) and prediction of dynamic aroma release and colour changes during "Chinese tea ceremony" with machine learning. *Food Chemistry, 464*, 141537. https://doi.org/10.1016/j.foodchem.2024.141537

Yao, S., Guo, W., Lu, Y., & Jiang, Y. (2005a). Flavor characteristics of lapsang souchong and smoked lapsang souchong, a special Chinese black tea with pine smoking process. *Journal of Agricultural and Food Chemistry, 53*(22), 8688–8693. https://doi.org/10.1021/jf058059i

Yao, X., Li, Y., Tang, J., Yu, J., Zhang, Y., Wan, X., et al. (2025b). Characterization of cooked off-flavor volatile sulfur-containing compounds in green tea and their thermal inhibition via (-)-epigallocatechin gallate. *Food Chemistry, 463*, 141143. https://doi.org/10.1016/j.foodchem.2024.141143

Ye, F., Qiao, X., Gui, A., Liu, P., Wang, S., Wang, X., et al. (2022). Characterization of roasting time on sensory quality, color, taste, and nonvolatile compounds of Yuan An Yellow Tea. *Molecules, 27*(13), 4119. https://doi.org/10.3390/molecules27134119

Yin, P., Kong, Y., Liu, P., Jiang, C., Sun, M., Guo, G., & Liu, Z. (2022a). Temporal variation of the non-volatile compounds and key odorants in Xinyang Maojian green teas during the spring and autumn seasons. *Agronomy, 12*(5), 1085. https://doi.org/10.3390/agronomy12051085

Yin, P., Wang, J., Kong, Y., Zhu, Y., Zhang, J., Liu, H., et al. (2022b). Dynamic changes of volatile compounds during the Xinyang Maojian Green tea manufacturing at an industrial scale. *Foods, 11*(17), 2682. https://doi.org/10.3390/foods11172682

Yin, X., Wei, Y., Li, T., Zhang, J., Zou, L., Cui, Q., et al. (2023). Heterocyclic compounds formation in large-leaf yellow tea induced by the Maillard reaction at different roasting temperatures. *LWT, 182*, 114856. https://doi.org/10.1016/j.lwt.2023.114856

Yong-Mei, X. U., Fang-Bin, Q., & Ji-Kun, H. (2022). Black tea markets worldwide: Are they integrated? *Journal of Integrative Agriculture, 21*(2), 552–565. https://doi.org/10.1016/S2095-3119(21)63850-9

Yu, J., Liu, Y., Zhang, S., Luo, L., & Zeng, L. (2021). Effect of brewing conditions on phytochemicals and sensory profiles of black tea infusions: A primary study on the effects of geraniol and *β*-ionone on taste perception of black tea infusions. *Food Chemistry, 354*, Article 129504. https://doi.org/10.1016/j.foodchem.2021.129504

Yu, J., Ho, C., Lin, Z., Zhu, Y., Feng, Z., Ni, D., et al. (2023). Sensomics-assisted characterization of key flowery aroma compounds in Lu'an Guapian Green Tea Infusion (*Camellia sinensis*). *Journal of Agricultural and Food Chemistry,* 3c–486c. https://doi.org/10.1021/acs.jafc.3c00486

Yue, C., Cao, H., Zhang, S., Hao, Z., Wu, Z., Luo, L., & Zeng, L. (2023). Aroma characteristics of Wuyi rock tea prepared from 16 different tea plant varieties. *Food Chemistry: X, 17*, Article 100586. https://doi.org/10.1016/j.fochx.2023.100586

Zeng, L., Zhou, Y., Gui, J., Fu, X., Mei, X., Zhen, Y., et al. (2016). Formation of volatile tea constituent indole during the oolong tea manufacturing process. *Journal of Agricultural and Food Chemistry, 64*(24), 5011–5019. https://doi.org/10.1021/acs.jafc.6b01742

Zeng, L., Fu, Y., Huang, J., Wang, J., Jin, S., Yin, J., & Xu, Y. (2022a). Comparative analysis of volatile compounds in Tieguanyin with different types based on HS-SPME-GC-MS. *Foods, 11*(11), 1530. https://doi.org/10.3390/foods11111530

Zeng, L., Jin, S., Fu, Y. Q., Chen, L. S., Yin, J. F., & Xu, Y. Q. (2022b). A targeted and untargeted metabolomics analysis of 'Oriental Beauty' oolong tea during processing. *Beverage Plant Research, 2*(1), 164–173.

Zeng, L., Fu, Y., Liu, Y., Huang, J., Chen, J., Yin, J., et al. (2023). Comparative analysis of different grades of Tieguanyin oolong tea based on metabolomics and sensory evaluation. *LWT, 174*, 114423. https://doi.org/10.1016/j.lwt.2023.114423

Zeng, L., Liu, Y. Y., Wang, F., Jin, S., Chen, L. S., Fu, Y. Q., et al. (2024). Effect of tea green leafhopper (*Empoasca onukii* Matsuda) sucking on the quality of Oriental Beauty. *Food Frontiers, 5*(4), 1571–1582. https://doi.org/10.1002/fft2.415

Zhai, X., Wang, J., Wang, H., Xue, M., Yao, X., Li, M., et al. (2022). Formation of dimethyl sulfide from the decomposition of S-methylmethionine in tea (*Camellia sinensis*) during manufacturing process and infusion brewing. *Food Research International, 162*, 112106. https://doi.org/10.1016/j.foodres.2022.112106

Zhai, X., Hu, Y., Pei, Z., Yu, J., Li, M., Zhang, L., et al. (2023). Insights into the Key Odorants in Large-Leaf Yellow Tea (*Camellia sinensis*) by Application of the Sensomics Approach. *Journal of Agricultural and Food Chemistry, 71*(1), 690–699. https://doi.org/10.1021/acs.jafc.2c05881

Zhang, S., Sun, L., Yan, F., Li, Q., Chen, R., Zhang, Z., et al. (2023). Metabolomics analysis reveals the patterns and mechanisms of volatile metabolite in oolong tea during storage. *LWT, 185*, 115173. https://doi.org/10.1016/j.lwt.2023.115173

Zhang, H., Zhang, J., Liu, S., Li, T., Wei, Y., Gu, Z., et al. (2024). Characterization of the key volatile compounds in longjing tea (*Camellia sinensis*) with different aroma types at different steeping temperatures by GC-MS and GC-IMS. *LWT, 200*, 116183. https://doi.org/10.1016/j.lwt.2024.116183

Zhao, M., Zhou, Y., Xia, H., Peng, M., Li, J., Jin, L., et al. (2022). Quantitative microbiome analysis reveals the microbial community assembly along with its correlation with the flavor substances during the manufacturing process of Qingzhuan brick tea at an industrial scale. *LWT, 167*, 113835. https://doi.org/10.1016/j.lwt.2022.113835

Zhao, R., Yao, H., Hou, Z., Zhou, Q., Zhao, M., Wu, C., et al. (2024). Sensomics-assisted analysis unravels the formation of the Fungus Aroma of Fu Brick Tea. *Food Chemistry, 458*, 140174. https://doi.org/10.1016/j.foodchem.2024.140174

Zheng, F., Gan, S., Zhao, X., Chen, Y., Zhang, Y., Qiu, T., et al. (2023). Unraveling the chemosensory attributes of Chinese black teas from different regions using GC-IMS combined with sensory analysis. *LWT, 184*, 114988. https://doi.org/10.1016/j.lwt.2023.114988

Zheng, X., Zeng, X., Lin, X., Chen, D., Li, Y., Huang, J., et al. (2024). Exploring aromatic components differences and composition regularity of 5 kinds of these 4 aroma types Phoenix Dancong tea based on GC-MS. *Scientific Reports, 14*(1), 2727. https://doi.org/10.1038/s41598-024-53307-6

Zheng, X., Wang, L., Yao, H., Wang, J., An, H., Li, Q., et al. (2025a). Unraveling the unique profile of Fu brick tea: Volatile components, analytical approaches and metabolic mechanisms of key odor-active compounds. *Trends in Food Science & Technology, 156*, 104879. https://doi.org/10.1016/j.tifs.2025.104879

Zheng, X., Xie, H., Fu, W., Wang, C., Hu, T., Ou, X., et al. (2025b). Perceptual interactions of key aroma compounds in Fu brick tea: Odor threshold, sensory, and E-nose analysis. *Current Research in Food Science, 10*, 101004. https://doi.org/10.1016/j.crfs.2025.101004

Zheng, Z., Wu, W., Wang, Z., Zhang, P., Ma, B., Huang, Y., & Sun, W. (2025c). Molecular sensory basis of the unique "cocoa aroma" in white tea: An integrated study combining HS-SPME-GC-MS, GC-O, molecular docking, and addition tests. *Food Research International, 211*, Article 116413. https://doi.org/10.1016/j.foodres.2025.116413

Zhu, Y., Kang, S., Yan, H., Lv, H., Zhang, Y., & Lin, Z. (2020). Enantiomeric distributions of volatile lactones and terpenoids in white teas stored for different durations. *Food Chemistry, 320*, Article 126632. https://doi.org/10.1016/j.foodchem.2020.126632

Zhu, J., Niu, Y., & Xiao, Z. (2021a). Characterization of the key aroma compounds in Laoshan green teas by application of odour activity value (OAV), gas chromatography-mass spectrometry-olfactometry (GC-MS-O) and comprehensive two-dimensional gas chromatography mass spectrometry (GC × GC-qMS). *Food Chemistry, 339*, Article 128136. https://doi.org/10.1016/j.foodchem.2020.128136

Zhu, J., Zhu, Y., Wang, K., Niu, Y., & Xiao, Z. (2021b). Characterization of key aroma compounds and enantiomer distribution in Longjing tea. *Food Chemistry, 361*, Article 130096. https://doi.org/10.1016/j.foodchem.2021.130096

Zhou, H., Liu, Y., Yang, J., Wang, H., Ding, Y., & Lei, P. (2022). Comprehensive profiling of volatile components in Taiping Houkui green tea. *LWT, 163*, Article 113523. https://doi.org/10.1016/j.lwt.2022.113523

Zhu, J., Chen, F., Wang, L., Niu, Y., Yu, D., Shu, C., et al. (2015). Comparison of aroma-active volatiles in oolong tea infusions using GC-olfactometry, GC-FPD, and GC-MS. *Journal of Agricultural and Food Chemistry, 63*(34), 7499–7510. https://doi.org/10.1021/acs.jafc.5b02358

Zhu, Y., Shao, C., Lv, H., Zhang, Y., Dai, W., Guo, L., et al. (2017). Enantiomeric and quantitative analysis of volatile terpenoids in different teas (*Camellia sinensis*). *Journal of Chromatography A, 1490*, 177–190. https://doi.org/10.1016/j.chroma.2017.02.013

Zhu, Y., Lv, H., Shao, C., Kang, S., Zhang, Y., Guo, L., et al. (2018). Identification of key odorants responsible for chestnut-like aroma quality of green teas. *Food Research International, 108*, 74–82. https://doi.org/10.1016/j.foodres.2018.03.026

Zhu, J., Liu, X., Lin, Y., Ke, Q., Niu, Y., Zhang, J., et al. (2024a). Unraveling the characteristic chestnut aroma compounds in MeiTanCuiYa green tea and their interaction mechanisms with broad-spectrum olfactory receptors using molecular docking. *LWT, 194*, 115785. https://doi.org/10.1016/j.lwt.2024.115785

Zhu, W., Zhou, S., Guo, H., Hu, J., Cao, Y., Xu, Y., et al. (2024b). Golden-flower fungus (*Eurotiwm Cristatum*) presents fungal flower aroma as well as accelerates the aging of white tea (Shoumei). *Food Chemistry, 451*, 139452. https://doi.org/10.1016/j.foodchem.2024.139452

Zhu, J., Liu, X., Sun, Z., Shen, T., Wang, Z., Yang, E., et al. (2025a). Reviewing the mechanisms of the interaction of characteristic aroma compounds of Longjing Tea with S-curve, molecular docking, and molecular dynamics simulation: with a focus on (+)-(1R, 2S)-Methyl Epijasmonate and β-Ionone as examples. *Journal of Agricultural and Food Chemistry, 73*(6), 3645–3656. https://doi.org/10.1021/acs.jafc.4c08754

Zhu, W., Feng, X., Pan, Y., Guo, H., Liu, Y., Lin, X., et al. (2025b). Flowering in aged white tea: Recovering umami taste and amplifying of stale aroma. *Food Chemistry, 465*, 141649. https://doi.org/10.1016/j.foodchem.2024.141649

Zou, L., Sheng, C., Xia, D., Zhang, J., Wei, Y., & Ning, J. (2024). Mechanism of aroma formation in white tea treated with solar withering. *Food Research International, 194*, Article 114917. https://doi.org/10.1016/j.foodres.2024.114917

Chapter 5
The Effects of Processing on Tea Aroma Quality

Yujie Wang

Abstract Aroma is an important quality attribute of tea. Through different processing techniques, tea leaves present diverse and fascinating aroma flavors. In recent years, numerous efforts have been invested to promote the production and accumulation of aroma volatiles in tea, which include enzymatic, thermal and microbial-induced metabolism. The volatile changes and aroma flavor formation induced by processing techniques, which have received much attention, are discussed in this chapter.

Keywords Aroma quality · Volatile components · Processing technologies · Volatile precursors · Carotenoids · Lipid pathway · Glycosides

5.1 Introduction

More than 700 volatile compounds have been identified in tea, and they are produced through four main pathways: carotenoid precursors, lipid precursors, glycoside precursors, and the Maillard reaction pathway (Feng et al., 2019). This section summarizes the changes in aroma volatiles at various stages of tea production.

5.2 Aroma Precursors in Fresh Leaves

Fresh tea leaves are virtually odorless, but they store numerous precursors of aroma volatiles, including carotenoids, lipids, and glycosides (Ho et al., 2015).

(a) *Carotenoids* Carotenoids, such as β-carotene, lutein, zeaxanthin, neoxanthin, lutein, and lycopene, are housed in the chloroplast cells of leaves, and are an essential component of plant photosynthesis. These carotenoids were

Y. Wang (✉)
College of Biosystems Engineering and Food Science, Zhejiang University, Hangzhou, China
e-mail: wangyjnet@163.com

X. Zhai (ed.), *Tea Chemistry and Health Benefits*,
https://doi.org/10.1007/978-3-032-11828-8_5

identified as aroma precursors of β-ionone, β-damascenone, and oxygenated thapsigarginone derivatives (Shi et al., 2023).

(b) *Lipids*: Lipids Particularly unsaturated fatty acids like α-linolenic, linoleic, oleic, and palmitoleic acids, were recognized as precursors of six- to ten-carbon aroma compounds such as (*E*)-2-hexanal, (*E*)-2-hexanol, and (*Z*)-3-hexanol, which impart a fresh or grassy aroma to the tea infusion (Ho et al., 2015).

(c) *Glycosides* Glycosides, sugar portions linked to functional groups through glycosidic bonds, are stored in the leaf vesicles. Under conditions of abiotic and biotic stress, they produced volatile compounds for chemical defense or biological communication signals through enzymatic or acidic hydrolysis. During the tea processing, enzymes were released from damaged tea tissues into cell walls or lumens to hydrolyze glycosidic bonds, thereby releasing volatile aromas such as monoterpene alcohols (linalool, linalool oxides, and geraniol) or aromatic alcohols (benzyl alcohol and phenylethanol) (Song et al., 2018).

5.3 Withering and Aroma Modulation

Withering, identified as the initial step in the processing of oolong, black, and white teas to augment flavor, was found to be positively modulated by sunlight, particularly enhancing the floral and fruity flavors of oolong tea. In general, there are three widely applied withering methods including sunlight withering, indoor (natural) withering, and warm-air withering (Fig. 5.1).

During the withering process, the fresh tea leaves underwent water depletion stress and experienced changes in aroma substances, among which the content of fatty acid-derived volatiles was significantly impacted. Specifically, (*Z*)-3-hexenal and (*E*)-2-hexenal, known for their grassy odor and being the most abundant volatiles in fresh leaves, present a significant reduction in content after withering. Furthermore, terpene volatiles, including β-terpene, limonene, γ-terpinene, α-farnesene, and β-ocimene, were reported to be more abundant in sunlight withering than in indoor withering, indicating that sunlight withering may be more conducive to the development of high-quality oolong tea (Zhu et al., 2019).

Moreover, sunlight withering was shown to significantly enhance the floral and fruity aroma of black tea, while warm-air withering was associated with an increase in the sweet and roasted aroma of the same tea. The higher temperature and air circulation rate provided by both sunlight withering and the warm-air withering were credited with resulting in faster water dissipation and an enhanced activity of endogenous hydrolases. This, in turn, promoted enzymatic hydrolysis and oxygen utilization of proteins, polysaccharides, lipids, and other macromolecules, leading to the production of a significant number of volatiles.

Additionally, sunlight withering was found to effectively increase the aroma concentration of glycoside and fatty acid degradation products, thereby enhancing the floral and fruity aroma of black tea, while the enhancement of the baking aroma

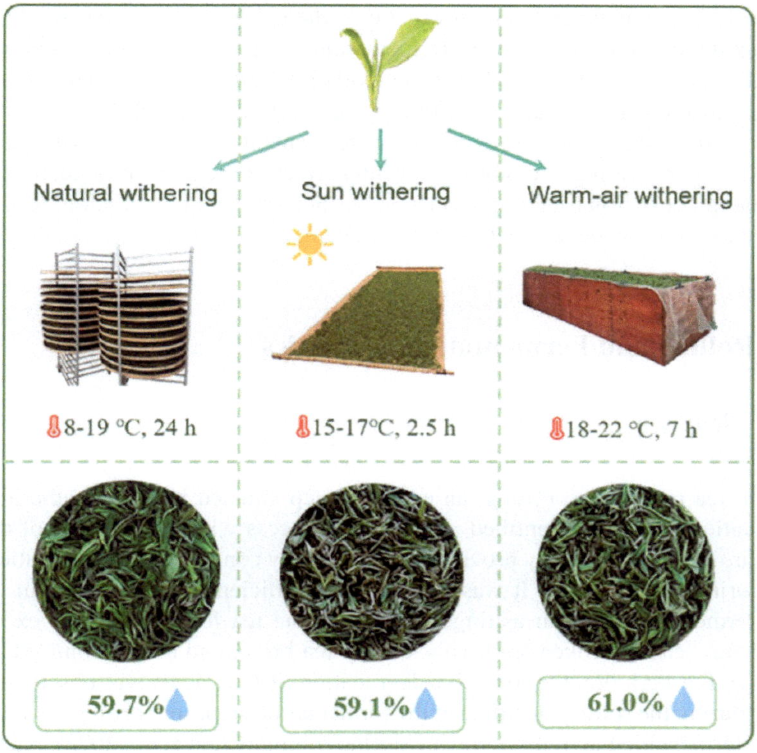

Fig. 5.1 Some typical withering methods used for processing black tea (Huang et al., 2023)

during warm-air withering was attributed to the degradation of amino acids (Huang et al., 2022).

In comparison with other tea varieties, white tea undergoes a unique processing method that consists solely of an extended period of withering and drying, without any enzyme inactivation or fermentation process. It was observed that sunlight withering enhances the aroma of white tea, while trough withering tends to enhance its grassy flavor. Nine compounds, including dimethyl sulfide, 2-methylbutyraldehyde, 1-penten-3-one, hexanal, (*Z*)-4-heptenal, β-ionone, linalool, geraniol, and β-ocimene, were identified as key aroma-active compounds that differentiate white tea produced by varying withering methods. Among these, linalool, geraniol, and β-ocimene were found to be the most significant contributors to the pronounced floral aroma characteristic of white tea subjected to sunlight withering. Additionally, volatile compounds such as β-ionone and phenylglyoxal, known for their floral aromas, were also identified as contributors to the floral aroma profile of white tea (Chen et al., 2019).

During the withering stage, sunlight was suggested to be a critical factor influencing the formation of floral aroma. Two pathways for the degradation of carotenoids, enzymatic and non-enzymatic oxidation, were discussed. While enzymatic oxidation is associated with fermentation or thermal drying, non-enzymatic oxidation is linked to photo-oxidation, autoxidation, and thermal degradation. It was indicated by low carotenoid cleavage dioxygenase activity and high β-ionone concentration during sunlight withering to low water content that the enzymatic reaction induced by solar wilting does not significantly affect carotenoids. Instead, the photo-oxidative process was found to accelerate the non-enzymatic oxidation, leading to an increase in β-ionone concentration (Kawakami et al., 2002; Shi et al., 2023).

5.4 Rolling and Fermentation Dynamics

5.4.1 Rolling

In black tea processing, rolling, an essential step situated between withering and fermentation, has been identified as a critical process wherein the cells of the tea leaves are disrupted, thereby providing the necessary conditions for enzymatic reactions during fermentation. It was found that insufficient rolling results in inadequate fermentation and an astringent flavor in the tea leaves, whereas excessive rolling may lead to broken tea leaves, cloudy tea broth, and a compromised aroma and taste. Furthermore, the oxidative degradation of fatty acids was observed to be greater during the rolling step than in the fermentation step, supporting the notion of intense fermentation activity occurring during rolling (Chen et al., 2022). Additionally, lipoxygenase (LOX) activity was maintained at a high level, and the degradation of linolenic and linoleic acids, unsaturated fatty acids, was found to be greater than that of other fatty acids, underscoring the propensity of unsaturated fatty acids for more rapid degradation compared to saturated fatty acids.

5.4.2 Fermentation in Black Tea

Fermentation, a pivotal process determining the flavor of black tea, was identified to differ between the dark and black teas. The former was found to undergo post-fermentation, dominated by microorganisms, while the latter was pre-fermented, predominantly influenced by endogenous enzymes. Through the fermentation process in black tea, it was observed that the aroma gradually transitioned from a robust grassy flavor to a sweet floral or fruity aroma (Chen et al., 2022). Fatty acids, encompassing linolenic, linoleic, palmitic, and oleic acids, were established as well-known aroma precursors in plants. Upon oxidation, these molecules were found to form C6–C9 volatile alcohols and aldehydes, contributing to green, grassy, or fresh odors. A significant role was attributed to many carotenoid derivatives in determining the aroma quality of black tea, with β-ionone, β-damascenone, geranylacetone, and α-ionone, being volatile compounds that contribute to the floral and fruity aroma of black tea (Chen et al., 2022). Most glycosides were found to be stored in the vesicles, with glycosidases being most abundant in the cytoplasm and cell wall. A significant decrease in the number of glycosides post-rolling was recorded, with further reduction to basal levels during fermentation (Hu et al., 2018).

5.4.3 *Pile-Fermentation in Dark Tea*

Eurotium cristatum, the primary fungus in the production of brick dark tea, was found to exert the most significant impact on the aroma characteristics of dark tea, especially during mid-fermentation. The pile-fermentation process facilitated by *Eurotium cristatum* was observed to enhance the level of volatiles exuding aged and floral aromas. Sensory evaluation revealed that the green aroma's intensity considerably diminished as fermentation progressed, supplanted by floral, minty, woody, and fungal floral notes. Hydrolytic enzymes secreted by *Eurotium cristatum* were found to catalyze a series of biochemical reactions contributing to the formation of flavor metabolites (Xiao et al., 2022). The transformation of Pu-erh tea from floral and fruity aroma to an aged and woody flavor was noted to occur through pile-fermentation. Microbial activity was identified as a crucial factor in the production of the stale (musty) aroma of Pu-erh tea during the pile fermentation process, while the humidity-heat effect was primarily responsible for the flavor characteristics of plum aroma, as well as the sweetness of Pu-erh tea (Li et al., 2024).

5.5 Fixation and Thermal Effects

Fixation, a critical step in green tea production, was noted to involve the inactivation of enzymes through high temperatures, cessation of enzyme oxidation, emission of grassy flavors, and promotion of aroma release (Xia et al., 2024). The typical fixation methods for green tea includes electromagnetic roller-hot air coupling fixation (ROHF), electromagnetic roller-steam coupling fixation (ROSF), electromagnetic roller-hot air–steam coupling fixation (RHSF), electromagnetic single roller fixation (ROLF), and carding fixation (CDF) (Fig. 5.2). Fixation was determined to have the most significant influence on the volatile metabolites of green tea, outperforming drying and rolling. The selection of the fixation method, which varied in temperature, duration, and heating technique, was found to profoundly influence the formation of volatile compounds that shaped the unique aroma profile of tea.

Large amounts of low-boiling, grassy-air volatile metabolites were dissipated during the fixation process. The formation of esters, ketones, aldehydes, and other metabolites with chestnut, floral, and fruity aromas was facilitated (Wang et al., 2021).

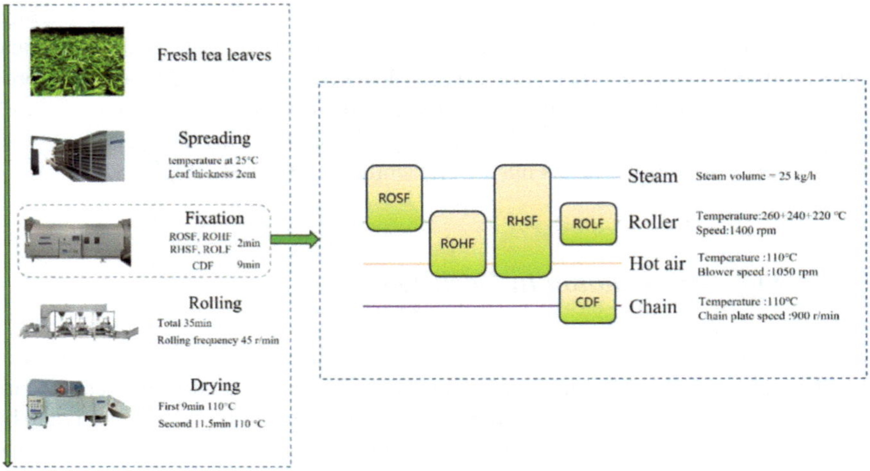

Fig. 5.2 Some typical fixation methods used for processing green tea (Zhu et al., 2025). ROHF, electromagnetic roller-hot air coupling fixation; ROSF, electromagnetic roller-steam coupling fixation; RHSF, electromagnetic roller-hot air–steam coupling fixation; ROLF, electromagnetic single roller fixation; CDF, carding fixation

The fixation process resulted in a more even heating of the tea sample and stimulated the occurrence of the Maillard reaction, leading to the production of pyrazine compounds. Roller hot air fixation was deemed as the most effective technique for generating floral aromas (Xia et al., 2024). Traditional wok fixation, which involved rapid frying at high temperatures to release the low-boiling volatiles in tea leaves, was observed to enhance the tea's high aroma. While this method typically resulted in tea retaining more of its natural aroma, it could also yield a burnt flavor if not appropriately controlled.

5.6 Shaking and Yellowing Processes

5.6.1 Shaking in Oolong Tea

Shaking is a key process in oolong tea processing, which promotes the characteristics of oolong tea's strong floral flavor. Under the dual stresses of mechanical damage and low temperature, JA was found to accumulate rapidly and regulate the formation of volatile compounds (Ma et al., 2018; Wu et al., 2023). Prolonged accumulation of JA resulted in the significant generation of key aroma compounds, including indole, jasmonolactone, and (*E*)-nerolidol, in response to the oolong tea processing. The effect of shaking on volatiles was observed to surpass that of sunlight withering. After shaking, the concentrations of hexyl esters, linalool oxides, aromatic alcohols, phenylacetaldehyde, sesquiterpenes, *cis*-jasmone, jasmolide, benzyl cyanide, and indole were markedly increased, favoring the formation of floral and fruity aroma of oolong tea. In-depth molecular biology experiments have facilitated the excavation of key genes that cause alterations in aroma volatiles. Methyl jasmonate (MeJA) is an important fatty acid derivative and contributes to the floral aroma of oolong tea. A recent study by Feng et al. (2021) focused on the first jasmonate carboxymethyl-transferase (JMT) gene. They found that the recombinant CsJMT protein was able to transfer methyl groups to jasmonic acid in vitro, producing orchid-like volatile 1R, 2R-MeJA. Further studies showed that mechanical damage (shaking) significantly induced the expression of CsJMT and the accumulation of MeJA during processing (Feng et al., 2021). Besides, the synthases related to (*E*)-nerolidol (Zhou et al., 2017) and β-ocimene (Chen et al., 2021) were successfully identified and functionally characterized.

5.6.2 Yellowing in Yellow Tea

The yellowing process, a unique step of yellow tea, involves the gradual yellowing of tea leaves under the influence of high temperature and high humidity. It was observed that the aroma of yellow tea is primarily sweet, fruity, and floral, distinguishing it from green tea, which is characterized mainly by a chestnut-like aroma. This differentiation is attributed to the addition of the yellowing process after the completion of fixation in yellow tea processing.

Following the yellowing process, it was found that aromatic compounds with high boiling points, low thresholds, and elegant aromas increased, contributing to yellow tea's high intensity, long-lasting sweetness, and fruity aroma. Concurrently, volatiles with low boiling points decreased, leading to the gradual diminishment of chestnut and grassy aromas (Wei et al., 2022). The yellowing process was shown to facilitate the hydrolysis of glycosides in tea, releasing part of the aroma substances, and to promote the decarboxylation, hydrolysis, and Maillard reaction of amino acids, as well as the degradation of carotenoids, all providing reaction substrates or precursors for aroma substances.

The non-enzymatic hydrolysis of glycoside precursors after the yellowing treatment was promoted, leading to the release of linalool, geraniol alcohol, and hexenol. Linalool and geraniol, identified as the compounds with the lowest thresholds in all samples, were found to play crucial roles in the aroma formation of yellow tea. The increase in linalool was linked to the long-lasting aroma of yellow tea. Geraniol, known for its mild sweet rose flavor, and its increase, contributed to the sweet aroma of yellow tea (Dong et al., 2023; Zhai et al., 2022). During the yellowing process, it was observed that the contents of phenylacetaldehyde, benzaldehyde, phenylethanol, benzyl alcohol, methyl salicylate, and geraniol increased. This enhancement and synergistic effect of these sweet aroma components resulted in the formation of a characteristic yellow tea aroma profile, which was dominated by a sweet floral aroma. A decrease in the content of (*Z*)-3-hexen-1-ol, 1-hexanol, heptanal, and 1-octanol

Fig. 5.3 Some typical drying methods used for processing green tea (Zhang et al., 2023)

led to a reduction in the grassy aroma. The decline in jasmone and β-ionone resulted in the loss of the chestnut-like aroma. These changes altered the aroma from the chestnut-like and fresh aroma characteristic of green tea to the sweet fruity aroma of yellow tea. However, it was noted that yellow tea exhibited a watery stuffiness after prolonged yellowing due to the presence of isoamyl alcohol, which was a result of the extended yellowing process (Dong et al., 2023).

5.7 Drying and Roasting Impacts

5.7.1 Drying Process

Drying, as the final process of tea processing, was found to produce different aroma profiles in tea leaves depending on the temperature used (Fig. 5.3). Traditionally, warm-air drying has been applied for a long history, while vacuum freeze-drying, as a new drying method, attracted tea factories very well.

It was noted that warm-air drying could increase the pyrroles, furans, pyrazines, thereby forming the aroma characteristics of dried black tea. On the other hand, vacuum freeze-drying was found to retain more of the original flavor, imparting the tea with a natural floral and fruity aroma, a unique attribute of oolong tea.

5.7.2 Roasting Process

Prolonged high-temperature roasting was shown to induce strong chemical reactions, such as the Maillard reaction, caramelization reaction, and the degradation of carotenoids, lipids, and glycosides. After the first or second round of roasting, oolong tea exhibited a distinct grassiness, a phenomenon referred to as "re-greening", which was largely eliminated by high-temperature roasting (Zhan et al., 2023). It was observed that roasting at low temperatures favored the formation of green tea aroma, whereas roasting at high temperatures led to the formation of black tea aroma (Wang et al., 2022). As the degree of roasting increased, the aroma of green tea was found to change from light to floral, fruity, and sweet, and then to roasted, nutty, or even burnt. The Maillard reaction was identified as one of the most important reactions in the roasting process, where the reaction of sugars with amino acids produced brown pigments with a typical roasting aroma. The proportion of Maillard reaction products (e.g., pyrazines, pyrroles, furans, and their derivatives) in the total volatile matter was seen to increase (Zhu et al., 2021). *N*,*O*-heterocycles, the important odorants of the Maillard reaction, were commonly associated with pleasant roasts, nuts, grains, licorice, coffee, and wood due to their strong odors and low odor thresholds (Yang et al., 2022). The characteristic crunchy rice flavor and the absence of a burnt taste were identified in the high-temperature baking of large-leaf yellow tea. In large-leaf yellow tea, the roasted and floral aroma was found to be provided by 2,3-diethyl-5-methylpyrazine and (*E*)-β-ionone, both of which exhibited high odor activity values (Guo et al., 2021). High-temperature baking, a characteristic process of *Lu'an Guapian* green tea, was shown to favor the formation and accumulation of (*E*)-β-ionone, γ-hexalactone, and dimethyl sulfide at higher roasting temperatures. Additionally, odorants produced from lipids, carotenoids, and glycosides were more likely to be affected by roasting temperatures than by roasting time (Yu et al., 2024).

5.8 Conclusion

Tea aroma arises from a complex interplay of volatiles shaped by processing. Strategic modulation of withering, rolling, fermentation, and thermal steps enable precise control over aroma profiles. Understanding these mechanisms is pivotal for optimizing tea quality and meeting diverse consumer preferences.

References

Chen, Q., Zhu, Y., Dai, W., Lv, H., Mu, B., Li, P., Tan, J., Ni, D., & Lin, Z. (2019). Aroma formation and dynamic changes during white tea processing. *Food Chemistry, 274*, 915–924. https://doi.org/10.1016/j.foodchem.2018.09.072

Chen, Q., Zhu, Y., Liu, Y., Liu, Y., Dong, C., Lin, Z., & Teng, J. (2022). Black tea aroma formation during the fermentation period. *Food Chemistry, 374*, 131640. https://doi.org/10.1016/j.foodchem.2021.131640

Chen, S., Xie, P., Li, Y., Wang, X., Liu, H., Wang, S., Han, W., Wu, R., Li, X., Guan, Y., Yang, Z., & Yu, X. (2021). New insights into stress-induced β-ocimene biosynthesis in tea (Camellia sinensis) leaves during oolong tea processing. *Journal of Agricultural and Food Chemistry, 69*, 11656–11664. https://doi.org/10.1021/acs.jafc.1c04378

Dong, R.-J., Sheng, X.-H., Xie, Q.-T., Huang, X., Yan, F., & Liu, S. (2023). Aroma formation and transformation during sealed yellowing process of Pingyang yellow tea. *Food Research International, 165*, 112535. https://doi.org/10.1016/j.foodres.2023.112535

Feng, Y., Wang, J., Zhao, Y., Zhang, M., Zhou, Z., Li, Y., Hu, Y., Wu, Y., Feng, Z., Schwab, W., Wan, X., & Song, C. (2021). Biosynthesis of orchid-like volatile methyl jasmonate in tea (Camellia sinensis) leaves in response to multiple stresses during the shaking process of oolong tea. *LWT, 143*, 111184. https://doi.org/10.1016/j.lwt.2021.111184

Feng, Z., Li, Y., Li, M., Wang, Y., Zhang, L., Wan, X., & Yang, X. (2019). Tea aroma formation from six model manufacturing processes. *Food Chemistry, 285*, 347–354. https://doi.org/10.1016/j.foodchem.2019.01.174

Guo, X., Ho, C.-T., Schwab, W., & Wan, X. (2021). Effect of the roasting degree on flavor quality of large-leaf yellow tea. *Food Chemistry, 347*, 129016. https://doi.org/10.1016/j.foodchem.2021.129016

Ho, C.-T., Zheng, X., & Li, S. (2015). Tea aroma formation. *Food Science and Human Wellness, 4*(1), 9–27. https://doi.org/10.1016/j.fshw.2015.04.001

Hu, C.-J., Li, D., Ma, Y.-X., Zhang, W., Lin, C., Zheng, X.-Q., Liang, Y.-R., & Lu, J.-L. (2018). Formation mechanism of the oolong tea characteristic aroma during bruising and withering treatment. *Food Chemistry, 269*, 202–211. https://doi.org/10.1016/j.foodchem.2018.07.016

Huang, W., Fang, S., Su, Y., Xia, D., Wu, Y., Liu, Q., Zou, L., Lu, C., Deng, W.-W., & Ning, J. (2023). Insights into the mechanism of different withering methods on flavor formation of black tea based on target metabolomics and transcriptomics. *LWT, 189*, 115537. https://doi.org/10.1016/j.lwt.2023.115537

Huang, W., Fang, S., Wang, J., Zhuo, C., Luo, Y., Yu, Y., Li, L., Wang, Y., Deng, W.-W., & Ning, J. (2022). Sensomics analysis of the effect of the withering method on the aroma components of Keemun black tea. *Food Chemistry, 395*, 133549. https://doi.org/10.1016/j.foodchem.2022.133549

Kawakami, M. (2002). Carotenoid-derived aroma compounds in Tea. https://doi.org/10.1021/bk-2002-0802.ch011

Li, T., Wei, Y., Lu, M., Wu, Y., Jiang, Y., Ke, H., Shao, A., & Ning, J. (2024). Exploring microbial and moist-heat effects on Pu-erh tea volatiles and understanding the methoxybenzene formation mechanism using molecular sensory science. *Food Chemistry: X, 23*, 101553. https://doi.org/10.1016/j.fochx.2024.101553

Ma, C., Li, J., Chen, W., Wang, W., Qi, D., Pang, S., & Miao, A. (2018). Study of the aroma formation and transformation during the manufacturing process of oolong tea by solid-phase micro-extraction and gas chromatography-mass spectrometry combined with chemometrics. *Food Research International, 108*, 413–422. https://doi.org/10.1016/j.foodres.2018.03.052

Shi, J., Wu, W., Zhang, Y., Baldermann, S., Peng, Q., Wang, J., Xu, L., Yang, G., Fu, J., Lv, H., & Lin, Z. (2023). Comprehensive analysis of carotenoids constituents in purple-colored leaves and carotenoid-derived aroma differences after processing into green, black, and white tea. *LWT, 173*, 114286. https://doi.org/10.1016/j.lwt.2022.114286

Song, C., Härtl, K., McGraphery, K., Hoffmann, T., & Schwab, W. (2018). Attractive but toxic: Emerging roles of glycosidically bound volatiles and glycosyltransferases involved in their formation. *Molecular Plant, 11*(10), 1225–1236. https://doi.org/10.1016/j.molp.2018.09.001

Wang, B., Qu, F., Wang, P., Zhao, L., Wang, Z., Han, Y., & Zhang, X. (2022). Characterization analysis of flavor compounds in green teas at different drying temperature. *LWT, 161*, 113394. https://doi.org/10.1016/j.lwt.2022.113394

Wang, H., Hua, J., Yu, Q., Li, J., Wang, J., Deng, Y., Yuan, H., & Jiang, Y. (2021). Widely targeted metabolomic analysis reveals dynamic changes in non-volatile and volatile metabolites during green tea processing. *Food Chemistry, 363*, 130131. https://doi.org/10.1016/j.foodchem.2021.130131

Wei, Y., Yin, X., Wu, H., Zhao, M., Huang, J., Zhang, J., Li, T., & Ning, J. (2022). Improving the flavor of summer green tea (*Camellia sinensis* L.) using the yellowing process. *Food Chemistry, 388*, 132982. https://doi.org/10.1016/j.foodchem.2022.132982

Wu, S., Gu, D., Chen, Y., Wang, F., Qian, J., Zeng, L., Tang, J., Yan, Y., Chen, C., & Li, J. (2023). Variations in oolong tea key characteristic floral aroma compound contents among tea (*Camellia sinensis*) germplasms exposed to postharvest stress. *Postharvest Biology and Technology, 197*, 112201. https://doi.org/10.1016/j.postharvbio.2022.112201

Xia, D., Zhang, J., Xiong, Z., Huang, W., Wei, Y., Feng, W., Huang, J., & Ning, J. (2024). Effects of different fixation methods on the aroma quality of Anjibai tea. *LWT, 204*, 116430. https://doi.org/10.1016/j.lwt.2024.116430

Xiao, Y., Huang, Y., Chen, Y., Zhu, M., He, C., Li, Z., Wang, Y., & Liu, Z. (2022). Characteristic fingerprints and change of volatile organic compounds of dark teas during solid-state fermentation with *Eurotium cristatum* by using HS-GC-IMS, HS-SPME-GC-MS, E-Nose and Sensory Evaluation. *LWT, 169*, 113925. https://doi.org/10.1016/j.lwt.2022.113925

Yang, Y., Xie, J., Chen, J., Deng, Y., Shen, S., Hua, J., Wang, J., Zhu, J., Yuan, H., & Jiang, Y. (2022). Characterization of N,O-heterocycles in green tea during the drying process and unraveling the formation mechanism. *Food Control, 139*, 109079. https://doi.org/10.1016/j.foodcont.2022.109079

Yu, J., Li, J., Lin, Z., Zhu, Y., Feng, Z., Ni, D., Zeng, S., Zeng, X., Wang, Y., Ning, J., Zhang, L., Wan, X., & Zhai, X. (2024). Dynamic changes and the effects of key procedures on the characteristic aroma compounds of Lu'an Guapian green tea during the manufacturing process. *Food Research International, 188*, 114525. https://doi.org/10.1016/j.foodres.2024.114525

Zhai, X., Zhang, L., Granvogl, M., Ho, C. T., & Wan, X. (2022). Flavor of tea (*Camellia sinensis*): A review on odorants and analytical techniques. *Comprehensive Reviews in Food Science and Food Safety, 21*(5), 3867–3909. https://doi.org/10.1111/1541-4337.12999

Zhan, S., Liu, Z., Su, W., Lin, C.-C., & Ni, L. (2023). Role of roasting in the formation of characteristic aroma of Wuyi rock tea. *Food Control, 147*, 109614. https://doi.org/10.1016/j.foodcont.2023.109614

Zhang, J., Xia, D., Li, T., Wei, Y., Feng, W., Xiong, Z., Huang, J., Deng, W.-W., & Ning, J. (2023). Effects of different over-fired drying methods on the aroma of Lu'an Guapian tea. *Food Research International, 173*, 113224. https://doi.org/10.1016/j.foodres.2023.113224

Zhou, Y., Zeng, L., Liu, X., Gui, J., Mei, X., Fu, X., Dong, F., Tang, J., Zhang, L., & Yang, Z. (2017). Formation of (E)-nerolidol in tea (Camellia sinensis) leaves exposed to multiple stresses during tea manufacturing. *Food Chemistry, 231*, 78–86. https://doi.org/10.1016/j.foodchem.2017.03.122

Zhu, C., Zhang, S., Fu, H., Zhou, C., Chen, L., Li, X., Lin, Y., Lai, Z., & Guo, Y. (2019). Transcriptome and phytochemical analyses provide new insights into long non-coding RNAs modulating characteristic secondary metabolites of oolong tea (*Camellia sinensis*) in solar-withering. *Frontiers in Plant Science, 10*, 1638. https://doi.org/10.3389/fpls.2019.01638

Zhu, Y.-M., Dong, J. J., Jin, J., Liu, J.-H., Zheng, X.-Q., Lu, J.-L., Liang, Y.-R., Ye, J.-H. (2021). Roasting process shaping the chemical profile of roasted green tea and the association with aroma features. *Food Chemistry, 353*, 129428. https://doi.org/10.1016/j.foodchem.2021.129428

Zhu, X., Yang, L., Ge, Z., Ouyang, W., Wang, J., Chen, M., Yu, Y., Wu, S., Qin, Y., Huang, C., Zhang, G., Zhang, Y., Yuan, H., Jiang, Y., & Hua, J. (2025). Non-volatile and volatile metabolite analyses and objective quantitative technique reveal the effect of fixation methods on the flavor quality and metabolites of green tea. *Current Research in Food Science, 10*, 101037. https://doi.org/10.1016/j.crfs.2025.101037

Chapter 6
Taste Characteristics in Six Types of Tea

Zisheng Han

Abstract This chapter systematically examines the taste characteristics of six major tea types, focusing on key nonvolatile compounds and their sensory impacts. Bitter compounds, primarily caffeine (threshold: 500 μM), catechins (e.g., EGCG), and anthocyanins-activate TAS2R receptors, with caffeine synergistically enhancing EGCG bitterness. Astringency arises from galloylated catechins, flavonol glycosides, and phenolic acids, notably 4-*O*-*p*-coumaroylquinic acid. Umami taste is dominated by L-theanine, acting via T1R1–T1R3 receptors, and augmented by Maillard reaction intermediates and γ-glutamyl peptides. Sweetness derives from soluble sugars, while sweet aftertaste is enhanced by hydrolyzed catechins (EGC/EC), hydrolysable tannins, and aroma-sweetness interactions. Critical findings include structural features dictate taste thresholds, compound interactions profoundly modulate sensory profiles, and tea type significantly influences taste compound abundance, with amino acids peaking in green/white teas and theaflavins governing black tea astringency. Unresolved taste characteristics and thresholds highlight research gaps.

Keywords Tea · Taste · Theanine · Catechins · Maillard reaction

6.1 Introduction

The taste of tea beverage is one of the most important and intuitive criteria for customers to assess the quality of tea and decide whether to enjoy it. According to chapter of book *Advances in Food and Nutrition Research* (Yu et al., 2023), the taste quality and threshold values of major taste active compounds in tea were summarized and shown in Table 6.1.

Z. Han (✉)
State Key Laboratory of Tea Plant Germplasm Innovation and Resource Utilization, Anhui Agricultural University, Hefei, Anhui, China
e-mail: hanzisheng@outlook.com

International Joint Laboratory on Tea Chemistry and Health Effects of Ministry of Education, Anhui Agricultural University, Hefei, Anhui, China

X. Zhai (ed.), *Tea Chemistry and Health Benefits*,
https://doi.org/10.1007/978-3-032-11828-8_6

Table 6.1 Taste qualities and thresholds of tea tastants in green tea, oolong tea, and black tea (Yu et al., 2023)

Tastant	Threshold (μmol/L)[a]	Green tea	Oolong tea	Black tea	References
Bitter-tasting alkaloid					
Caffeine	500	+	+	+	a, b, c, d, e, f, g
Puckering astringent-tasting flavan-3-ols					
Catechin	24	+	+		a, b, c, d, e, g
Catechin-3-gallate	250				a, b, c, g
Epicatechin	930				a, b, c, d, e, f, g
Epicatechin-3-gallate	260				a, b, c, d, e, f, g
Epigallocatechin	520				a, b, c, d, e, f, g
Epigallocatechin-3-gallate	190		+	+	a, b, c, d, e, f, g
Gallocatechin	540				a, b, c, f, g
Gallocatechin-3-gallate	390				a, b, c, g
Mouth-drying and rough astringent-tasting benzotropolones					
Theaflavic acid	410				d, e, f, g
Theaflavin	16				d, e, f, g
Theaflavin-3-gallate	15				d, e, f, g
Theaflavin-3′-gallate	15				d, e, f, g
Theaflavin-3,3′-digallate	13				f, g
Silky and velvety-like astringent-tasting flavonol glycosides					
Apigenin-8-*C*-[*α*-*L*-rhamnopyranosyl -(1 → 2)-*β*-*D*-glucopyranoside]	2.8				a, b, c, f, g
Kaempferol-3-*O*-*β*-*D*-galactopyranoside	6.7				a, b, c, d, e, f, g
Kaempferol-3-*O*-*β*-*D*-glucopyranoside	0.67			+	a, b, c, d, e, f, g

(continued)

Table 6.1 (continued)

Tastant	Threshold (μmol/L)[a]	Green tea	Oolong tea	Black tea	References
Kaempferol-3-*O*-[*α*-*L*-rhamnopyranosyl -(1 → 6)-*β*-*D*-glucopyranoside]	0.25			+	a, b, c, d, e, f, g
Kaempferol-3-*O*-[*β*-*D*-glucopyranosyl-(1 → 3) -*O*-*α*-*L*-rhamnopyranosyl-(1 → 6)-*O*-*β*-*D*-galactopyranoside]	5.8				a, b, c, f, g
Kaempferol-3-*O*-[*β*-*D*-glucopyranosyl -(1 → 3)-*O*-*α*-*L*-rhamnopyranosyl -(1 → 6)-*O*-*β*-*D*-glucopyranoside]	19.8				a, b, c, f, g
Myricetin-3-*O*-*β*-*D*-galactopyranoside	2.7			+	a, b, c, d, e, f, g
Myricetin-3-*O*-*β*-*D*-glucopyranoside	2.1			+	a, b, c, d, e, f, g
Myricetin-3-*O*-[*α*-*L*-rhamnopyranosyl -(1 → 6)-*β*-*D*-glucopyranoside]	10.5				a, b, c, f, g
Quercetin-3-*O*-*β*-*D*-galactopyranoside	0.43			+	a, b, c, d, e, f, g
Quercetin-3-*O*-*β*-*D*-glucopyranoside	0.65			+	a, b, c, d, e, f, g
Quercetin-3-*O*-[*α*-*L*-rhamnopyranosyl -(1 → 6)-*β*-*D*-glucopyranoside]	0.00115			+	a, b, c, d, e, f, g
Quercetin-3-*O*-[*β*-*D*-glucopyranosyl -(1 → 3)-*O*-*α*-*L*-rhamnopyranosyl -(1 → 6)-*O*-*β*-*D*-galactopyranoside]	1.36			+	a, b, c, f, g
Quercetin-3-*O*-[*β*-*D*-glucopyranosyl -(1 → 3)-*O*-*α*-*L*-rhamnopyranosyl -(1 → 6)-*O*-*β*-*D*-glucopyranoside]	18.4				a, b, c, f, g
Other astringent-tasting compounds					
Gallic acid	200		+		a, b, c, d, e
L-Theanine	600		+	+	a, b, c, d, e, f, g

(continued)

Table 6.1 (continued)

Tastant	Threshold (μmol/L)[a]	Green tea	Oolong tea	Black tea	References
Theogallin	90	+			a, b, c
Umami-tasting amino acids					
L-Aspartic acid	7500				a, b, c, d, e
L-Glutamic acid	2000	+			a, b, c, d, e
Umami enhancers					
Gallic acid	n.m.[b]		+		a, b, c, d, e
Succinic acid	900				a, b, c
L-Theanine	24,000		+		a, b, c, d, e
Theogallin	n.m.[b]	+			a, b, c
Sour-tasting organic acids					
Oxalic acid	5600				f, g
Salty-tasting inorganic compounds					
Magnesium	4000				a, c
Potassium	15,000	+			a, c

[a] Thresholds were obtained from following cited references. [+]: the tastant was present with DoT factors ≥ 1 in the tea infusion, indicating the compound as key tastant of tea

[b] n.m.: Thresholds were not mentioned in the references

References: a: (Yu et al., 2014); b: (Zhang et al., 2020); c: (Kaneko et al., 2006); d: (Wang et al., 2016); e: (Dou et al., 2007); f: (Scharbert & Hofmann, 2005); g: (Scharbert et al., 2004)

6.2 Bitter Compounds in Tea

Bitterness and astringency are the two most intense taste qualities in almost all teas, while a highly perceived intensity of bitterness commonly results in a low consumption of tea. The major bitter compounds include

(a) caffeine,
(b) catechins, and
(c) anthocyanins.

6.2.1 *Caffeine*

Caffeine, the most abundant alkaloid in tea, has been verified as an important purely bitter-tasting compound for tea taste (Scharbert & Hofmann, 2005; Scharbert et al., 2004; Yu et al., 2014). It remains relatively stable during processing of the six major tea types, and has a bitterness threshold of 500 μM (Ye et al., 2022; Zhang et al., 2020). In addition to caffeine, tea alkaloids also include theobromine, and theophylline, while theobromine and theophylline occur at substantially lower concentrations. Theobromine exhibits a bitterness threshold of 800 μM, while the threshold for theophylline remains unreported, likely due to its low abundance and limited research focus.

6.2.2 *Catechins*

The primary bitter compounds in tea also include catechins and anthocyanins (Ye et al., 2022; Zhang et al., 2020). Upon ingestion of foods or medicines rich in bitter compounds, the perception of bitterness is mediated by the binding of these compounds to bitter taste receptors (TAS2Rs) in the oral cavity (Narukawa & Misaka, 2018). Catechins, the predominant polyphenols in tea, comprise both epicatechins (epigallocatechin gallate (EGCG), epicatechin gallate (ECG), epigallocatechin (EGC), epicatechin (EC)) and non-epicatechins (gallocatechin gallate (GCG), catechin gallate (CG), gallocatechin (GC), catechin (C)). They constitute a major source of bitterness in tea. Specific TAS2Rs for several catechins have been identified: Tas2r125 (EGCG); hTAS2R39, hTAS2R14, Tas2r125 (ECG); TAS2R4, TAS2R5, TAS2R39 (EC) (Zhang et al., 2020). The content of epicatechins significantly exceeds that of non-epicatechins in tea, particularly EGCG, which is especially prominent in non-fermented green tea and lightly fermented teas such as yellow tea, oolong tea, and white tea (Wang et al., 2019). Despite being key bitter compounds, systematic data on the bitterness thresholds of individual catechins remain unreported.

The bitterness of tea infusion results from the combined effect of multiple compounds and their interactions. Notably, caffeine synergistically enhances the bitterness and astringency of EGCG when co-present (Yin et al., 2014). Research on tea aftertaste enhancement revealed that EGC and EC-hydrolysis products of ester-type catechins (EGCG, ECG)—potentiate the lingering sweet aftertaste intensity (Zhang et al., 2020).

6.2.3 *Anthocyanins*

Beyond catechins and alkaloids, anthocyanins are significant bitter compounds. Over 10 distinct anthocyanins have been identified in tea, though their taste thresholds remain uncharacterized. Purple-leaf cultivars (e.g., 'Zijuan' tea) exhibit elevated anthocyanin levels, preliminarily quantified via colorimetric analysis at 1.65–35.57 mg/g dry weight (Jiang et al., 2013); however, systematic data on their precise concentrations across tea types are lacking. Additionally, theaflavins contribute to tea bitterness. While their bitterness thresholds are unreported, their cognate receptors (hTAS2R39 and hTAS2R14) have been identified (Zhang et al., 2020).

6.3 Astringent Compounds in Tea

Astringency is a complicated sensation, which can be induced by certain physical effects, such as friction force, drying, drawing, velvety, and puckering, but can also result from the physiological response of trigeminal nerves. The primary astringent compounds in tea are attributed to catechins, theaflavins, flavonol glycosides, and phenolic acids.

6.3.1 *Catechins and Theaflavins*

A puckering astringency, mouth-drying, and roughness taste in tea beverages derive from eight catechins (i.e., epigallocatechin and epigallocatechin-3-gallate) and five theaflavins (i.e., theaflavin-3-gallate and theaflavin-3′-gallate). The human sensory recognition thresholds determined in water using the half-tongue test showed catechins with relatively high taste thresholds (190–930 μmol/L), while the theaflavins revealed significantly lower thresholds (13–26 μmol/L) (Scharbert & Hofmann, 2005). Significant differences in astringency intensity exist among catechin stereoisomers, with galloylated catechins exhibiting lower astringency thresholds than their non-galloylated counterparts. This indicates that structural variations substantially modulate their protein-binding affinity with salivary proline-rich proteins (Ye et al., 2022).

6.3.2 Flavonol Glycosides

Contrary to the high amounts of catechins and theaflavins in tea, the flavonol glycosides imparting silky and velvety-like astringent oral sensation were present at low concentrations. However, due to very low threshold concentrations (0.001–19.8 μmol/L), the flavonol glycosides are believed to significantly contribute to the taste of black tea (Table 6.1) (Scharbert & Hofmann, 2005; Scharbert et al., 2004).

6.3.3 Phenolic Acids

Phenolic acids constitute significant astringent compounds in tea, exemplified by chlorogenic acid, 4-*O*-*p*-coumaroylquinic acid, quinic acid, gallic acid, and theogallin (Zhang et al., 2020). To identify key astringent contributing compounds in black tea, liquid chromatography separation coupled with turbidimetric assay (measuring protein precipitation) and sensory evaluation revealed 4-*O*-*p*-coumaroylquinic acid as the predominant astringent agent, exhibiting an astringency threshold of 38 μM (Wen et al., 2022). Although substantial quantities of quinic acid and chlorogenic acid (particularly quinic acid) are present in tea, their specific astringency thresholds remain uncharacterized in the literature.

Very recently, Zhang et al. reviewed researches on tea tastants as well as respective taste receptors (Zhang et al., 2020). Many attempts to sensory analysis revealed that not only a correlation between single taste compounds and respective taste qualities, but also interactions between different taste compounds take place. For example, purified L-theanine exhibiting a velvety astringent sensation to the oral space is able to enhance the umami intensity of sodium L-glutamate (Kaneko et al., 2006). Besides, sensory analysis of a taste profile of the taste recombinate containing the tea volatiles while wearing/not wearing a nose clamp demonstrated olfactory responses to tea volatiles are modifying the perception of taste-active tea nonvolatiles (Kaneko et al., 2006).

6.4 Umami Compounds in Tea

The umami taste described in tea shares sensory characteristics with monosodium glutamate (MSG), imparting a distinct savory and refreshing sensation (Zhang et al., 2020). The primary umami compounds in tea are amino acids, particularly L-theanine, a unique type of amino acid in tea.

Among the six major tea categories, green tea, white tea, and yellow tea contain higher amino acid levels, especially in spring harvests. Amino acid content correlates positively with plucking tenderness-higher concentrations are found in younger shoots and buds (Wang et al., 2019). Free amino acids constitute 2–5% of tea dry weight, with L-theanine accounting for approximately 50% of this fraction. As the dominant amino acid, L-theanine represents approximately 2% of tea dry weight and exhibits an umami recognition threshold of 24 mM (Kaneko et al., 2006). Its perception is mediated by the T1R1–T1R3 heterodimeric receptor (Narukawa et al., 2014). Umami thresholds and receptors for other amino acids remain unreported, likely due to their lower concentrations.

During thermal processing of teas, such as roasting and drying, amino acids (notably L-theanine) will undergo Maillard reactions with reducing sugars (e.g., glucose) to form Maillard reaction intermediates (MRIs). These MRIs contribute umami taste and have been detected across all tea types (Han et al., 2022a, 2022b). An MRI formed during processing of oolong tea, *N*-(1-carboxyethyl)-6-(hydroxymethyl) pyridinium-3-ol (alapyridaine), were found to surprisingly increase the perceived umami and sweetness intensity of oolong tea infusions (Feng et al., 2024). Wang et al. (2024) observed a significant reduction in L-theanine and galacturonic acid content following thermal processing during tea drying. Concurrently, they identified two novel MRIs derived from these precursors in processed tea samples. Using an in vitro thermal simulation system, they confirmed that one L-theanine-derived MRI, 2-(2-formyl-1*H*-pyrrol-1-yl)-L-theanine, exhibits umami characteristics with a detection threshold of 0.18 mg/mL. Through comprehensive analysis on green tea samples, researchers also found that γ-glutamylglutamine (γ-Glu-Gln) and γ-glutathione (γ-Glu-Cys-Gly) were major umami contributors, as well as enhancers in green tea infusions. Beyond amino acids, succinic acid also imparts umami taste, with a reported threshold of 900 μM, though its cognate receptor has not been identified (Zhang et al., 2020).

6.5 Sweet Compounds in Tea

The primary sweet-tasting compounds in tea are saccharides, notably glucose. While total carbohydrates constitute approximately 25% of tea dry weight, the content of soluble sugars is comparatively low (Zhang et al., 2020). Quantitative analysis via liquid chromatography during black tea processing revealed elevated levels of *D*-mannose, *D*-glucuronic acid, *D*-glucose, *D*-galactose, and *L*-arabinose during the withering and fermentation stages. These monosaccharides undergo significant reduction following drying, with D-glucose emerging as one of the most abundant monosaccharides in black tea (Wen et al., 2021). Comparative analysis across the six major tea categories demonstrated higher glucose concentrations in black tea and oolong tea, ranging from 4–6 mg/g dry weight (Han et al., 2022a, 2022b). While the sweetness recognition threshold for glucose is 90.0 mM, and fructose exhibits a lower sweetness threshold (52.0 mM), thresholds for other monosaccharides remain unreported (Scharbert & Hofmann, 2005).

In addition to saccharides, Wei et al. studied the sweet taste of *Camellia nanchuanica* black tea, and results showed that phloridzin, dulcitol, L-alanine, and sucrose were the major sweetness contributing compounds. While phloridzin was recognized as the key sweet taste compound with the recognition threshold of 0.01 mg/mL, a synergistic effects between phloridzin, dulcitol, L-alanine and sucrose on the sweetness intensity of *Camellia nanchuanica* black tea was also observed (Wei et al., 2023). A similar research also confirmed the sweetness enhancing role of D-turanose, D-kojibiose, L-phenylalanine, and L-leucine during tea processing (Wei et al., 2025a, 2025b). Despite taste active compounds, aroma compounds are also capable of offering sweetness to tea infusions. Over 10 aromatic compounds, including (*E*)-β-damascenone, linalool, geraniol, dimethyl sulfide, (*E*,*E*)-2,4-heptadienal, (*E*,*Z*)-2,6-nonadienal, (*E*)-linalool oxide, dihydroactinidiolide, γ-nonalactone, and (*E*)-β-ionone significantly increased the sweet intensity of sucrose ($p < 0.05$). They enhanced the sweetness of tea infusion possibly by reducing the binding energy of sucrose to sweet taste receptors through the formation of new hydrogen bonds and hydrophobic interactions (Wei et al., 2025a, 2025b).

Different from direct sweetness, the sweet aftertaste of tea has always been a critical aspect regarding the quality assessment of tea and tea products. By adding tannase to hydrolyze EGCG and ECG, researchers found that the obtained non-galloylated catechins, EGC and EC were able to bring sweet aftertaste with a synergistic effect (Cao et al., 2019). In a study of green teas with different grades, hydrolysable tannins, including monogalloyl glucose, digalloyl glucose, trigalloyl glucose and galloyl-hexahydroxydiphenoyl-glucose were found to not only correlate with grade, but also associate with the sweet aftertaste of tea infusions (Han et al., 2022a, 2022b). The sweet aftertaste of raw Pu-erh tea was also investigated through sensory evaluation, mass spectrometry, and taste addition experiments, and results indicated that theophylline was significantly linked to the sweet aftertaste of tea infusions (Yuan et al., 2025).

References

Cao, Q. Q., Zou, C., Zhang, Y. H., Du, Q. Z., Yin, J. F., Shi, J., Xue, S., & Xu, Y. Q. (2019). Improving the taste of autumn green tea with tannase. *Food Chemistry, 277*, 432–437. https://doi.org/10.1016/j.foodchem.2018.10.146

Dou, J., Lee, V. S., Tzen, J. T., & Lee, M. R. (2007). Identification and comparison of phenolic compounds in the preparation of Oolong tea manufactured by semifermentation and drying processes. *Journal of Agricultural and Food Chemistry, 55*(18), 7462–7468. https://doi.org/10.1021/jf0718603

FAOSTAT statistics database, Food and Agriculture Organization of the United Nations; crops-world production quantity (tea). Retrieved 2021, from https://www.fao.org/faostat/

Feng, X., Wang, H., Zhu, Y., Ma, J., Ke, Y., Wang, K., Liu, Z., Ni, L., Lin, C.-C., Zhang, Y., & Liu, Y. (2024). New insights into the umami and sweet taste of oolong tea: Formation of enhancer N-(1-carboxyethyl)-6-(hydroxymethyl) pyridinium-3-ol (alapyridaine) in roasting via Maillard reaction. *Journal of Agricultural and Food Chemistry, 72*(15), 8760–8773. https://doi.org/10.1021/acs.jafc.3c09011

GB/T 23776-2018. *Methodology for sensory evaluation of tea*. Chinese National Standard. Retrieved 2021, from http://www.foodmate.net

Han, Z., Jiang, Z., Zhang, H., Qin, C., Rong, X., Lai, G., Wen, M., Zhang, L., Wan, X., & Ho, C.-T. (2022a). Amadori reaction products of theanine and glucose: Formation, structure, and analysis

in tea. *Journal of Agricultural and Food Chemistry, 70*(37), 11727–11737. https://doi.org/10.1021/acs.jafc.2c04560
Han, Z., Wen, M., Zhang, H., Zhang, L., Wan, X., & Ho, C.-T. (2022b). LC-MS based metabolomics and sensory evaluation reveal the critical compounds of different grades of Huangshan Maofeng green tea. *Food Chemistry, 374*, 131796. https://doi.org/10.1016/j.foodchem.2021.131796
Jiang, L., Shen, X., Shoji, T., Kanda, T., Zhou, J., & Zhao, L. (2013). Characterization and activity of anthocyanins in Zijuan tea (Camellia sinensis var. kitamura). *Journal of Agricultural and Food Chemistry, 61*(13), 3306–3310. https://doi.org/10.1021/jf304860u
Kaneko, S., Kumazawa, K., Masuda, H., Henze, A., & Hofmann, T. (2006). Molecular and sensory studies on the umami taste of Japanese green tea. *Journal of Agriculture and Food Chemistry, 54*(7), 2688–2694. https://doi.org/10.1021/jf0525232
Narukawa, M., & Misaka, T. (2018). Tas2r125 functions as the main receptor for detecting bitterness of tea catechins in the oral cavity of mice. *Biochemical and Biophysical Research Communications, 503*(4), 2301–2305. https://doi.org/10.1016/j.bbrc.2018.06.152
Narukawa, M., Toda, Y., Nakagita, T., Hayashi, Y., & Misaka, T. (2014). L-Theanine elicits umami taste via the T1R1 + T1R3 umami taste receptor. *Amino Acids, 46*(6), 1583–1587. https://doi.org/10.1007/s00726-014-1713-3
Scharbert, S., & Hofmann, T. (2005). Molecular definition of black tea taste by means of quantitative studies, taste reconstitution, and omission experiments. *Journal of Agriculture and Food Chemistry, 53*(13), 5377–5384. https://doi.org/10.1021/jf050294d
Scharbert, S., Holzmann, N., & Hofmann, T. (2004). Identification of the astringent taste compounds in black tea infusions by combining instrumental analysis and human bioresponse. *Journal of Agriculture and Food Chemistry, 52*(11), 3498–3508. https://doi.org/10.1021/jf049802u
Wang, H., Yang, Y., Chen, L., Xu, A., Wang, Y., Xu, P., & Liu, Z. (2024). Identifying the structures and taste characteristics of two novel Maillard reaction products in tea. *Food Chemistry, 431*, 137125. https://doi.org/10.1016/j.foodchem.2023.137125
Wang, Y., Kan, Z., Thompson, H. J., Ling, T., Ho, C.-T., Li, D., & Wan, X. (2019). Impact of six typical processing methods on the chemical composition of tea leaves using a single Camellia sinensis cultivar, Longjing 43. *Journal of Agricultural and Food Chemistry, 67*(19), 5423–5436. https://doi.org/10.1021/acs.jafc.8b05140
Wei, F., Luo, L., & Zeng, L. (2023). Characterization of key sweet taste compounds in Camellia nanchuanica black tea. *Food Science & Technology, 182*, 114858. https://doi.org/10.1016/j.lwt.2023.114858
Wei, Q., Liu, X., Long, S., Zhu, L., Yang, H., Song, X., Peng, Y., Chen, T., Liu, J., Wen, B., & Li, M. (2025a). Exploration of the substances and key processing steps related to the sweetness of Niangniang tea. *Food Chemistry: X, 28*, 102556. https://doi.org/10.1016/j.fochx.2025.102556
Wei, Y., Yu, Y.-Y., Li, Y.-C., Zhong, X.-Y., Zou, C., Ning, J., Dong, W.-J., Wu, K., & Xu, Y.-Q. (2025b). Aroma compounds with enhanced sweet perception in tea infusions: Screening, characterization, and sweetening mechanism. *Journal of Advanced Research*. https://doi.org/10.1016/j.jare.2025.05.044
Wen, M., Cui, Y., Dong, C.-X., & Zhang, L. (2021). Quantitative changes in monosaccharides of Keemun black tea and qualitative analysis of theaflavins-glucose adducts during processing. *Food Research International, 148*, 110588. https://doi.org/10.1016/j.foodres.2021.110588
Wen, M., Han, Z., Cui, Y., Ho, C.-T., Wan, X., & Zhang, L. (2022). Identification of 4-O-p-coumaroylquinic acid as astringent compound of Keemun black tea by efficient integrated approaches of mass spectrometry, turbidity analysis and sensory evaluation. *Food Chemistry, 368*, 130803. https://doi.org/10.1016/j.foodchem.2021.130803
Wang, C., Lv, S., Wu, Y., Gao, X., Li, J., Zhang, W., & Meng, Q. (2016). Oolong tea made from tea plants from different locations in Yunnan and Fujian China showed similar aroma but different taste characteristics. *SpringerPlus, 5*(1). https://doi.org/10.1186/s40064-016-2229-y
Ye, J. H., Ye, Y., Yin, J. F., Jin, J., Liang, Y. R., Liu, R. Y., Tang, P., & Xu, Y. Q. (2022). Bitterness and astringency of tea leaves and products: Formation mechanism and reducing strategies. *Trends in Food Science & Technology, 123*, 130–143.

Yin, J.-F., Zhang, Y.-N., Du, Q.-Z., Chen, J.-X., Yuan, H.-B., & Xu, Y.-Q. (2014). Effect of Ca^{2+} concentration on the tastes from the main chemicals in green tea infusions. *Food Research International, 62*, 941–946. https://doi.org/10.1016/j.foodres.2014.05.016

Yu, P., Yeo, A. S., Low, M. Y., & Zhou, W. (2014). Identifying key non-volatile compounds in ready-to-drink green tea and their impact on taste profile. *Food Chemistry, 155*, 9–16. https://doi.org/10.1016/j.foodchem.2014.01.046

Yu, J. Y., Zhang, K. Y., Wang, Y. J., Zhai, X. T, & Wan, X. C. (2023). Flavor perception and health benefits of tea. In F. Toldrá (Ed.), *Advances in food and nutrition research* (Vol. 106, pp. 129–218). Elsevier.

Yuan, X., Chen, X., Chai, C., Feng, M., Hu, Y., Yi, Z., Gu, Y., Ruan, L., & Yi, L. (2025). Identifying key contributors to the sweet aftertaste of raw Pu-erh tea through analytical and sensory methods. *Food Chemistry, 481*, 144067. https://doi.org/10.1016/j.foodchem.2025.144067

Zhang, L., Cao, Q.-Q., Granato, D., Xu, Y.-Q., & Ho, C.-T. (2020). Association between chemistry and taste of tea: A review. *Trends in Food Science & Technology, 101*, 139–149. https://doi.org/10.1016/j.tifs.2020.05.015

Chapter 7
The Effects of Processing on Tea Taste Quality

Yujie Wang

Abstract The taste of tea is formed by the contribution of non-volatile metabolites. The prevailing view is about the bitterness of tea originates from alkaloids (caffeine) and tea polyphenols, astringency from tea polyphenols (catechins), and umami taste from free amino acids, especially theanine. In addition to volatile metabolites, the content of these non-volatile metabolites changes significantly during tea processing, resulting in a diversity of tea flavors. In this chapter, we focus on the effects of key processing steps on taste quality and metabolites in different tea types, and some novel processing methods are also described.

Keywords Taste quality · Non-volatile components · Processing technologies · Tea polyphenols · Amino acids

7.1 Introduction

The composition and proportion of primary flavor components in tea are influenced by a complex interplay of factors, including the genetic variety of tea plants, cultivation conditions (e.g., soil composition, climate), and the specific biochemical transformations induced during processing. Among these, tea processing—conducted under controlled environmental parameters such as temperature, humidity, and oxygen levels—plays a dominant role in shaping taste profiles through enzymatic reactions, hygrothermal effects, and microbial activity. These biochemical changes are not merely incidental but are systematically regulated to achieve desired flavor outcomes. For instance, the oxidation of polyphenols, hydrolysis of glycosides, and Maillard reactions are meticulously manipulated across processing stages to balance bitterness, astringency, sweetness, and umami. The unique sensory characteristics of tea—whether the brisk freshness of green tea, the floral mellowness of oolong, or the robust sweetness of black tea—are thus the direct result of carefully orchestrated processing techniques.

Y. Wang (✉)
College of Biosystems Engineering and Food Science, Zhejiang University, Hangzhou, China
e-mail: wangyjnet@163.com

X. Zhai (ed.), *Tea Chemistry and Health Benefits*,
https://doi.org/10.1007/978-3-032-11828-8_7

Tea is categorized into six major types based on manufacturing processes, each defined by critical steps that directly influence taste chemistry (Zhang et al., 2019):

(a) *Green Tea*: Spreading and fixation (enzyme inactivation via high heat).
(b) *White Tea*: Extended withering (natural enzymatic and non-enzymatic transformations).
(c) *Oolong Tea*: Shaking (mechanical stress to induce partial oxidation).
(d) *Yellow Tea*: Yellowing (controlled oxidation under high humidity).
(e) *Black Tea*: Full fermentation (enzymatic oxidation of polyphenols).
(f) *Dark Tea*: Pile-fermentation (microbial-dominated aging).

These unique processing techniques were observed to directly influence the changes in the tea leaf's internal components, bestowing upon the six tea types their unique flavor characteristics. The initial post-harvest step—spreading or withering—significantly impacts flavor development.

7.2 Spreading and Fixation in Green Tea

For green tea, short spreading (3–5 h) under low temperature (20–25 °C) and moderate humidity (60–70% RH) inhibits the downregulation of flavanol synthesis genes and polyphenol oxidase (PPO) activity, delaying oxidative hydrolysis of polyphenols. Concurrently, this process was found to promote the up-regulation of the expression of certain amino acid synthesis genes and the activity of peptidases, thereby increasing the content of amino acids. This series of changes effectively enhanced the freshness and concentration of the green tea flavor while reducing its bitterness and astringency (Yu, 2020). Low-temperature spreading combined with yellow light or carbon dioxide further elevates amino acid content compared to natural withering, improving green tea taste (Yu et al., 2020).

Furthermore, the high-temperature treatment involved in fixation was identified as facilitating the hydrolysis, oxidation, and isomerization of metabolites, along with other rapid thermochemical reactions. These reactions drive the rapid conversion of amino acids, flavonoids, and phenolic acids in green tea, marking a key process in the formation of green tea's taste quality (Wang et al., 2021). Previous studies demonstrated that higher fixation temperatures could effectively reduce the bitterness and astringency induced by tea polyphenols, enhancing the freshness. This effect was attributed to the reduction of epicatechin (EC) and catechin dimers (theaflavin-3′-gallate and theaflavin-3,3′-gallate), which significantly reduced the bitterness (Yin et al., 2023). In the current era, a variety of fixation methods are available, among which light wave fixation has been found to uniformly kill green tea in a short time span.

Compared with traditional roller fixation and steam fixation, light wave fixation was shown to produce a more mellow taste and was proven to promote the degradation of proteins and the conversion ofγ-glutamyl peptide. Furthermore, it was observed that this method led to the accumulation of substances that contribute to a fresh flavor, sweetness, and a strong and coordinated taste. In contrast, roller fixation was found to stimulate the oxidation of proteins and the accumulation of peptides, flavonoids, trans-catechins, and other bitter and astringent substances (Xue et al., 2023).

7.3 Extended Withering in White Tea

In white tea, prolonged withering shifts the flavor from bitter and astringent to fresh, sweet, and mellow taste. The changes in flavonoid glycosides (FGs) were primarily observed during the withering process, with the content of 90% difference FGs significantly decreasing, and the total FGs dropping from 7.02 mg/g to 4.35 mg/g (Wang et al., 2024). FGs, compounds that impart a dry and velvety astringent taste, were not present in high amounts but were characterized by a low threshold, thereby having a substantial impact on the astringency of tea broth (Scharbert & Hofmann, 2005). Earlier studies revealed that the highest amount of FGs was retained in green tea, followed by white tea, with the lowest amount retained in black tea. This was because FGs were hydrolyzed and oxidized by enzymes during the rolling and fermentation process of black tea, and the fixation process of green tea destroyed the enzyme activity, thereby retaining a large amount of FGs (Fang et al., 2019).

Compared to natural indoor withering, sunlight withering resulted in a stronger bitter flavor, as described in Fig. 7.1. This was attributed to sunlight-induced heat and light stress, which led to the upregulation of differential genes in the mangiferolic acid, phenylpropane, and flavonoid biosynthesis pathways, thereby increasing the concentration of bitter-tasting flavonoids (Zou et al., 2022). Moreover, the withering process was accompanied by the decomposition of other substances, such as the degradation of RNA into guanosine 5'-monophosphate (GMP), adenosine monophosphate (AMP), and cytidine monophosphate (CMP), and the breakdown of proteins into free peptides, both of which made significant contributions to the fresh flavor of tea broth (Chen et al., 2020; Zhao et al., 2019). The subtly sweet characteristic of white tea was potentially linked to the conversion of starch into soluble sugars and a noticeable increase in the levels of certain sweet amino acids, including glycine, proline, alanine, serine, and threonine, during the withering stage (Chen et al., 2019).

7.4 Shaking and Mechanical Stress Effects in Oolong Tea

The shaking process of oolong tea, which involves mechanical damage to the fresh leaves, was found to significantly mitigate the quality defects commonly associated with summer and fall teas, such as the pronounced bitterness and astringency, lack of freshness, and weak tea aroma. This process has been applied to various types of tea to enhance their flavor and aroma; for example, shaking has been utilized to render the withering of white and green teas more mellow and refreshing, to make

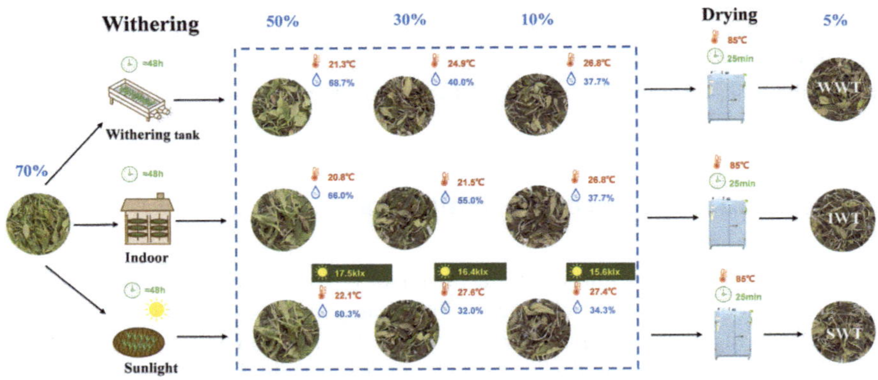

Fig. 7.1 Some typical withering methods used for processing white tea (Wu et al., 2024)

the flavors of black tea more rounded, and to diminish the bitterness and astringency of yellow tea (Liu et al., 2024).

The processing of oolong tea, characterized by shaking, as well as the rolling and fermentation of black tea, involves enzymatic reactions, given that enzymatic activity is not eliminated at high temperatures. In the production of oolong tea, withering and shaking are collectively referred to as enzyme-catalyzed processes (ECPs). As tea processing advances, light stress, dehydration stress, and mechanical damage lead to a gradual increase in the content of enzymes involved in sugar-degrading metabolism, thereby enhancing carbon flux and the subsequent synthesis of amino acids or secondary metabolites to adapt to unfavorable conditions. This process also accelerates the degradation of tea proteins and the oxidation of catechins. The resultant overall increase in the abundance of free amino acids leads to a reduction in the bitter taste-related catechins and flavonol glycosides, culminating in the characteristic mellow taste of oolong tea (Lin et al., 2024; Wu et al., 2022a, 2022b).

In contrast to the environmental stress experienced during withering, mechanical damage induced during shaking was identified as the primary stimulus for the development of oolong tea's unique flavor (Wu et al., 2022a, 2022b).

7.5 Rolling and Fermentation in Black Tea

The initial stage of rolling and fermentation, serves a dual purpose; it not only shapes the tea leaves into tightly curled strips but also promotes comprehensive contact between enzymes and substrates in the cell sap due to the crushing and deformation of the rolled leaves, thereby facilitating the subsequent enzyme-promoted reactions (Chiang et al., 2022).

7.5.1 Rolling

Rolling machines were widely employed in tea production, with the duration of rolling and the amount of pressure applied considered critical factors. It was demonstrated that a 100-min rolling period, with alternating light and heavy pressure, produced black tea with a sweet and mellow flavor (Wu et al., 2022a, 2022b). Furthermore, black tea produced using light pressure was found to exhibit a sweeter and mellower flavor compared to those produced under conditions of no pressure or frequent alternating light and heavy pressure (Zhang et al., 2023). Additionally, the practice of double rolling during the fermentation process was found to enhance the flavor of black tea by reducing astringency and increasing mellowness (Chen et al., 2024).

7.5.2 Fermentation

Fermentation has been recognized as a crucial process in the development of flavor substances and functional components in black tea, with the extension of fermentation time leading to a gradual reduction in the astringency and freshness of black tea. Conversely, bitterness, freshness, and sweetness are maintained, bestowing upon black tea a distinctive sweetness, mellowness, and refreshing flavor. It was observed that the relative content of astringent characteristic constituents, such as epigallocatechin gallate (EGCG), catechin gallate (CG), epigallocatechin (EGC), and theaflavins (TFs), significantly decreased after fermentation (Dong et al., 2024). Oxygen, an important substrate in the enzymatic oxidation of metabolites during fermentation, was found to enhance the process. Oxygen-enriched fermentation was shown to promote the hydrolysis of low-threshold flavonoid glycosides and the reduction of phenolic compounds, including caffeic acid and chlorogenic acid. This process effectively reduced the bitterness and astringency of the tea broth, thereby improving the taste of black tea (Chen et al., 2021).

7.6 Yellowing in Yellow Tea

Yellow tea, when compared with green tea, was noted to undergo an additional yellowing process. The withering and fixation processes were primarily found to regulate the bitterness and astringency of tea, while the yellowing process predominantly affected the sweetness of tea. It was revealed that the reduced forms of key compounds, *trans*-4-hydroxy-L-proline and glutathione, were mainly metabolized through arginine and proline. The metabolic pathways and ATP-binding cassette (ABC) transport proteins were significantly increased, resulting in an enhancement of the sweetness and a reduction of bitterness and astringency in yellow tea (Sun et al.,

2024). Soluble sugars, *L*-glutamine, *L*-theanine, gallic acid, catechin oxidized polymers, and catechin methylation products were also found to be positively correlated with the enhancement of sweetness in yellow tea (Wei et al., 2023).

7.7 Microbial Contributions in Dark Tea

Dark tea, post-fermented with green tea, exhibited a taste significantly different from that of green tea, characterized by a sweet and mellow flavor. This difference was attributed to the change in the ratio of the contents of tea polyphenols (especially ester catechins), amino acids and soluble sugars, and the formation of new substances such as theaflavins and thearubigins (Li et al., 2022).

The pile-fermentation process of dark tea was found to be accompanied by the rapid proliferation of microorganisms. Among these, the "golden flower fungus" of Fuzhuan brick tea, a beneficial *Aspergillus*, was identified to play a pivotal role in the formation of Fuzhuan brick tea's unique flavor. Research indicated that as the presence of *Aureobasidium* in the samples increased, the astringency gradually diminished and the taste became more mellow. This process was associated with a deeper biotransformation of catechins (particularly EGCG and ECG), amino acids (especially theanine), dimeric catechins, flavonol glycosides, aromatic proximate glycosides, lipids, nucleotides, and nucleosides, leading to the production of higher levels of macromolecular pigments, theaflavins, and catechin A- and B-ring derivatives such as EPSFs, Fuzhuanins, and degallocatechin like EGC, EC, and GC (Li et al., 2023a, 2023b). Among these, Fuzhuanins were found to contribute significantly to the unique flavor and quality of Fuzhuan brick tea (Zhu et al., 2015). The enzymes produced by microorganisms were also demonstrated to play a crucial role in accelerating oxidative degradation through the flavonoid metabolism pathway and lipid metabolite pathway, contributing to the development of Pu-erh tea's unique quality. Additionally, the heat and humidity present during the stacking process were found to be integral to the formation of the aged and rich flavor of Pu-erh tea (Li et al., 2023a, 2023b).

7.8 Drying and Final Flavor Stabilization

Drying, recognized as the final step in tea processing, serves dual purposes. On one hand, it evaporates the moisture from the tea, inhibits mildew growth, and facilitates subsequent storage; on the other hand, it plays a critical role in the formation and stabilization of the final quality. Temperature has been identified as a crucial factor in regulating the drying process. For black tea, the drying process is segmented into two stages. Initially, a higher temperature is required for the first drying phase to deactivate enzyme activity and halt enzymatic reactions. Subsequently, the second phase, known as the foot-fire, aims to further reduce moisture content, employing a generally lower temperature than that used in the initial phase. This strategy (Wang et al., 2020) allows black tea to retain a higher content of amino acids and soluble sugars, fostering the development of sweetness and freshness in the tea. Conversely, higher temperatures during the foot-fire phase were shown to increase the levels of flavonoids, phenolic acids, organic acids, tannins, amino acid derivatives, nucleotides, and alkaloids, while reducing the levels of amino acids, lignans, and sugars. This results in a darker brown appearance, broth color, and leaf underside, alongside the formation of bitter and acidic flavors in black tea (Hua et al., 2024).

However, for large-leaf yellow tea, it was found that a high-temperature pulling and aging process increased the gallocatechin gallate (GCG) content in the tea leaves but decreased the levels of certain polyphenols, caffeine, free amino acids, total sugars, and *L*-theanine. Sensory evaluations confirmed that the pulling and aging process yielded a more intense and richer flavor (Guo et al., 2021). After the application of the pulling old fire process, significant increases in the sweetness and aftertaste scores of large-leaf yellow tea were observed, a phenomenon that may be attributed to the production of caramelized substances and the presence of residual soluble sugar. During this process, an increase in populin was also noted. It has been suggested that populin, along with quercetin—both of which are present in relatively high concentrations in large-leaf yellow tea—may substitute for the lower levels of glutamic acid and theanine. These components, in conjunction with high levels of gallic acid, were found to activate taste receptors in the cells of the taste buds more efficiently, thereby enhancing the sweetness of large-leaf yellow tea (Li et al., 2024).

Additionally, the palatability of stored tea leaves experienced enhancement through a re-drying process. It was demonstrated that the aroma of stored single-form tea became diminished and was accompanied by a strong bitter and astringent flavor. Conversely, tea that underwent humidification and rehydration, followed by steaming and low-temperature drying after being shaped into a ball and exposed to air for 8 h, exhibited a reduction in bitter and astringent tastes and an increase in sweetness. This change in flavor profile was associated with a decrease in the tannin fractions responsible for bitter and astringent tastes and an increase in soluble sugars (Zeng et al., 2023).

7.9 Enzymatic Regulation and Industrial Advancements

Enzymatic reactions, essential for the synthesis of flavor substances during tea processing, are determined by the specific enzymes present in various stages such as withering (including salicylic acid carboxymethyl transferase, carotenoid cleavage dioxygenase, α-amylase, and β-amylase), shaking (tanninase and β-glucosidase), tumbling (alcohol dehydrogenase, lipoxygenase, and alkene synthase), fermentation (polyphenol oxidase, peroxidase, jasmonic acid carboxymethyl transferase, cellulase, and pectinase), moistening (polyphenol oxidase, peroxidase, cellulase, pectinase, jasmonic acid carboxymethyl transferase, and β-prime myrosinase), and torrefaction (polyphenol oxidase, peroxidase, cellulase, and pectinase). It has been demonstrated that the quality of tea can be enhanced by manipulating these enzymatic activities, which can be achieved through the careful modulation of key enzyme genes via the application of low temperatures, light waves, mechanical damage, and pest attacks, thereby regulating the synthesis and transformation of flavor compounds (Hu et al., 2024).

7.10 Conclusion

Concurrently, advancements and innovations in tea processing technology have introduced more stringent requirements for tea production, including the automation of equipment, standardization of processes, industrial scaling, and cleaner production practices. It has also been recognized that the control over specific parameters such as temperature, humidity, and processing time need to be more meticulously managed to facilitate the more efficient production of high-quality tea.

References

Chen, Q., Zhu, Y., Dai, W., Lv, H., Mu, B., Li, P., & Lin, Z. (2019). Aroma formation and dynamic changes during white tea processing. *Food Chemistry, 274*, 915–924. https://doi.org/10.1016/j.foodchem.2018.09.072

Chen, Q., Shi, J., Mu, B., Chen, Z., Dai, W., & Lin, Z. (2020). Metabolomics combined with proteomics provides a novel interpretation of the changes in nonvolatile compounds during white tea processing. *Food Chemistry, 332*, Article 127412. https://doi.org/10.1016/j.foodchem.2020.127412

Chen, L., Liu, F., Yang, Y., Tu, Z., Lin, J., Ye, Y., & Xu, P. (2021). Oxygen-enriched fermentation improves the taste of black tea by reducing the bitter and astringent metabolites. *Food Research International, 148*, Article 110613. https://doi.org/10.1016/j.foodres.2021.110613

Chen, Q., Fu, Y., Heng, W., Yu, S., Xie, F., Dong, F., & Fu, H. (2024). Re-rolling treatment in the fermentation process improves the taste and liquor color qualities of black tea. *Food Chemistry: X, 21*, Article 101143. https://doi.org/10.1016/j.fochx.2024.101143

Chiang, S., Yang, K., Wang, S., & Chen, C. (2022). Enzymatic treatment in black tea manufacturing processing: Impact on bioactive compounds, quality, and bioactivities of black tea. *LWT, 163*, Article 113560. https://doi.org/10.1016/j.lwt.2022.113560

Dong, H., Li, Y., Lai, X., Hao, M., Sun, L., Li, Q., & Liu, X. (2024). Effects of fermentation duration on the flavour quality of large leaf black tea based on metabolomics. *Food Chemistry, 444*, Article 138680. https://doi.org/10.1016/j.foodchem.2024.138680

Fang, Z., Song, C., Xu, H., & Ye, J. (2019). Dynamic changes in flavonol glycosides during production of green, yellow, white, oolong and black teas from Camellia sinensis L. (cv. Fudingdabaicha). *International Journal of Food Science & Technology, 54*(2), 490–498. https://doi.org/10.1111/ijfs.13961

Guo, X., Ho, C.-T., Schwab, W., & Wan, X. (2021). Effect of the roasting degree on flavor quality of large-leaf yellow tea. *Food Chemistry, 347*, Article 129016. https://doi.org/10.1016/j.foodchem.2021.129016

Hu, J., Feng, X., Song, H., Hao, Z., Ma, S., Hu, H., & Ch, Q. (2024). Enzymatic reactions throughout cultivation, processing, storage and post-processing: Progressive sculpture of tea quality. *Trends in Food Science & Technology, 143*, Article 104294. https://doi.org/10.1016/j.tifs.2023.104294

Hua, J., Ouyang, W., Zhu, X., Wang, J., Yu, Y., Chen, M., & Jiang, Y. (2024). Objective quantification technique and widely targeted metabolomic reveal the effect of drying temperature on sensory attributes and related non-volatile metabolites of black tea. *Food Chemistry, 439*, Article 138154. https://doi.org/10.1016/j.foodchem.2023.138154

Li, Y., Hao, J., Zhou, J., He, C., Yu, Z., Chen, S., & Ni, D. (2022). Pile-fermentation of dark tea: Conditions optimization and quality formation mechanism. *LWT, 166*, Article 113753. https://doi.org/10.1016/j.lwt.2022.113753

Li, T., Wei, Y., Feng, W., Lu, M., Ke, H., Li, Y., & Ning, J. (2023a). Exploring the mysterious effect of piling fermentation on Pu-erh tea quality formation: Microbial action and moist-heat action. *LWT, 185*, Article 115132. https://doi.org/10.1016/j.lwt.2023.115132

Li, Y., Zhou, H., Tian, T., Hou, Y., Chen, D., Zhou, J., & Zhou, T. (2023b). Nontargeted and targeted metabolomics analysis for evaluating the effect of "golden flora" amount on the sensory quality, metabolites, and the alpha-amylase and lipase inhibitory activities of Fu brick tea. *Food Chemistry, 416*, Article 135795. https://doi.org/10.1016/j.foodchem.2023.135795

Li, Y., Luo, Q., Qin, M., Xu, W., Wang, X., Zhou, J., & Ni, D. (2024). Study on color, aroma, and taste formation mechanism of large-leaf yellow tea during an innovative manufacturing process. *Food Chemistry, 438*, Article 138062. https://doi.org/10.1016/j.foodchem.2023.138062

Lin, J., Lin, H., Li, C., Liao, N., Zheng, Y., Yu, X., & Wu, L. (2024). Unveiling characteristic metabolic accumulation over enzymatic-catalyzed process of Tieguanyin oolong tea manufacturing by DESI-MSI and multiple-omics. *Food Research International, 181*, Article 114136. https://doi.org/10.1016/j.foodres.2024.114136

Liu, B., Liang, G., Li, Z., Yang, J., Wang, Y., & Zhao, Y. (2024). Research progress on the application of oolong tea withering techniques in tea processing. *Modern Agricultural Science and Technology, 05*, 156–160. (In Chinese).

Scharbert, S., & Hofmann, T. (2005). Molecular definition of black tea taste by means of quantitative studies, taste reconstitution, and omission experiments. Journal of Agricultural and *Food Chemistry, 53(13)*, 5377–5384.

Sun, L., Wen, S., Zhang, S., Li, Q., Cao, J., Chen, R., & Sun, S. (2024). Study on flavor quality formation in green and yellow tea processing by means of UPLC-MS approach. *Food Chemistry: X, 22*, Article 101342. https://doi.org/10.1021/jf050294d

Wang, J., Yuan, H., Hua, J., Jiang, Y., Dong, C., Deng, Y., & Yang, Y. (2020). The influence of full-fire process parameters on the hot air drying characteristics and quality of Gongfu black tea. *Transactions of the Chinese Society of Agricultural Engineering, 36*(10), 287–296. (In Chinese).

Wang, H., Hua, J., Yu, Q., Li, J., Wang, J., Deng, Y., & Jiang, Y. (2021). Widely targeted metabolomic analysis reveals dynamic changes in non-volatile and volatile metabolites during green tea processing. *Food Chemistry, 363*, Article 130131. https://doi.org/10.1016/j.foodchem.2021.130131

Wang, Z., Gao, C., Zhao, J., Zhang, J., Zheng, Z., Huang, Y., & Sun, W. (2024). The metabolic mechanism of flavonoid glycosides and their contribution to the flavor evolution of white tea during prolonged withering. *Food Chemistry, 439*, Article 138133. https://doi.org/10.1016/j.foodchem.2023.138133

Wei, Y., Yin, X., Zhao, M., Zhang, J., Li, T., Zhang, Y., & Ning, J. (2023). Metabolomics analysis reveals the mechanism underlying the improvement in the color and taste of yellow tea after optimized yellowing. *Food Chemistry, 428*, Article 136785. https://doi.org/10.1016/j.foodchem.2023.136785

Wu, L., Wang, Y., Liu, S., Sun, Y., Li, C., Lin, J., & Wei, S. (2022a). The stress-induced metabolites changes in the flavor formation of oolong tea during enzymatic-catalyzed process: A case study of Zhangping Shuixian tea. *Food Chemistry, 391*, Article 133192. https://doi.org/10.1016/j.foodchem.2022.133192

Wu, S., Yu, Q., Shen, S., Shan, X., Hua, J., Zhu, J., & Li, J. (2022b). Non-targeted metabolomics and electronic tongue analysis reveal the effect of rolling time on the sensory quality and nonvolatile metabolites of congou black tea. *LWT, 169*, Article 113971. https://doi.org/10.1016/j.lwt.2022.113971

Wu, H., Sheng, C., Lu, M., Ke, H., Li, T., Wei, Y., Shen, S., Yin, X., Lu, C., Wang, Y., & Ning, J. (2024). Identification of the causes of aroma differences in white tea under different withering methods by targeted metabolomics. *Food Bioscience, 59*, Article 104020. https://doi.org/10.1016/j.fbio.2024.104020

Xue, J., Liu, P., Feng, L., Zheng, L., Gui, A., Wang, X., & Zheng, P. (2023). Insights into the effects of fixation methods on the sensory quality of straight-shaped green tea and dynamic changes of key taste metabolites by widely targeted metabolomic analysis. *Food Chemistry: X, 20*, Article 100943. https://doi.org/10.1016/j.fochx.2023.100943

Yin, X., Wei, Y., Li, T., Zhang, J., Zou, L., Cui, Q., & Ning, J. (2023). Heterocyclic compounds formation in large-leaf yellow tea induced by the Maillard reaction at different roasting temperatures. *LWT, 182*, Article 114856. https://doi.org/10.1016/j.lwt.2023.114856

Yu, X. (2020). *Research on the influence of fresh leaf spreading method on the metabolism of color, aroma and taste quality components of green tea.* Central China Agricultural University. (In Chinese).

Yu, X., Li, Y., He, C., Zhou, J., Chen, Y., Yu, Z., & Ni, D. (2020). Nonvolatile metabolism in postharvest tea (Camellia sinensis L.) leaves: Effects of different withering treatments on nonvolatile metabolites, gene expression levels, and enzyme activity. *Food Chemistry, 327*, 126992. https://doi.org/10.1016/j.foodchem.2020.126992

Zeng, Y., Liang, J., Zhang, J., Chen, W., Hu, D., Xia, H., & Qiao, X. (2023). Improvement of the sensory characteristics of Dancong tea using a dry-heating post-treatment approach. *LWT, 185*, Article 115118. https://doi.org/10.1016/j.lwt.2023.115118

Zhang, L., Ho, C.-T., Zhou, J., Santos, J. S., Armstrong, L., & Granato, D. (2019). Chemistry and biological activities of processed camellia Sinensis teas: a comprehensive review. *Comprehensive Reviews in Food Science and Food Safety, 18*(5), 1474–1495. https://doi.org/10.1111/1541-4337.12479

Zhang, S., Wu, S., Yu, Q., Shan, X., Chen, L., Deng, Y., & Li, J. (2023). The influence of rolling pressure on the changes in non-volatile compounds and sensory quality of congou black tea: The combination of metabolomics, E-tongue, and chromatic differences analyses. *Food Chemistry: X, 20*, Article 100989. https://doi.org/10.1016/j.fochx.2023.100989

Zhao, F., Ye, N., Qiu, X., Qian, J., Wang, D., Yue, W., & Chen, M. (2019). Identification and comparison of oligopeptides during withering process of White tea by ultra-high pressure liquid chromatography coupled with quadrupole-orbitrap ultra-high resolution mass spectrometry. *Food Research International, 121*, 825–834. https://doi.org/10.1016/j.foodres.2019.01.002

Zhu, Y., Chen, J., Ji, X., Hu, X., Ling, T., Zhang, Z., & Wan, X. (2015). Changes of major tea polyphenols and production of four new B-ring fission metabolites of catechins from post-fermented Jing-Wei Fu brick tea. *Food Chemistry, 170*, 110–117. https://doi.org/10.1016/j.foodchem.2014.08.075

Zou, L., Shen, S., Wei, Y., Jia, H., Li, T., Yin, X., & Ning, J. (2022). Evaluation of the effects of solar withering on nonvolatile compounds in white tea through metabolomics and transcriptomics. *Food Research International, 162*, Article 112088. https://doi.org/10.1016/j.foodres.2022.112088

Chapter 8
Technologies for the Flavor Quality Improvement in Tea-Beverage Processing

Jing Wang and Xiaoting Zhai

Abstract Tea beverage, a kind of tea comprehensive processing product, is widely consumed due to its diverse flavors, convenience, and health benefits. Their quality and stability were influenced by many factors such as raw material, processing technique, and storage conditions. Understanding the interactions between these factors and processing methods is crucial for enhancing product stability, preserving flavor, and increasing consumer acceptance. This chapter reviews the key flavor compounds in tea beverages and the critical factors affecting their flavor qualities. It further explored the impacts of major processing technologies, including extraction, clarification, and sterilization, on the flavor development and overall quality, with a special emphasis on flavor improvement methods. Additionally, global innovations and advancements in tea beverage processing were discussed, highlighting their potential for improving sensory attributes and extending shelf life. Future research should focus on optimizing processing parameters and developing innovative technologies to ensure the production of high-quality tea beverages.

Keywords Tea-beverage · Flavor · Quality stability · Processing technology

8.1 Introduction

Tea-beverage, as a liquid drink instead of traditional tea-leaf product, is primarily derived from water extracts and concentrates of tea leaves, or instant tea powder according to GB/T 21,733–2008 (Standard, 2008). The production process involves

J. Wang
Key Laboratory of Tea Science of Ministry of Education, Hunan Agricultural University, Changsha 410128, China

X. Zhai (✉)
State Key Laboratory of Tea Plant Germplasm Innovation and Resource Utilization, School of Tea Science, Anhui Agricultural University, Hefei, China
e-mail: xiaotingzhai@ahau.edu.cn

International Joint Laboratory On Tea Chemistry and Health Effects of Ministry of Education, Anhui Agricultural University, Hefei, China

X. Zhai (ed.), *Tea Chemistry and Health Benefits*,
https://doi.org/10.1007/978-3-032-11828-8_8

the following steps: extraction, filtration, blending, canning, and sterilization. The final product contains a stable concentration of tea active-compounds, resulting in a beverage with characteristic tea flavor (Fig. 8.1). In recent years, the consumption of tea-beverages has seen a surge in global popularity, fueled by consumer desires for diverse flavors, health benefits, and cultural appeal. Countries with a long history of traditional tea products, such as China, Japan, and India, still lead in both production and innovation of the rising tea-beverage products, while new markets in Europe and North America are expanding rapidly (Yin et al., 2021). The global tea-beverage market remains robust with a market value of $ 404.4 billion in 2017. Among them, China was the largest consumption market (Liang et al., 2022a).

With customers' growing quest for health, convenience, and flavor diversity, the tea-beverage market is evolving, blending traditional tea with innovative products. Tea-beverages are often classified into two types: liquid tea-beverage (such as ready-to-drink tea, *kombucha*) and solid tea-beverage (e.g., instant tea powder, milk tea

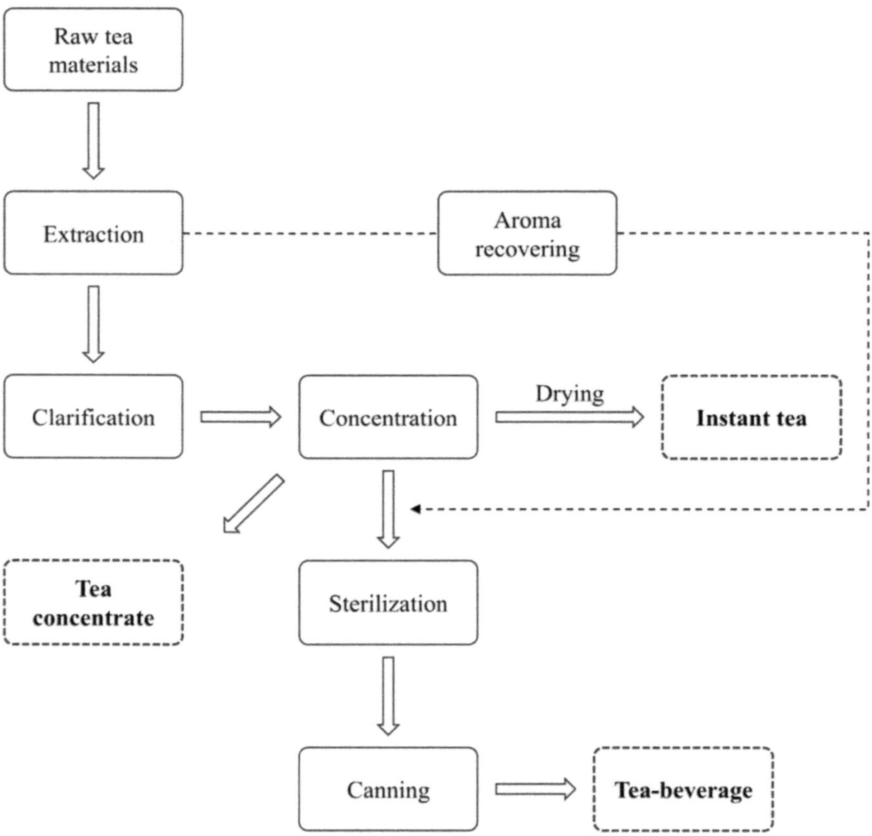

Fig. 8.1 Main production process of tea-beverages

powder), catering to a wide range of tastes and health-conscious demands across demographic groups (Liang et al., 2021). The flavor quality of tea-beverage plays a crucial role in their market appeal, influencing taste preference and brand loyalty. Tea-beverages are particularly susceptible to issues such as browning, turbidity, sedimentation, and aroma deterioration throughout processing and storage (Li et al., 2018; Liu et al., 2018), which would seriously compromise overall quality and flavor stability. Thus, tea factories should prioritize enhancing flavor quality, ensuring consumer' satisfaction. So far, controlled fermentation (Liang et al., 2022b), microbial inoculation (Rigling et al., 2021), and extraction procedures (Rehder et al., 2021) have been commonly employed to improve and enhance the flavor of tea drinks in order to fulfill the diverse needs of consumers.

This chapter explores the key factors influencing the flavor quality of tea-beverage and the modern technologies employed to enhance it, covering the entire production process from raw materials to post-processing. By examining the interactions between these factors and processing techniques, this chapter aims to provide insights into optimizing tea beverage quality and guiding industry innovation to meet evolving consumer preferences.

8.2 Factors Affecting the Quality of Tea-Beverages

In the tea market, there are three major types of tea-beverage products including green tea-beverage, black tea beverage, and oolong tea-beverage. Their quality is primarily determined by color, taste, aroma, and other aspects of the tea infusion. During tea-beverage production, the changes of all the ingredients contained in tea leaves such as the umami-tasting theanine (Thea), bitter-tasting caffeine, astringent and bitter-tasting tea polyphenols (TPs) would contribute to the formation of excellent quality (Li et al., 2018). However, the chemical properties of these components are not stable. Even when stored at room temperature, tea leaves were prone to generating precipitation, browning in color, degradation of aroma, and other challenges that have plagued researchers and factories for many years in terms of limiting shelf-life. Since the 1980s, researchers in the field of tea flavor have conducted extensive studies on the various quality variables that influence tea-beverages and their regulation. The goal of these studies is to identify appropriate solutions to the quality concerns that arise during production and storage.

8.2.1 Factors Influencing the Color of Tea-Beverages

Tea-beverage exhibits distinct colors based on its types: green tea-beverage presents typical yellow-green and bright color; black tea-beverage presents a red and bright color, and oolong tea-beverage shows a characteristic orange-yellow and bright color.

The color in tea-beverage arises from a variety of non-volatile compounds, predominantly catechins and flavonols, and their glycosides. Additional pigments including anthocyanins, chlorophyll, and carotenoids also contribute to their appearance. Exposure to external factors such as temperature, oxygen, and light can alter these pigments over time, leading to color shifts toward yellow, red, or brown. This sensitivity underscores the importance of controlling processing and storage conditions in preserving the desired color of tea-beverage product.

The following list shows the condition factors influencing on tea-beverage during processing and storage:

(a) *raw materials,*
(b) *water quality,*
(c) *temperature, and*
(d) *light.*

It has to be remembered that variable results concerning tea-beverage have to be taken into account: the abovementioned factors can remarkably influence the solubility of ingredients and other bioactive principles. The following sections discuss in detail the effects of abovementioned factors on the color of tea-beverage concerning four groups: raw materials, water quality, temperature, and light.

Raw Materials

The quality of tea-beverage is strongly influenced by the characteristics of the raw tea leaves, such as variety, grade, and leaf tenderness, which serve as the foundation for the quality of the final product. Generally, large-leaf varieties with high content of polyphenols and low content of chlorophyll have a yellow-green color. Therefore, they are typically applied in black tea-beverage production. While medium/small-leaf varieties are more suitable for green tea-beverage production due to their dark green leaves with high contents of chlorophyll and amino acid, but low contents of polyphenols (Luo et al., 2022). The tea plant varieties with a high concentration of anthocyanins are unsuitable for green tea production, as they cause browning and bitterness of the tea infusion (Zhao et al., 2020). Although flavor quality fluctuates during sterilization and canning processes, tender tea leaves are associated with high contents of ester catechins, which can enhance the final products' flavor (Chen et al., 2000). The application of low-grade raw materials can improve the green color and aroma stability of green tea-beverages. Therefore, low- to medium-grade raw materials with low contents of polyphenol and caffeine, as well as high content of ester catechin are often selected to extend shelf life while retaining beverage quality (Chen et al., 2001a).

Water Quality

The quality of water used in tea-beverages, particularly its pH and metal ion composition, has notable impacts on color stability and overall quality of tea-beverage. highly acidic tea-beverages exhibit more stable color profiles and significantly reduce browning during storage compared to their low-acid counterparts (Chen et al., 2021). In addition, metal ions such as Fe^{2+}, Ca^{2+}, and Al^{3+} significantly influence tea infusion's color and taste. For instance, Fe^{2+} catalyzed the oxidation of catechins, which diminished brightness and green color while increasing yellow coloration and imparted a bitter taste (Jiang, 2010; Yasuda et al., 2012). Ca^{2+}, complexed with pectins present in cell walls, can reduce the leaching of phenols, and when the concentration of Ca^{2+} in brewing water was > 40 mg/L, the green tea infusion became more turbid (Xu et al., 2013). In contrast, Al^{3+} intensifies green color but decreases brightness (Mossion et al., 2008). Other ions, such as Mg^{2+} and K^{+}, primarily affect the tea's taste profile (Zhou et al., 2009). Li (2001) found that tea-beverages extracted with mineral-rich water were more prone to browning, underscoring the critical role of water quality in preserving the desirable color and flavor characteristics of tea-beverages.

Temperature

During the tea-beverage processing, high temperatures are involved in the extraction and sterilization, thus impacting the color and brightness. High-temperature extraction often reduces brightness, but it substantially improves the extraction yield (Jiang, 2010). To balance quality and efficiency, extraction temperature and duration should be minimized while achieving optimal yields. Besides, lowering the ratio of tea to water can improve the brightness of the infusion, while high ratios intensify red and yellow color (Yu et al., 2021). Sterilization is commonly conducted at high temperatures to eliminate microorganisms. With the increase in temperature, the browning of the infusion becomes more and more serious. The reason was due to accelerated chlorophyll degradation, causing a noticeable rise in redness and yellowness (He et al., 2019). Kim's study demonstrated that sterilizing green tea infusions at 85–120 °C reduced brightness progressively while deepening redness and yellowness. However, there was no significant color difference between the unheated and the 85 °C heated beverage. Therefore, to preserve color integrity, tea-beverage extraction and pasteurization should ideally be carried out at 85 °C or lower temperature (Kim et al., 2007).

Light

TPs, anthocyanins, and chlorophylls in tea-beverages are light-sensitive and prone to photo-oxidation. Light accelerates the oxidation of TPs and induces the photodegradation of chlorophyll, therefore leading to the formation of pheophytin and the relevant epimers (Petrović et al., 2017). When tea-beverages are exposed to blue light illumination, photolysis of catechins occurs, further resulting in a color change from colorless to yellowish (Yuann et al., 2021). Packaging materials such as tinplate cans and Tetra Pak, which offer effective light protection, helping mitigation of the abovementioned light effects, whereas, when exposing transparent PET (polyethylene terephthalate) bottles to light for prolonged periods, oxidation of phenolics and vitamin C would be promoted and thus deepen the tea infusion's color (Luo et al., 2022).

8.2.2 Factors Influencing the Aroma of Tea-Beverage

Tea odorants are a complex blend of volatile compounds with low-concentrations. During tea-beverage preparation, only a small part of these compounds was retained, imparting a distinctive aroma to the infusion. Studies on aroma extraction and analysis reveal a significant effect of volatile compounds' changes that occur during processing of the raw material. Commonly, volatile oxidation and hydrolysis result in volatile compounds reduction and further sensory quality changes in tea-beverages during processing and storage (Liang et al., 2022a). According to their high aroma intensity perceived in gas chromatography–olfactometry (GC-O), odor activity values (OAV), or flavor dilution (FD) factors, key aroma-active compounds were characterized from the volatile extract of tea-beverage. These compounds contribute to the final aroma profile of the beverage product (Ho et al., 2015; Wang et al., 2022). Currently, aldehydes, alcohols, ketones, and terpenes were considered key contributors to tea's aromatic profile. For example, β-damascenone and linalool were identified as primary odorants in several tea-beverages (Liang et al., 2022a). According to sensory panelists' descriptions, aldehydes and alcohols evoke green, citrus-like, fatty, and sweet notes; terpenes are described as flowery, fruity, and sweet notes; and ketones are responsible for the typical nutty aroma.

Traditional dried tea leaves with a distinct taste are commonly used as an excellent basis of tea-beverages, ensuring the formation of a good combination of flavors and a unique memory point. For example, lemon-flavored tea is often more popular in the beverage market of Europe and the United States. Aroma extraction analysis reveals that the key contributor to the lemon flavor of this kind of product is terpenoids from citrus oils, as well as volatile constituents from tea leaves, such as (Z)-3-hexenol, hexanal, and 4-vinylguaiacol (Yuan et al., 2016).

8.2.3 Factors Influencing the Taste of Tea-Beverage

During the tea-beverage processing, internal components, water quality, and processing techniques collectively contribute to the development of desirable qualities (Fang et al., 2000). The main taste-active compounds in tea-beverages include

(a) *TPs, for both bitterness and astringency;*
(b) *caffeine, for bitterness;*
(c) *amino acids, for umami and sweetness; and*
(d) *carbohydrates, for sweetness.*

TPs

Among the polyphenols, simple catechins such as epicatechin (EC) and epigallocatechin (EGC) contribute to umami, whereas complex catechins like epicatechin gallate (ECG) and epigallocatechin gallate (EGCG) impart stronger bitter and astringent notes. During the processing of tea-beverages, catechin retention declines significantly, dropping to 40–60%, as thermal intensity and duration increase (Chen et al., 2001b). Furthermore, after one month of storage at room temperature, at least 50% of catechins degrade (Xu et al., 2003). Among the four major catechins, EGCG and EGC were equally unstable, whereas EC and ECG were relatively stable (Zhu et al., 1997). Despite extensive study, the mechanisms underlying catechin degradation during processing and storage are not fully elucidated. Xu et al. (2003) concluded that isomerization and thermal degradation were responsible for the loss of catechins during processing, whereas Wang et al. (2000) proposed that catechins with multiple hydroxyl groups were easily oxidized, with oxidation and polymerization were the dominant pathway for catechin degradation in green tea beverages.

Caffeine

Caffeine is another important taste and bioactive component of tea-beverages, contributing to bitterness and offering stimulant effects. During tea processing, caffeine can interact with tea polyphenols to form complexes that reduce both the astringency and bitterness, thereby enhancing the overall flavor. However, caffeine can also bind with polyphenols, theaflavins, thearubigins, and proteins to form insoluble sediment, tea cream, leading to reduced caffeine content and diminished clarity and brightness of the beverage (Ma et al., 2014). Despite these interactions, caffeine remains relatively stable throughout processing and storage. It is minimally affected by factors such as temperature, pH, and antioxidants, with retention rates generally exceeding 90% (Kanaya, 1991).

Amino Acids

Amino acids contribute significantly to the umami and sweetness of tea and are positively correlated with the taste quality of green tea-beverages (Zhang et al., 2020). They also play a role in aroma formation. Free amino acids in beverages are prone to degradation through the Strecker reaction, and can undergo Maillard reactions with reducing sugar under heat treatment, leading to reduced amino acid content. For instance, sterilization at 121 °C results in a 9–12% reduction in amino acids in oolong tea-beverages (Luo, 1996). In green tea beverages sterilized at 121 °C for 7 min, contents of glutamic acid (Glu), arginine (Arg), and Thea decreased by 10%, 20%, and 25%, respectively, while glutamine (Gln), aspartic acid (Asp), and serine (Ser) declined by 10%, 5%, and 5%, respectively (Kinugasa et al., 1992). Therefore, during thermal processing, especially sterilization, caffeine content remains largely stable, while amino acids and simple catechins decrease; in contrast, levels of ester catechins increase. Consequently, the pleasant umami and sweet characteristics of tea beverages are compromised, yielding a less desirable bitter and astringent profile.

Carbohydrates

Soluble sugars in tea infusion, including fructose, glucose, sucrose, maltose, and small amounts of starch and water-soluble pectin, constitute approximately 4% of the total dry weight and are important contributors to the flavor of tea beverages. Monosaccharides and disaccharides enhance sweetness, reduce astringency, and improve overall palatability. Soluble pectin contributes to mouthfeel by increasing viscosity and imparting a thick and mellow sensation to the beverage (Lu et al., 2001). Luo (1996) reported a 12–28% reduction in soluble sugar content in oolong tea-beverages following sterilization. Under thermal conditions, sugars undergo Maillard reactions and may also be transformed into caramel-like/roasty aroma compounds (Baptista et al., 1998). Additionally, the soluble fraction of pectin enhances flavor perception (Sun, 2004). Zhang et al. (2003) further noted that sugar degradation reduces bitterness and astringency while increasing sweetness, thereby improving the mellow quality of the tea beverage.

8.2.4 The Main Components and Formation Mechanism of Tea Sediment

Sediment, known as tea cream, is formed easily during the processing and storage of tea-beverage and tea concentrates. Tea sediment not only decreases the exterior quality of tea-beverages, but also their flavor and shelf life. It has been an important technological challenge during the production of instant tea-beverages, particularly after low-temperature storage (Qin et al., 2022; Yu et al., 2022). Tea sediment is often generated from the interaction of multiple compounds in tea infusion, commonly observed in black, green, and oolong teas (Chao et al., 1999b; Liang et al., 2002).

The following list shows the main contributors to generating tea-beverage sediment during processing and storage:

(a) *TPs*,

(b) *caffeine,*
(c) *proteins, and*
(d) *mineral elements.*

The following sections discuss in detail the contribution of the abovementioned sediment ingredients.

TPs

TPs, including catechins, phenolic acids, and flavonoid glycosides, are the primary contributors to the formation of green tea cream (Kim et al., 2012). TPs represent the largest proportion of green tea cream, accounting for 29.86–78.66% of the total sedimentation content. Among these, catechins make about 12.8–42.5% of the tea cream (Yin et al., 2009). Catechins account for 30% of oolong tea cream (Chao et al., 1999a), with ester catechins being more involved in the sedimentation due to more hydroxyl groups (Chao et al., 1999b). In black tea, oxidized catechins, such as theaflavins and thearubigins, play a key role in the production of cold sedimentation (Liang et al., 2002).

Caffeine

Caffeine is also a significant ingredient in the formation of tea cream. In green tea-beverage, caffeine accounts for 88% of the total tea cream content (Dong et al., 2021), and their concentrations can predict the content of cold sedimentation (Xu et al., 2011). In semi-fermented teas that filtered out tea cream, the addition of caffeine induces re-sedimentation in the pellucid infusion (Chao et al., 1999b). Similarly, external addition of caffeine and gallated catechins significantly promotes the formation of tea cream in black tea infusion, while de-caffeination remarkably reduces turbidity and sedimentation (Penders et al., 1998). Studies on instant tea have shown a strong positive correlation between turbidity and caffeine concentration (Yue, 1999). In simulated systems investigating the interaction between TPs and caffeine, EGCG and caffeine were identified as key contributors to turbidity. The accumulation of these compounds was associated with the low solution transmittance, large particle sizes, and high precipitation levels (Ma et al., 2014).

Protein

Protein is colloidal in tea infusion, accounting for a large proportion of all types of tea sedimentation. Further, protein is one of the top three main components of the tea cream in semi-fermented tea and accounts for about 16% of total tea cream (Chao et al., 1999a). Besides, certain amino acids, such as proline and histidine, in the protein peptide chains can bind with polyphenols, thus forming macromolecular complexes that contribute to protein sedimentation. Yu et al. (2020) found that the interaction between proteins and TPs in tea cream significantly lowers the sensory quality and shelf life of tea-beverage. Therefore, reducing the contents of gallated catechins and proteins can effectively reduce sedimentation.

Mineral Elements

Mineral elements, including Ca^{2+}, Mg^{2+}, K^{+}, Mn^{2+}, and Zn^{2+}, also play a key role in the formation of tea cream by neutralizing charges on the surface of colloidal particles, thus facilitating rapid sedimentation (Yin et al., 2009). Ca^{2+} can form complexes with oxalic acid, contributing to the generation of tea cream and sediment (Xu et al., 2013).

Other components, such as amino acids (e.g., aspartic acid and glutamic acid), also participated in the formation of tea cream, although they were not the primary contributors (Xu et al., 2016). Flavonoids, while involved in the sedimentation of green tea beverage, represent a minor fraction (Xu et al., 2010). Additionally, certain saccharides, including pectin and polysaccharides, are believed to play a role in tea cream formation in green tea (Yin et al., 2009).

To summarize, tea sediment is a complex of various chemical components in the tea beverage. It can be formed by intermolecular forces such as hydrogen bonding and hydrophobic interaction. The main components include TPs, proteins, caffeine, with smaller amounts of carbohydrates, free amino acids, organic acids, and minerals. Tea cream contains both reversible precipitates, which can be redissolved upon heating, and irreversible sediment formed from metal ions and organic acids (Xu et al., 2012).

8.3 Technologies for Flavor Quality Improvement

8.3.1 *Pre-harvest Technologies*

Raw tea materials are crucial to the production of high-quality tea-beverages. China's vast tea-producing regions yield a wide variety of teas, each with distinct qualities. The following pre-harvest technologies are widely used for improving the quality of raw tea leaf materials.

(a) *Blending.* The unique characteristics of different finished teas can be leveraged for blending, which is an effective method to improve the flavor of tea-beverages (Wu, 2012).
(b) *Processing Technologies.* Appropriate processing techniques, such as roasting and drying, can improve the flavor durability and color of tea-beverages, thus ensuring a bright green without cooked off-flavors (Fu et al., 2020).

(c) *Exogenous Additives.* The addition of exogenous tannase into the rolled tea leaves could reduce bitterness and astringency in tea infusions while increasing sweetness by hydrolyzing gallated catechins (Cao et al., 2019). Adding 2% and 4% of sucrose during the raw tea processing could improve the beverage's freshness, sweetness, and aroma concentration (Wang et al., 2020).

(d) *Variety Selection.* Proper selection of tea plant varieties is equally important. The color difference in raw tea infusion can serve as a screening criterion, with a color difference (-a/b) greater than 0.475 recommended for selecting optimal materials for tea-beverages (Yuan et al., 2018).

8.3.2 Processing Technologies

Extraction

Extraction is the first step in the production of tea-beverage. Methods and conditions applied during extraction significantly affect the final sensory quality. Therefore, optimizing the extraction process is very important for beverage quality development. Conventional tea beverage processing typically relies on a high-temperature extraction method, which results in significant leaching of TPs, amino acids, and other soluble substances (Dai, 2013). However, this method often has low extraction efficiency while also causing low aroma intensity, dark coloration, and inferior taste. Therefore, the tea beverage industry has increasingly turned to nonthermal-extraction techniques, such as ultrasonic-assisted, solvent, enzymatic, and microwave-assisted extractions (Raghunath et al., 2020; Yang et al., 2005). These methods effectively preserve the original flavor and aroma of raw materials, while minimizing color changes and other undesirable reactions. Additionally, combining low-temperature extraction with techniques such as superfine grinding, ultra-high pressure, and vacuum distillation can prevent thermal degradation of sensitive compounds like vitamins, catechins, and pigments (Cai et al., 2018; Xie et al., 2020). Modern extraction methods, such as ethanol and ultrasonic-assisted extraction as well as dynamic extraction, enhance the concentration of solids, TPs, and free amino acids, thereby improving the taste and enriching the flavor profile of tea-beverages (Xu et al., 2018; Zhang et al., 2017).

Clarification

Tea-beverages are prone to sedimentation during processing and storage, which can negatively impact their appearance and quality. Therefore, filtration and clarification are essential steps in tea beverage processing. Initially, filtration removes larger solid particles and suspended solids from the tea extract, followed by clarification to further refine the beverage. Current clarification methods for reducing precipitation in tea-beverages can be categorized into physical, chemical, and biological approaches.

(1) Physical Clarification Methods

The interaction between TPs, their oxidation products, caffeine, and proteins leads to the formation of tea cream. While the tea cream can be removed using adsorbents such as activated carbon (AC), polyvinylpolypyrrolidone (PVPP), and chitosan to improve the stability of tea-beverages. AC can adsorb both caffeine and catechins, but it will affect the sensory and bioactivity of tea-beverages (Shiono et al., 2017). PVPP can quantitatively adsorb some of the tea polyphenols in the green tea extract under the appropriate conditions by hydrogen bonding with phenolic hydrogens (Oliva et al., 2007), and it retains the flavor of tea-beverages and prolongs the storage period (Kong et al., 2011). Chitosan can be used as an effective absorption enhancer that can form complexes with proteins and metal ions so that the tea broth can remain clear for a long period of time (Ding et al., 2021; Zohri et al., 2011). Montmorillonite (MMT) and saponite can effectively adsorb caffeine in the aqueous solution, but the binding with catechins is not significant. Meanwhile, Ca^{2+} in MMT may react with oxalic acid to form calcium oxalate precipitation, leading to an increase in the turbidity of decaffeinated tea. By exchanging MMT interlayer ions with Na^+, K^+, Mg^{2+}, or NH_4^+, the formation of calcium oxalate precipitation could be inhibited, thus avoiding turbidity of decaffeinated green tea (Shiono et al., 2018). Membrane filtration technology offers an effective approach for removing tea cream while preserving the flavor and color of tea-beverages. Early applications, such as the use of acetate ultrafiltration membranes in 1987 (Luo et al., 2005), were now been further studied extensively in black tea and oolong tea-beverages (Ye et al., 1995; Zhang, 1997). By selecting appropriate membrane materials and optimizing process parameters, proteins, pectin, and other large molecules can be efficiently removed, while retaining key compounds such as tea polyphenols, caffeine, and amino acids. This significantly enhances the stability and anti-sedimentation properties of tea-beverages (Zhang, 1997). However, membrane fouling is still the key bottleneck limiting its large-scale industrial application, highlighting the need for further advancements in membrane materials, cleaning strategies, and regeneration technologies (Luo et al., 2005).

(2) Chemical Clarification Methods

(a) The addition of fructosyl-containing sugars during the rolling process of green tea can reduce precipitation and enhance stability by forming glycosylated polyphenols and glycoproteins (Xu et al., 2017).
(b) β-Cyclodextrin (β-CD) can be used as an embedding agent, encapsulating catechins and preventing their interaction with other components, effectively inhibiting the formation of tea cream in the tea beverage (Zhang et al., 2019).
(c) The solubilization method inhibits tea cream formation by introducing alkaline compounds (such as NaOH, $NaHCO_3$, and KOH), which disrupt hydrogen bonding between TPs, caffeine, and proteins. Simultaneously, the alkaline solution forms water-soluble salts, further enhancing stability (Zou, 1990).

(d) Adding complexing agents such as ethylene diamine tetraacetic acid (EDTA) allows stable complex formation with metal ions or polyphenols, effectively preventing tea cream formation while maintaining clarity, transparency, and sensory quality (Zhao, 1998).

(3) Biological Clarification Methods

Enzymatic treatment using tannase, pectinase, amylase, or enzyme combinations can effectively reduce tea cream formation in tea-beverages. Tannase hydrolyzes the ester bond between gallic acid and catechins, releasing free gallic acid, which subsequently complexes with caffeine to form water-soluble compounds. This reduces the aggregation of macromolecules and inhibits precipitation formation (Xu et al., 2015).

Sterilization

Sterilization is essential for ensuring the safety and extending the shelf life of tea-beverages (Liang et al., 2021). Tea-beverages are highly susceptible to microbial contamination due to their high nutritional content, including proteins, carbohydrates, amino acids, and vitamins. Therefore, the microbial contamination accelerates product degradation. Conventional sterilization methods include autoclave sterilization and ultra-high temperature (UHT) treatment (Yin, 2006). Autoclave sterilization conducted at 121 °C for 7–15 min causes tea color darkening and flavor degradation. In contrast, UHT treatment performed at 135–150 °C for 2–8 s eliminates microorganisms and preserves the tea's original flavor effectively (Yin, 2006). In addition, non-thermal sterilization techniques such as high-voltage pulsed electric field sterilization were gaining attention. This method applies high-voltage pulses to disrupt bacterial cell membranes, achieving sterilization without heat and thereby retaining the beverage's nutritional and sensory qualities (Takeo et al., 1994). Additionally, modern membrane separation technology effectively separates suspended particles, colloids, and molecules of varying sizes while retaining microorganisms. This approach minimizes flavor loss and maximizes the preservation of tea's original quality (Takeo et al., 1994; Wang et al., 2009).

To mitigate the adverse effects of heat extraction and sterilization on the flavor quality of tea-beverages, various protective technologies and auxiliary methods have been developed. One effective approach is the addition of antioxidants, such as *L*-ascorbic acid, *D*-ascorbic acid, and their sodium salts. They can inhibit the oxidation of TPs and vitamin C, thereby preventing browning and aroma deterioration. Another method involves microencapsulation using β-CD and other macromolecules to encapsulate flavor compounds. They can effectively reduce flavor loss and minimize the development of undesirable "cooked" note caused by sterilization. The integrated application of these techniques plays a crucial role in enhancing the processing quality and improving the storage stability of tea-beverages (Yin, 2006).

Canning

Tea-beverages are packaged in various formats, including glass bottles, tinplate cans, aluminum cans, and polyethylene terephthalate (PET) bottles (Yin et al., 2002). Initially, tinplate cans were the preferred choice due to their adaptability, low equipment investment, and good preservation properties. Whereas glass bottles were gradually phased out due to their fragility and high transportation costs. In recent years, heating-resistant PET bottles have become the dominant packaging form due to their low cost, simple processing, and ease of quality control. Their core technology is the hot canning process, which originated in Japan. It involves UHT sterilization followed by hot filling and sealing to ensure product quality. However, PET bottles may release acetaldehyde during heating and affect flavor. The emergence of heating-resistant BOPP bottles offers a potential solution to this issue (Liu et al., 2001). Additionally, the emergence of cold canning technology not only mitigates the adverse effects of heat sterilization on the color and quality of tea beverages but also reduces packaging costs, thus making it a promising innovation. Aluminum cans are particularly suitable for carbonated tea beverages and can be nitrogen-filled to stabilize catechins, reduce oxygen content, slow oxidation, and thereby preserve flavor, color, and shelf life (Qian, 2019).

8.3.3 Post-processing Technologies

The flavor quality of processed tea-beverages can be enhanced through various post-processing techniques. Rapid cooling minimizes oxidation, preserves the fresh taste of tea, and prevents aging. Low-temperature storage (e.g., refrigeration at 4 °C) further delays flavor and color deterioration. Aroma recovery technology extracts volatile compounds at low temperatures and reintegrates them into instant tea. It helps maintain the authentic tea aroma (Pintauro, 1977). Additionally, aroma enhancers such as maltol, ethyl maltol, and monosodium glutamate can improve flavor perception by lowering the olfactory threshold, thereby enhancing the sensory experience of tea-beverages (Feng, 2002).

References

Baptista, J. A. B., Tavares, JFd. P., & Carvalho, R. C. B. (1998). Comparison of catechins and aromas among different green teas using HPLC/SPME-GC. *Food Research International, 31*(10), 729–736. https://doi.org/10.1016/S0963-9969(99)00052-6

Cai, Y., Jie, G., & Che, K. (2018). A method to produce instant tea with cold solubility by cold extraction in the whole process. China Patent.

Cao, Q. Q., Zou, C., Zhang, Y. H., Du, Q. Z., Yin, J. F., Shi, J., Xue, S., & Xu, Y. Q. (2019). Improving the taste of autumn green tea with tannase. *Food Chemistry, 277*, 432–437. https://doi.org/10.1016/j.foodchem.2018.10.146

Chao, Y. C., & Chiang, B. H. (1999a). Cream formation in a semifermented tea. *Journal of Science and Food Agriculture, 79*(13), 1767–1774. https://doi.org/10.1002/(SICI)1097-0010(199910)79:13%3c1767::AID-JSFA433%3e3.0.CO;2-8

Chao, Y. C., & Chiang, B. H. (1999b). The roles of catechins and caffeine in cream formation in a semi-fermented tea. *Journal of Science and Food Agriculture, 79*(12), 1687–1690. https://doi.org/10.1002/(SICI)1097-0010(199909)79:12%3c1687::AID-JSFA422%3e3.0.CO;2-L

Chen, Y. Q., Ni, D. J., & Zhang, J. N. (2000). Effect of additives on the quality of canned green tea. *Journal of Food Science, 21*(09), 31–34.

Chen, Z.-Y., Zhu, Q. Y., Tsang, D., & Huang, Y. (2001). Degradation of green tea catechins in tea drinks. *Journal of Agricultural and Food Chemistry, 49*(1), 477–482. https://doi.org/10.1021/jf000877h

Chen, J.D., Gao, Y.X. (2021), Quality changes of different tea beverages during storage (in Chinese). *Science and Technology of Food Industry, 42*(11), 281–289. https://doi.org/10.13386/j.issn1002-0306.2020090291

Chen, Y.Q., Zhang, J.F. (2001a), Effects of raw material on quality of canned green tea beverage (in chinese). *Journal of Tea Science* (02), 120–123.

Dai, Z. (2013). An overview of tea leaf extraction technologies for rtd tea beverages (in chinese). *Beverage Industry, 16*(09), 39–45.

Ding, P., Li, G.-Y., Chen, C.-M., & Huang, K.-L. (2021). Kinetics and mechanism of chelating reaction between chitosan derivatives with ca(ii). *Journal of Coordination Chemistry, 64*(8), 1333–1343. https://doi.org/10.1080/00958972.2011.568615

Dong, Y.R., Cao, Y.Y., Lu, J.C., Mei, S.F., He, R., Chen, P. (2021), Crystallographic study on the complex of catechins and caffeine. *Journal of Tea Science, 41*(01), 80–89. https://doi.org/10.13305/j.cnki.jts.20201127.001

Fang, Y.C., Mei, C.X., Yang, C.W. (2000). The mechanism of the change of the color, fragrance, taste of the tea beverage and the method of improving the quality (in chinese). *Beverage & Fast Frozen Food Industry* (04), 19–21.

Feng, X.H. (2002), An overview of several factors affecting the flavor of tea beverages (in chinese). *The Food Industry,* (02), 27–29.

Fu, Y. Q., Wang, J. Q., Chen, J. X., Wang, F., Yin, J. F., Zeng, L., Shi, J., & Xu, Y. Q. (2020). Effect of baking on the flavor stability of green tea beverages. *Food Chemistry, 331*, Article 127258. https://doi.org/10.1016/j.foodchem.2020.127258

He, S. Y., Zhang, N., & Jing, P. (2019). Insights into interaction of chlorophylls with sodium caseinate in aqueous nanometre-scale dispersion: Color stability, spectroscopic, electrostatic, and morphological properties. *RSC Advances, 9*(8), 4530–4538. https://doi.org/10.1039/C8RA09329F

Ho, C.-T., Zheng, X., & Li, S. M. (2015). Tea aroma formation. *Food Science and Human Wellness, 4*(1), 9–27. https://doi.org/10.1016/j.fshw.2015.04.001

Jiang, C. L. (2010). *Research on the quality of tea beverages extracted and blended by different water property*. Agronomy, Fujian Agricultural University, Fujian.

Kanaya, Y. (1991). Proceedings of the international symposium on tea science. *Shizuoka Japan, 8*(26–19), 571–575.

Kim, Y., & Talcott, S. T. (2012). Tea creaming in nonfermented teas from camellia sinensis and ilex vomitoria. *Journal of Agricultural and Food Chemistry, 60*(47), 11793–11799. https://doi.org/10.1021/jf303555f

Kim, E. S., Liang, Y. R., Jin, J., Sun, Q. F., Lu, J. L., Du, Y. Y., & Lin, C. (2007). Impact of heating on chemical compositions of green tea liquor. *Food Chemistry, 103*(4), 1263–1267. https://doi.org/10.1016/j.foodchem.2006.10.031

Kinugasa, H., Takeo, C., & Fukumoto, K. (1992). Effects of high-pressure treatment on sterilization efficiency and component changes in tea beverages. *Journal of the Agricultural Chemical Society of Japan, 66*(4), 707–712.

Kong, Q.L., Sun, J.H., Chen, X.Q., Tu, Y.F., Yang, X.F. (2011), Application of polyvinyl polypyrrolidone (pvpp) in deposition control of tea beverage (in chinnese). *China Tea Processing,* (02), 29–32. https://doi.org/10.15905/j.cnki.33-1157/ts.2011.02.013

Li, J., Dong, X.Y., Feng, Y., Wang, X.J., Yang, X.P. (2018). Research progress and development prospect of liquid tea beverage (in chinese). *China tea processing*, (2), 23–27+30. https://doi.org/10.15905/j.cnki.33-1157/ts.2018.02.009

Li, X.M. (2001), Reserch on the influence of different water quality on the quality of green tea beverage. *China Tea Processing,* (02), 28–30. https://doi.org/10.15905/j.cnki.33-1157/ts.2001.02.011

Liang, S., Granato, D., Zou, C., Gao, Y., Zhu, Y., Zhang, L., Yin, J. F., Zhou, W. B., & Xu, Y. Q. (2021). Processing technologies for manufacturing tea beverages: From traditional to advanced hybrid processes. *Trends in Food Science & Technology, 118*, 431–446. https://doi.org/10.1016/j.tifs.2021.10.016

Liang, S., Gao, Y., Fu, Y. Q., Chen, J. X., Yin, J. F., & Xu, Y. Q. (2022a). Innovative technologies in tea-beverage processing for quality improvement. *Current Opinion in Food Science, 47*, Article 100870. https://doi.org/10.1016/j.cofs.2022.100870

Liang, S., Wang, F., Chen, J. X., Granato, D., Li, L., Yin, J. F., & Xu, Y. Q. (2022b). Optimization of a tannase-assisted process for obtaining teas rich in theaflavins from camellia sinensis leaves. *Food Chemistry: X, 13*, Article 100203. https://doi.org/10.1016/j.fochx.2022.100203

Liang, Y.R., Lu, J.L., Zhang, L.Y. (2002), Comparative study of cream in infusions of black tea and green tea [camellia sinensis (l.) o. Kuntze]. *International Journal of Food Science & Technology, 37*(6):627–634. https://doi.org/10.1046/j.1365-2621.2002.00589.x

Liu, J.Q., Gong, J.S., Wu, W.H., Chen, W.P., Qi, D.M. (2001), A study on the penetration of oxygen during shelf-life for beverages packed in bopp bottles (in chinese). *Beverage Industry,* (02):7–11+16.

Liu, P.P., Xu, Y.Q., Zou, C., Gao, Y., Wang, F., Chen, J.X., Yin, J.F. (2018). Studies on the quality change of tea infusion beverages during sterilization and storage (in chinese). *Journal of Chinese Institute of Food Science and Technology, 18*(1):202–210. https://doi.org/10.16429/j.1009-7848.2018.01.026

Lu, S., & Shi, Z. (2001). *Tea evaluation and inspection.* China Agriculture Press.

Luo, R., Shao, W., Wu, H. (2005). The deposition of tea beverages and application of technology of clarification. *Chinese Agricultural Science Bulletin,* (12):95–98.

Luo, X. L., & Gao, Y. X. (2022). Research progress of color substances, color degradation mechanism and color protection technology for tea beverages (in chinese). *China Food Additives, 02,* 218–229.

Luo, L.X. (1996), Quality changes during storage of liquid tea (in chinese). *China Tea,* (04), 11–13.

Ma, M. J., Luo, Y. L., Li, S., & Zeng, L. (2014). Effects of different concentrations of caffeine and tea polyphenolic compounds on the formation of tea cream (in chinese). *Food Science, 35*(13), 15–19.

Mossion, A., Potin-Gautier, M., Delerue, S., Le Hécho, I., & Behra, P. (2008). Effect of water composition on aluminium, calcium and organic carbon extraction in tea infusions. *Food Chemistry, 106*(4), 1467–1475. https://doi.org/10.1016/j.foodchem.2007.05.098

Oliva, J., Payá, P., Cámara, M. A., & Barba, A. (2007). Removal of famoxadone, fluquinconazole and trifloxystrobin residues in red wines: Effects of clarification and filtration processes. *Journal of Environmental Science and Health, 42*(7), 775–781. https://doi.org/10.1080/03601230701550964

Penders, M. H. G. M., Scollard, D. J. P., Needham, D., Pelan, E. G., & Davies, A. P. (1998). Some molecular and colloidal aspects of tea cream formation. *Food Hydrocolloids, 12*(4), 443–450. https://doi.org/10.1016/S0268-005X(98)00060-5

Petrović, S., Zvezdanović, J., & Marković, D. (2017). Chlorophyll degradation in aqueous mediums induced by light and uv-b irradiation: An uhplc-esi-ms study. *Radiation Physics and Chemistry, 141*, 8–16. https://doi.org/10.1016/j.radphyschem.2017.05.024

Pintauro, N.D. (1977). Tea and soluble tea products manufacture. Food technology review; no. 38. Noyes Data Corp., Park Ridge, N.J.

Qian, W. (2019). Preparation method of tea beverages. CN110074221A.

Qin, C. Y., Lian, L., Xu, W., Jiang, Z. D., Wen, M. C., Han, Z. S., & Zhang, L. (2022). Comparison of the chemical composition and antioxidant, anti-inflammatory, α-amylase and α-glycosidase inhibitory activities of the supernatant and cream from black tea infusion. *Food & Function, 13*(11), 6139–6151. https://doi.org/10.1039/D2FO00707J

Raghunath, S., & Mallikarjunan, K. (2020). Optimization of ultrasound-assisted extraction of cold-brewed black tea using response surface methodology. *Journal of Food Process Engineering, 43*(11), Article e13540. https://doi.org/10.1111/jfpe.13540

Rehder, A. P., Silva, P. B., Xavier, A. M. F., & Barrozo, M. A. S. (2021). Optimization of microwave-assisted extraction of bioactive compounds from a tea blend. *Journal of Food Measurement and Characterization, 15*(2), 1588–1598. https://doi.org/10.1007/s11694-020-00750-4

Rigling, M., Liu, Z. B., Hofele, M., Prozmann, J., Zhang, C., Ni, L., Fan, R., & Zhang, Y. Y. (2021). Aroma and catechin profile and in vitro antioxidant activity of green tea infusion as affected by submerged fermentation with wolfiporia cocos (fu ling). *Food Chemistry, 361*, Article 130065. https://doi.org/10.1016/j.foodchem.2021.130065

Shiono, T., Yamamoto, K., Yotsumoto, Y., Kawai, J., Imada, N., Hioki, J., Naganuma, H., Eguchi, T., Kurihara, M., Yoshida, A., & Deuchi, K. (2017). Selective decaffeination of tea extracts by montmorillonite. *Journal of Food Engineering, 200*, 13–21. https://doi.org/10.1016/j.jfoodeng.2016.12.015

Shiono, T., Yamamoto, K., Yotsumoto, Y., & Yoshida, A. (2018). Effects of interlayer ion in montmorillonite on appearance of decaffeinated tea beverage. *Food Science and Technology Research, 24*(2), 215–221. https://doi.org/10.3136/fstr.24.215

Standard tCN. (2008). Tea beverages. In vol GB/T 21733-2008. pp. Standardization Administration of China, China.

Sun, Q. (2004). *A study on maintenance of volatile compounds and related processing parameters of ready-to-drink green tea.* Zhejiang University.

Takeo, T., Kinugasa, H., Ishihara, M., Fukumoto, K., Shinno, T. (1994). Application of sterilization technique by hydrostatic high pressure for green tea drink. In: Yano T, Matsuno R & Nakamura K (eds) Developments in food engineering: Proceedings of the 6th international congress on engineering and food (pp 864–86). Springer US, Boston, MA.

Wang, L.-F., Kim, D.-M., & Lee, C. Y. (2000). Effects of heat processing and storage on flavanols and sensory qualities of green tea beverage. *Journal of Agricultural and Food Chemistry, 48*(9), 4227–4232. https://doi.org/10.1021/jf0003597

Wang, J., Li, M. R., Wang, H., Huang, W. J., Li, F., Wang, L. L., Ho, C.-T., Zhang, Y. Y., Zhang, L., Zhai, X. T., & Wan, X. C. (2022). Decoding the specific roasty aroma wuyi rock tea (camellia sinensis: Dahongpao) by the sensomics approach. *Journal of Agricultural and Food Chemistry, 70*(34), 10571–10583. https://doi.org/10.1021/acs.jafc.2c02249

Wang, Z.L., Li, S.K., Xu, Y.Q., Yin, J.F. (2009). Overview of tea beverage sterilization technology (in chinese). *Newsletter of Sericulture and Tea,* (03):31–34.

Wang, J.Q., Yang, Y., Wang, F., Zeng, L., Xu, Y.Q., Yin, J.F. (2020). Effect of sucrose on the physicochemical composition and flavor stability of green tea beverages (in chinese). *Journal of Chinese Institute of Food Science and Technology, 20*(04), 142–151. https://doi.org/10.16429/j.1009-7848.2020.04.019

Wu, B. C. (2012). *Feasibility study on blending among multi-varieties of Jingshan tea.* Zhejiang Universuty, Zhejiang.

Xie, Z., Wang, Y., Wang, Y., Liu, R., Zhang, J., Liu, D., Wang, Z., Liang, J., Han, J. (2020), A method to produce cold-extracted tea concentrate. China Patent.

Xu, J. Z., Leung, L. K., Huang, Y., & Chen, Z.-Y. (2003). Epimerisation of tea polyphenols in tea drinks. *Journal of Science and Food Agriculture, 83*(15), 1617–1621. https://doi.org/10.1002/jsfa.1597

Xu, Y. Q., Chen, S. Q., Shen, D. Y., & Yin, J. F. (2011). Effects of chemical components on the amount of green tea cream. *Agricultural Sciences in China, 10*(6), 969–974. https://doi.org/10.1016/S1671-2927(11)60083-7

Xu, Y. Q., Chen, G. S., Wang, Q. S., Yuan, H. B., Feng, C. H., & Yin, J. F. (2012). Irreversible sediment formation in green tea infusions. *Journal of Food Science, 77*(3), C298–C302. https://doi.org/10.1111/j.1750-3841.2011.02583.x

Xu, Y. Q., Zhong, X. Y., Yin, J. F., Yuan, H. B., Tang, P., & Du, Q. Z. (2013). The impact of ca2+ combination with organic acids on green tea infusions. *Food Chemistry, 139*(1), 944–948. https://doi.org/10.1016/j.foodchem.2013.01.025

Xu, Y. Q., Hu, X. F., Zou, C., Shi, J., Du, Q. Z., Teng, B. T., & Yin, J. F. (2017). Effect of saccharides on sediment formation in green tea concentrate. *LWT, 78*, 352–360. https://doi.org/10.1016/j.lwt.2017.01.003

Xu, Y.-Q., Ji, W.-B., Yu, P., Chen, J.-X., Wang, F., & Yin, J.-F. (2018). Effect of extraction methods on the chemical components and taste quality of green tea extract. *Food Chemistry, 248*, 146–154. https://doi.org/10.1016/j.foodchem.2017.12.060

Xu, Y.Q., Yin, J.F. (2010). Study on the characteristics of cream formation in green tea infusion (in chinese). *Journal of Tea Science, 30*(S1), 527–532. https://doi.org/10.13305/j.cnki.jts.2010.s1.005

Xu, Y.Q., Yin, J.F. (2016). Review on tea sediment formation and its controlling methods (in chinese). *Journal of Tea Science, 36*(4):337–346. https://doi.org/10.13305/j.cnki.jts.2016.04.001

Xu, Y.Q., Hu, X.F., Chen, J.X., Chen, G.S., Wang, F, Yin, J.F. (2015). Re-dissolution and reclaim of green tea sediment based on the treatment of tannase (in chinese). *Journal of Tea Science, 35*(06), 589–595. https://doi.org/10.13305/j.cnki.jts.2015.06.011

Yang, W., Gao, Y. (2005). Study on improving extraction quality of tea soup by enzyme. *Science and Technology of Food Industry,* (04), 61–63+67. https://doi.org/10.13386/j.issn1002-0306.2005.04.010

Yasuda, M., Matsuda, C., Ohshiro, A., Inouye, K., & Tabata, M. (2012). Effects of metal ions (cu2+, fe2+ and fe3+) on hplc analysis of catechins. *Food Chemistry, 133*(2), 518–525. https://doi.org/10.1016/j.foodchem.2012.01.018

Ye, W.Y., Fu, K.R., Yang, J., Chen, L.Q. (1995). Clarification of oolong tea beverage by ultrafitration (in chinese). *Journal of Fujian Agriculture and Forestry University,* (01):99–104.

Yin, J. F., Xu, Y. Q., Yuan, H. B., Luo, L. X., & Qian, X. J. (2009). Cream formation and main chemical components of green tea infusions processed from different parts of new shoots. *Food Chemistry, 114*(2), 665–670. https://doi.org/10.1016/j.foodchem.2008.10.004

Yin, J.F., Lin, Z. (2002). Progress in the processing technology of tea drink (in chinese). *Journal of Tea Science,* (01):7–13+29.

Yin, J.F., Xu, Y.Q., Yuan, H. (2021). Development trends and technical requirements of tea beverage in china mainland (in chinese). *Journal of Tea Science, 30*(S1), 588–592. https://doi.org/10.13305/j.cnki.jts.2010.s1.014

Yin, J.F. (2006). Sterilization technology in tea beverages (in chinese). *China Tea,* (03), 17–18.

Yu, J. Y., Liu, Y., Zhang, S. R., Luo, L. Y., & Zeng, L. (2021). Effect of brewing conditions on phytochemicals and sensory profiles of black tea infusions: A primary study on the effects of geraniol and β-ionone on taste perception of black tea infusions. *Food Chemistry, 354*, Article 129504. https://doi.org/10.1016/j.foodchem.2021.129504

Yu, X., Cai, X., Li, S., Luo, L., Wang, J., Wang, M., & Zeng, L. (2022). Studies on the interactions of theaflavin-3,3′-digallate with bovine serum albumin: Multi-spectroscopic analysis and molecular docking. *Food Chemistry, 366*, Article 130422. https://doi.org/10.1016/j.foodchem.2021.130422

Yu, X., Cai, X.H., Luo, L.Y., Wang, J., Ma, M.J., Wang, M., Zeng, L. (2020). Influence of tea polyphenol and bovine serum albumin on tea cream formation by multiple spectroscopy methods and molecular docking. *Food Chemistry, 333*. https://doi.org/10.1016/j.foodchem.2020.127432

Yuan, F., He, F., Qian, Y., Zheng, J., & Qian, M. C. (2016). Aroma stability of lemon-flavored hard iced tea assessed by chirality and aroma extract dilution analysis. *Journal of Agricultural and Food Chemistry, 64*(28), 5717–5723. https://doi.org/10.1021/acs.jafc.6b01776

Yuan, H., Deng, Y., Hua, J., Li, J., Dong, C., Yang, Y., Wang, J., Yin, J., Jiang, Y. (2018), A method based on c&r decision tree for primary screening of tea beverage raw materials for tea beverage (in chinese). *Food Science, 39*(17):67–72. https://link.cnki.net/urlid/11.2206.TS.20180112.1641.002

Yuann, J. M. P., Lee, S. Y., Yang, M. J., Huang, S. T., Cheng, C. W., & Liang, J. Y. (2021). A study of catechin photostability using photolytic processing. *Processes, 9*(2), 293. https://doi.org/10.3390/pr9020293

Yue, P.X. (1999). The effect of caffeine concentration on the turbidity of instant tea (in chinese). *Beverage Industry,* (02):33–36.

Zhang, Q.-A., Zhang, X.-L., Yan, Y.-Y., & Fan, X.-H. (2017). Antioxidant evaluation and composition analysis of extracts from Fuzhuan brick tea and its comparison with two instant tea products. *Journal of AOAC INTERNATIONAL, 100*(3), 653–660. https://doi.org/10.5740/jaoacint.16-0403

Zhang, Y.-H., Chen, G.-S., Chen, J.-X., Liu, Z.-Q., Yu, L.-Y., Yin, J.-F., & Xu, Y.-Q. (2019). *Effects of β-Cyclodextrin and Sodium Ascorbate on the Chemical Compositions and Sensory Quality of Instant Green Tea Powder during Storage, 2019*(1), 5618723. https://doi.org/10.1155/2019/5618723

Zhang, L., Cao, Q.-Q., Granato, D., Xu, Y.-Q., & Ho, C.-T. (2020). Association between chemistry and taste of tea: A review. *Trends in Food Science & Technology, 101*, 139–149. https://doi.org/10.1016/j.tifs.2020.05.015

Zhang, L., Liang, Y., Sun, Q., Sun, Q., Lu, J., Mamati, E.G. (2003). Effect of heat sterilization and storage on quality attributes of fresh green tea beverages (in chinese). *Journal of Tea Science,* (02):171–176.

Zhang, C.B. (1997). Study on the production of canned black tea beverage by ultrafiltration method (in chinese). *Science and Technology of Food Industry,* (06):22–23.

Zhao, Y. L., Lai, W. Y., Xu, A. A., Jin, J., Wang, Y. F., & Xu, P. (2020). Characterizing relationships among chemicals, sensory attributes and in vitro bioactivities of black tea made from an anthocyanins-enriched tea cultivar. *LWT, 132*, Article 109814. https://doi.org/10.1016/j.lwt.2020.109814

Zhao, L. (1998). Study on color protection and stability of green tea beverages (in chinese). *Food Research and Development,* (01), 29–31.

Zhou, D. R., Chen, Y. Q., & Ni, D. J. (2009). Effect of water quality on the nutritional components and antioxidant activity of green tea extracts. *Food Chemistry, 113*(1), 110–114. https://doi.org/10.1016/j.foodchem.2008.07.033

Zhu, Q. Y., Zhang, A., Tsang, D., Huang, Y., & Chen, Z.-Y. (1997). Stability of green tea catechins. *Journal of Agricultural and Food Chemistry, 45*(12), 4624–4628. https://doi.org/10.1021/jf9706080

Zohri, M., Nomani, A., Gazori, T., Haririan, I., Mirdamadi, S. S., Sadjadi, S. K., & Ehsani, M. R. (2011). Characterization of chitosan/alginate self-assembled nanoparticles as a protein carrier. *Journal of Dispersion Science and Technology, 32*(4), 576–582. https://doi.org/10.1080/01932691003757314

Zou, Y.G. (1990). Methods and principles of solving cloudy precipitation of tea beverages. *Food Science,* (07), 14–19.

Chapter 9
Antioxidant Effect of Tea

Shanshan Hu

Abstract This chapter comprehensively examines the biological mechanisms of oxidative stress and its health impacts, with an in-depth analysis of the antioxidant effects of tea bioactive compounds. Oxidative stress arises from an imbalance between free radical production and elimination, triggered by endogenous pathways or exogenous factors, leading to lipid peroxidation, DNA damage, mitochondrial dysfunction, and disrupted signaling cascades. These alterations promote inflammatory responses, cellular senescence, and chronic diseases. Critical antioxidant components in tea—including polyphenols (notably catechins), alkaloids, amino acids, and flavonoids—exhibit structure–activity relationships dependent on phenolic hydroxyl group count/spatial configuration, esterification/glycosylation modifications. Mechanistically, tea polyphenols not only directly scavenge free radicals but also activate the Nrf2/ARE pathway to upregulate antioxidant enzymes, while modulating AKT-JNK and MAPK-NF-κB signaling axes to suppress inflammation and apoptosis. Collectively, tea constituents mitigate oxidative damage through multi-target interactions, establishing a molecular basis for preventing oxidative stress-related pathologies.

Keywords Oxidative stress · Tea · Tea polyphenols · Structure–activity relationship · Antioxidant mechanisms

S. Hu (✉)
State Key Laboratory of Tea Plant Germplasm Innovation and Resource Utilization, Anhui Agricultural University, Hefei, Anhui, China
e-mail: Hshan1008@163.com

International Joint Laboratory on Tea Chemistry and Health Effects of Ministry of Education, Anhui Agricultural University, Hefei, Anhui, China

X. Zhai (ed.), *Tea Chemistry and Health Benefits*,
https://doi.org/10.1007/978-3-032-11828-8_9

9.1 Occurrence and Harm of Oxidative Stress

9.1.1 Occurrence of Oxidative Stress

Oxidative stress is a complex biochemical process that involves an imbalance between free radicals and the antioxidant defense system within the biological organism. Under normal physiological conditions, free radicals, such as reactive oxygen species (ROS) and reactive nitrogen species (RNS), are naturally produced during cellular metabolism and play important roles in signal transduction, cell division, and immune responses. Free radicals are highly reactive molecules or atoms within the biological organism that possess unpaired electrons, thereby exhibiting strong oxidative properties. The generation of free radicals within the organism is a complex biochemical process involving various enzymatic and non-enzymatic reactions. During normal cellular metabolism, the mitochondrial electron transport chain (ETC) is the primary source of ROS. At complexes I and III of the ETC, electron leakage occurs and reacts with oxygen molecules to generate superoxide anion (O_2^-) (Sabharwal & Schumacker, 2014). Additionally, the NADPH oxidase (NOX) family, cytochrome P450 enzyme system, and certain activated inflammatory cells (such as neutrophils and macrophages) can also produce large amounts of ROS (Cipriano et al., 2023). When the production of free radicals exceeds the body's capacity for clearance, oxidative stress ensues, leading to damage to cells and tissues. The damage to cells caused by oxidative stress is mainly manifested in the following aspects:

(a) *Lipid Peroxidation* ROS attack polyunsaturated fatty acids in cell membranes, leading to the formation of lipid peroxides. These substances further decompose to produce secondary products such as malondialdehyde (MDA) (Ayala et al., 2014). These products can react with proteins and nucleic acids, resulting in damage to cellular structure and function.

(b) *Protein Oxidation* Free radicals can attack amino acid residues in proteins, leading to oxidative modifications such as carbonylation of proteins. Such oxidative modifications alter the structure and function of proteins, thereby affecting their normal biological activities (Akagawa, 2021).

(c) *Nucleic Acid Damage* ROS can induce oxidative damage to DNA, forming oxidized products such as 8-hydroxy-2′-deoxyguanosine (8-OHdG) (Trosko, 2021). If these damages are not promptly repaired, they may lead to DNA mutations and even trigger carcinogenesis.

(d) *Mitochondrial Dysfunction* mitochondria are both the primary site of ROS production and a major target for ROS attack. Oxidative damage to mitochondrial DNA (mtDNA) can impair mitochondrial function, leading to disturbances in energy metabolism and cellular apoptosis (Cojocaru et al., 2023).

(e) *Interference with Cellular Signal Transduction* ROS can function as signaling molecules involved in intracellular signal transduction (Sies & Jones, 2020). However, excessive ROS can interfere with normal signaling pathways, resulting in abnormalities in cell proliferation, differentiation, and apoptosis.

The occurrence of oxidative stress mainly includes two pathways: endogenous and exogenous. The endogenous pathway involves the inevitable production of free radicals during cellular metabolism, such as electron leakage from the mitochondrial respiratory chain. As the cell's energy powerhouse, the mitochondrial electron transport chain (ETC) produces a small amount of reactive oxygen species (ROS) as a byproduct while generating ATP (Suomalainen & Nunnari, 2024). Under normal conditions, these ROS are neutralized by the antioxidant systems in the mitochondria and cytoplasm. However, when mitochondrial function is impaired or metabolic demands increase, ROS production may rise, thereby triggering oxidative stress. The exogenous pathway includes exposure to environmental pollution, ultraviolet radiation, tobacco smoke, and certain chemicals, all of which can increase the body's free radical load and place additional strain on the antioxidant system (García-Caparrós et al., 2021).

9.1.2 Harm of Oxidative Stress

The effects of oxidative stress on health are multifaceted. First, oxidative stress can promote inflammatory responses. Inflammation is a natural response of the body to injury and infection, involving the activation of immune cells and the release of inflammatory mediators (Xu et al., 2023). However, chronic oxidative stress can lead to the overproduction of inflammatory mediators, resulting in a state of chronic inflammation, which is closely associated with the development of various diseases. Second, oxidative stress can accelerate the cellular aging process. The imbalance between DNA damage and repair mechanisms within cells, as well as the impact of oxidative stress on cell cycle and apoptosis pathways, can lead to cellular dysfunction and aging (Xie et al., 2023). Additionally, oxidative stress is closely related to the development of various chronic diseases, such as cardiovascular diseases, neurodegenerative diseases, cancer, and diabetes. These diseases are often associated with cell damage, inflammatory responses, and metabolic disorders caused by oxidative stress (Jomova et al., 2023). Therefore, maintaining the balance between oxidative stress and antioxidant defense is crucial for cellular homeostasis and overall health.

9.2 Antioxidant Active Ingredients in Tea

In recent years, with the advancement of science and technology, numerous experimental studies have demonstrated the potent antioxidant properties of tea. The key active components in tea are of significant importance for the prevention and improvement of diseases related to oxidative stress. The medicinal history of tea dates back to ancient China (Pan et al., 2022). According to the "*Shennong Bencao Jing*" (The Classic of Materia Medica), Shennong tasted hundreds of herbs and encountered seventy-two poisons daily, which were neutralized by tea. This reflects the earliest medicinal value of tea. In the "*Chajing*" (The Classic of Tea), Lu Yu provided a detailed discussion on the types, origins, harvesting, processing, and consumption methods of tea, emphasizing its health benefits. Ancient medical literature described tea as having properties such as clearing heat and detoxifying, invigorating the mind, and aiding digestion. Over time, the medicinal value of tea has gradually been recognized and accepted worldwide.

The chemical composition of tea is complex and diverse, forming the basis for its unique flavor and health benefits. The main chemical components of tea include tea polyphenols, alkaloids, amino acids, flavonoids, vitamins, etc. (Bag et al., 2022; Fang et al., 2025; Shi et al., 2023). Among these, polyphenols, especially catechins, are one of the most important antioxidant components in tea. Catechins are a class of phenolic compounds based on gallic acid, including epicatechin, epicatechin gallate, and epigallocatechin gallate. These compounds possess strong antioxidant activity, capable of scavenging free radicals, inhibiting lipid peroxidation, and protecting cells from oxidative damage. Alkaloids in tea, such as caffeine, theophylline, and theobromine, also exhibit certain antioxidant effects. Caffeine can reduce oxidative stress and protect nerve cells. The antioxidant effects of theophylline and theobromine are relatively weaker but can enhance the antioxidant efficacy of caffeine. Amino acids, such as theanine, which is unique to tea, can scavenge free radicals, promote the release of neurotransmitters, and protect cells from oxidative damage. In addition, flavonoids (quercetin and kaempferol) and vitamins (vitamin C and vitamin E) in tea also play antioxidant roles by regulating cellular signal transduction and inhibiting inflammatory responses. It is evident that tea, with its unique chemical composition and biological activities, has become a hot topic in modern scientific research.

9.3 Antioxidant Mechanisms of Tea Polyphenols

Antioxidant components in tea, especially tea polyphenols and their related compounds, exert complex and sophisticated antioxidant effects within the human body. This section will elaborate on the mechanisms of tea's antioxidant actions, focusing on free radical scavenging, regulation of antioxidant enzyme activity, and modulation of signaling pathways.

9.3.1 Free Radical Scavenging and Structure–Activity Relationship of Catechins

Tea polyphenols, particularly catechins, possess strong free radical scavenging abilities. Tea polyphenols mainly include catechins, flavonoids, anthocyanins, and phenolic acids. These compounds can react with free radicals, neutralizing them through electron transfer or hydrogen atom transfer, thereby preventing cellular damage (Braicu et al., 2013; Yan et al., 2020). Catechins are the most abundant components of tea polyphenols, including epigallocatechin (EGC), epigallocatechin gallate (EGCG), epicatechin gallate (ECG), and epicatechin (EC). The key to their antioxidant activity lies in the multiple phenolic hydroxyl groups in their molecular structures.

Research has shown that the structure–activity relationship of catechins is primarily reflected in the following aspects:

(a) number of phenolic hydroxyl Groups,
(b) position of phenolic hydroxyl Groups,
(c) molecular configuration, and
(d) esterification and glycosylation.

The following sections discuss in detail about the structure–activity relationship of catechins.

Number of Phenolic Hydroxyl Groups

The number of phenolic hydroxyl groups directly affects the antioxidant capacity of tea polyphenols (Chen et al., 2020). These hydroxyl groups (–OH) are functional groups capable of donating hydrogen atoms or electrons, which can neutralize free radicals and reduce oxidative damage to cells (Rana et al., 2022). Studies have demonstrated that the antioxidant activity of catechins is positively correlated with the number of phenolic hydroxyl groups. Additionally, the chemical environment of the phenolic hydroxyl groups also influences their antioxidant efficacy. The presence of a galloyl group enhances the antioxidant capacity of the phenolic hydroxyl groups. In EGCG, the galloyl group increases the antioxidant activity of the phenolic hydroxyl groups.

Position of Phenolic Hydroxyl Groups

The distribution of phenolic hydroxyl groups on the molecular structure also influences the antioxidant activity of tea polyphenols. In terms of chemical structure, the pattern of phenolic hydroxyl group distribution determines the polarity, solubility, and the ability of the molecule to interact with other molecules. These factors collectively impact the antioxidant potential of tea polyphenols. When phenolic hydroxyl groups are concentrated in a specific region of the molecule, the polarity of that region is enhanced, which facilitates the interaction of tea polyphenols with free radicals or oxidizing agents, thereby increasing antioxidant activity. Conversely, if the phenolic hydroxyl groups are dispersed, it may reduce the local polarity of the molecule, affecting its reaction efficiency with free radicals (Ishii et al., 2010). Furthermore, the distribution of phenolic hydroxyl groups affects the solubility of tea polyphenols, which in turn influences their interaction with other biomolecules

(Dudek & Day, 2019). The presence of phenolic hydroxyl groups increases the solubility of the molecule in water or other solvents, aiding the distribution and absorption of tea polyphenols in biological systems. For example, catechin compounds, such as epicatechin and epicatechin gallate, exhibit good solubility in water due to the distribution of their phenolic hydroxyl groups. Enhanced water solubility can improve the binding capacity of catechins with proteins, enzymes, or other large biomolecules, contributing to their stability and transport within the biological system and ultimately enhancing their antioxidant effects (Intra & Kuo, 2007).

Molecular Configuration

The stereochemistry of tea polyphenols also affects their interaction with free radicals. The stereochemical configuration of tea polyphenols, which refers to the spatial arrangement of their molecules, significantly impacts their interaction with free radicals (Muzolf-Panek et al., 2012; Yang et al., 2018). Stereochemistry determines the spatial orientation of phenolic hydroxyl groups, thereby influencing the efficiency and selectivity of the reaction between tea polyphenols and free radicals. In chemistry, stereochemistry is typically defined by the relative positions of atoms within a molecule and the rotational freedom of chemical bonds. Firstly, if phenolic hydroxyl groups in tea polyphenol molecules are in adjacent positions, their synergistic effect can enhance the molecule's antioxidant capacity. For example, the phenolic hydroxyl groups on the B and D rings of catechin compounds, due to their adjacent stereochemical configuration, can more effectively donate hydrogen atoms or electrons, thereby more efficiently neutralizing free radicals (Colon & Nerín, 2016; Ma et al., 2025). Secondly, stereochemistry also affects the solubility and bioavailability of tea polyphenols (Jakobek, 2015; Li & Wiele, 2023). The stereochemical configuration of a molecule can influence its distribution and absorption within the biological system. Some specific stereochemical configurations may be more readily recognized and metabolized by the body's enzyme systems, releasing antioxidant-active metabolites. For example, the stereochemical configuration of epigallocatechin gallate (EGCG) enhances its solubility in the gastrointestinal tract. The alteration of its stereochemical configuration by gut microbiota facilitates its absorption and metabolism, thereby exerting antioxidant effects (Zagury et al., 2019).

Esterification and Glycosylation

Esterification and glycosylation of tea polyphenols can alter their solubility and bioavailability, thereby affecting their antioxidant activity in vivo. Esterification refers to the process where phenolic hydroxyl groups in tea polyphenol molecules form ester bonds with organic acids (such as gallic acid). This modification enhances the lipophilicity of tea polyphenols, thereby improving their absorption and distribution within the biological system. For example, the esterified structure of EGCG not only increases its absorption rate in the gastrointestinal tract but also enhances its antioxidant activity in vivo (Zhuang et al., 2024). Studies have shown that the esterified structure of EGCG results in a higher absorption rate in the gastrointestinal tract compared to unesterified epicatechin (EC). Esterification facilitates easier passage through cell membranes, thereby exerting more effective antioxidant actions within the body. Glycosylation refers to the process where phenolic hydroxyl groups in tea polyphenol molecules form glycosidic bonds with sugar molecules. Glycosylation can improve the water solubility of tea polyphenols, thereby increasing their solubility and bioavailability in vivo (Xu et al., 2016; Zhang et al., 2023). For example, glycosylated derivatives of tea polyphenols, such as tea polyphenol-3-O-glucoside, exhibit better water solubility, which aids their distribution and absorption within the body. Moreover, glycosylation can protect tea polyphenols from degradation by enzymes within the body, prolonging their effective duration in vivo.

9.4 Antioxidant Enzyme System of Tea

9.4.1 Antioxidant Effects by Enzyme System

In addition to directly scavenging free radicals, bioactive compounds in tea, especially tea polyphenols, enhance the body's defense capabilities by modulating the intracellular antioxidant enzyme system. Antioxidant enzymes such as superoxide dismutase (SOD), glutathione peroxidase (GPx), and catalase (CAT) play a crucial role in the cellular defense system.

Nrf2/ARE Signaling Pathway

One of the primary mechanisms by which tea polyphenols activate antioxidant enzymes is through the Nrf2/ARE (nuclear factor erythroid 2-related factor 2/antioxidant response element) signaling pathway. Nrf2 is a basic transcription factor that regulates the expression of various antioxidant and stress-related enzymes (Horie et al., 2021). Under normal conditions, Nrf2 is sequestered in the cytoplasm by binding to Keap1 (Kelch-like ECH-associated protein 1), which restricts its nuclear translocation. Tea polyphenols can alter this equilibrium by forming Michael adducts with the thiol groups of Keap1, leading to the dissociation of Nrf2 from Keap1 and its subsequent translocation to the nucleus (Li et al., 2023; Zhao et al., 2024). Once in the nucleus, Nrf2 binds to the antioxidant response element (ARE) and activates the transcription of a series of antioxidant enzyme genes, including but not limited to superoxide dismutase (SOD), glutathione peroxidase (GPx), catalase (CAT), and glutathione S-transferase (GST). The up-regulation of these enzymes helps to mitigate oxidative stress-induced cellular damage.

Influencing Post-translational Modifications and Stability

Tea polyphenols not only enhance the expression of antioxidant enzymes but may also modulate their activity by influencing post-translational modifications and stability (Liu et al., 2024). For instance, by interacting with the active sites or auxiliary regions of antioxidant enzymes, tea polyphenols may alter their three-dimensional structure, thereby increasing the thermal stability and protease resistance of these enzymes. This structural protection prolongs the active lifespan of antioxidant enzymes within the cell. Additionally, tea polyphenols may be involved in regulating the post-translational modification processes of antioxidant enzymes, such as phosphorylation, acetylation, and methylation. These modifications can alter enzyme activity, stability, and subcellular localization. For example, tea polyphenols may inhibit the activity of certain protein kinases (including the MAPK/NF-κB signaling pathway, NF-κB/Notch signaling pathway, etc.), reducing the phosphorylation-mediated inhibition of antioxidant enzymes, thereby maintaining or enhancing their activity (Hu et al., 2021; Krajka-Kuźniak et al., 2017; Zhao et al., 2022a, 2022b).

Inhibiting the Feneration of Free Radicals

In addition to enhancing antioxidant enzyme levels, tea components can exert antioxidant effects by inhibiting the generation of free radicals. For example, tea polyphenols can inhibit the activity of oxidases such as xanthine oxidase and NADPH oxidase, which catalyze the production of free radicals in the body. By inhibiting the activity of these enzymes, tea polyphenols can reduce the generation of free radicals, thereby lowering oxidative stress levels.

9.4.2 Signal Transduction Pathway Regulation

Nrf2-Keap1-ARE Pathway

Nrf2 is a fundamental transcription factor that controls cellular antioxidant responses and metabolic detoxification processes. Under non-oxidative stress conditions, Nrf2 is bound to its inhibitory protein Keap1 and retained in the cytoplasm. Keap1 acts as a negative regulator of Nrf2, promoting its degradation via the ubiquitination pathway. When cells are exposed to oxidative stress, tea polyphenols such as epigallocatechin gallate (EGCG) can alter the conformation of Keap1, preventing its binding to Nrf2 and thereby inhibiting the ubiquitination and degradation of Nrf2. The released Nrf2 translocates to the nucleus, binds to the antioxidant response element (ARE), and activates the expression of a series of antioxidant enzymes and related genes.

Activation of Nrf2 leads to increased expression of antioxidant enzymes such as GST, GPx, SOD, and heme oxygenase-1 (HO-1). Nrf2-mediated HO-1 expression also helps regulate the activity of the p38 MAPK and extracellular signal-regulated kinase (ERK) 1/2 signaling pathways, linking Nrf2 to the MAPK signaling cascade (Hong et al., 2010). This ultimately reduces the production of ROS in animals and alleviates oxidative damage to cells and organs caused by ROS. Additionally, circadian rhythm disruption, which leads to decreased expression of Bmal1, can also result in reduced Nrf2 pathway expression, and is one of the key factors contributing to oxidative stress (Xiong et al., 2016). EGCG can activate the Nrf2 oxidative defense pathway in the liver and HepG2 cells, increasing the expression levels of enzymes such as HO-1 and NAD(P)H:quinone oxidoreductase 1 (NQO1) to reduce oxidative stress (Qi et al., 2017a; Qi et al., 2018). HO-1 helps enhance cellular antioxidant capacity, while NQO1 reduces the production of ROS in cells. The protective effect of tea polyphenols against oxidative stress in animals has been demonstrated.

AKT-JNK Pathway

AKT, also known as protein kinase B (PKB), is a serine/threonine kinase that is involved in the regulation of cell survival, proliferation, metabolism, and apoptosis. Under conditions of oxidative stress, the activation of AKT can promote cell survival (Fakhri et al., 2021; Kumar & Bansal, 2022). AKT exerts its effects by phosphorylating downstream target proteins, such as mTOR (mammalian target of rapamycin) and GSK-3β (glycogen synthase kinase 3β), thereby inhibiting their activities and reducing cell damage induced by oxidative stress. JNK (c-Jun N-terminal kinase) is a member of the MAPK family and is involved in cellular responses to various types of stress, including oxidative stress, ultraviolet irradiation, and inflammatory stimuli. Activation of JNK can promote the phosphorylation of c-Jun, which in turn activates AP-1 (activator protein 1) and regulates the expression of genes associated with cell apoptosis. After tea consumption, key bioactive compounds in tea, such as EGCG and ECG, enhance the activity of PI3K (phosphoinositide 3-kinase) within cells (Wu et al., 2023; Zhu et al., 2021). This increase in PI3K activity promotes the phosphorylation and activation of AKT, while inhibiting the activation of JNK, thereby enhancing cell survival signals. The activation of AKT can upregulate the expression of antioxidant enzymes such as SOD and GPx, while the inhibition of JNK can reduce the suppression of antioxidant enzymes induced by oxidative stress, collectively enhancing the cellular antioxidant capacity. Furthermore, tea polyphenols can modulate the expression of apoptosis-related proteins (such as Bcl-2 and Bax) by regulating the activities of AKT and JNK, thereby inhibiting cell apoptosis caused by oxidative damage (Ma et al., 2022; Qi et al., 2017b). Additionally, the activation of AKT is involved in the regulation of cellular metabolism, such as promoting glucose uptake and glycolysis. Tea polyphenols may provide energetic support through this pathway, helping cells to resist oxidative stress.

MAPK-NF-κB Pathway

The MAPK signaling pathway is a key cellular response to external stress, including ERK, JNK, and p38 MAPK. The MAPK pathway is involved in the regulation of cell growth, differentiation, apoptosis, and the modulation of inflammatory and immune responses (Kwon & Lee, 2020). Oxidative stress is closely related to the MAPK pathway. The accumulation of ROS activates ERK/p38 MAPK, leading to the oxidation of intracellular proteins and lipid peroxidation of cell membranes. NF-κB is a core transcription factor that regulates cellular inflammatory responses, and its activity is inhibited by IκB proteins (Blaser et al., 2016; Shanmugam et al., 2016). NF-κB also plays an important role in cellular stress and antioxidant responses by regulating the expression of antioxidant enzymes and cell-protective proteins. Tea polyphenols can reduce cell damage caused by oxidative stress by inhibiting the activation of the MAPK pathway (Zhao et al., 2022a, 2022b). For example, tea polyphenols may inhibit the phosphorylation of JNK and p38, thereby reducing the production of inflammatory factors (Wang et al., 2022). Additionally, tea polyphenols can inhibit the nuclear translocation of NF-κB by blocking the degradation of IκB proteins, thereby reducing the expression of NF-κB-dependent inflammatory genes. Studies have shown that overexpression of IκB in cells can significantly inhibit NF-κB and increase the activation of ARE and NQO1genes (Downton et al., 2023). NF-κB activation has a negative regulatory effect on Nrf2 expression. After tea consumption, key components in tea inhibit the MAPK-NF-κB signaling pathway, further enhancing the activation of the Nrf2/HO-1 signaling pathway and the activity of antioxidant enzymes such as SOD and GSH-Px, significantly improving oxidative stress-mediated intestinal inflammation.

References

Akagawa, M. (2021). Protein carbonylation: Molecular mechanisms, biological implications, and analytical approaches. *Free Radical Research, 55*(4), 307–320. https://doi.org/10.1080/10715762.2020.1851027

Ayala, A., Muñoz, M. F., & Argüelles, S. (2014). Lipid peroxidation: Production, metabolism, and signaling mechanisms of malondialdehyde and 4-hydroxy-2-nonenal. *Oxidative Medicine and Cellular Longevity, 2014*, 360438. https://doi.org/10.1155/2014/360438

Bag, S., Mondal, A., Majumder, A., & Banik, A. (2022). Tea and its phytochemicals: Hidden health benefits and modulation of signaling cascade by phytochemicals. *Food Chemistry, 371*, 131098. https://doi.org/10.1016/j.foodchem.2021.131098

Blaser, H., Dostert, C., Mak, T. W., & Brenner, D. (2016). TNF and ROS crosstalk in inflammation. *Trends in Cell Biology, 26*(4), 249–261. https://doi.org/10.1016/j.tcb.2015.12.002

Braicu, C., Ladomery, M. R., Chedea, V. S., Irimie, A., & Berindan-Neagoe, I. (2013). The relationship between the structure and biological actions of green tea catechins. *Food Chemistry, 141*(3), 3282–3289. https://doi.org/10.1016/j.foodchem.2013.05.122

Chen, J., Yang, J., Ma, L., Li, J., Shahzad, N., & Kim, C. K. (2020). Structure-antioxidant activity relationship of methoxy, phenolic hydroxyl, and carboxylic acid groups of phenolic acids. *Scientific Reports, 10*(1), 2611. https://doi.org/10.1038/s41598-020-59451-z

Cipriano, A., Viviano, M., Feoli, A., Milite, C., Sarno, G., Castellano, S., & Sbardella, G. (2023). NADPH oxidases: From molecular mechanisms to current inhibitors. *Journal of Medicinal Chemistry, 66*(17), 11632–11655. https://doi.org/10.1021/acs.jmedchem.3c00770

Cojocaru, K.-A., Luchian, I., Goriuc, A., Antoci, L.-M., Ciobanu, C.-G., Popescu, R., Vlad, C. E., Blaj, M., & Foia, L. G. (2023). Mitochondrial dysfunction, oxidative stress, and therapeutic strategies in diabetes, obesity, and cardiovascular disease. *Antioxidants, 12.*

Colon, M., & Nerín, C. (2016). Synergistic, antagonistic and additive interactions of green tea polyphenols. *European Food Research and Technology, 242*(2), 211–220. https://doi.org/10.1007/s00217-015-2532-9

Downton, P., Bagnall, J. S., England, H., Spiller, D. G., Humphreys, N. E., Jackson, D. A., Paszek, P., White, M. R., & Adamson, A. D. (2023). Overexpression of IκBα modulates NF-κB activation of inflammatory target gene expression. *Frontiers in Molecular Biosciences, 10*, 1187187. https://doi.org/10.3389/fmolb.2023.1187187

Dudek, M. K., & Day, G. M. (2019). Explaining crystallization preferences of two polyphenolic diastereoisomers by crystal structure prediction. *CrystEngComm, 21*(13), 2067–2079. https://doi.org/10.1039/C8CE01783B

Fakhri, S., Iranpanah, A., Gravandi, M. M., Moradi, S. Z., Ranjbari, M., Majnooni, M. B., Echeverría, J., Qi, Y., Wang, M., Liao, P., Farzaei, M. H., & Xiao, J. (2021). Natural products attenuate PI3K/Akt/mTOR signaling pathway: A promising strategy in regulating neurodegeneration. *Phytomedicine, 91*, 153664. https://doi.org/10.1016/j.phymed.2021.153664

Fang, Q., Qin, C., Peng, Y., Zhang, L., Li, D., & Hu, S. (2025). Navigating the gut-liver axis: The potential of tea to alleviate inflammatory bowel disease and associated metabolic syndromes. *Trends in Food Science & Technology, 157*, 104893. https://doi.org/10.1016/j.tifs.2025.104893

García-Caparrós, P., De Filippis, L., Gul, A., Hasanuzzaman, M., Ozturk, M., Altay, V., & Lao, M. T. (2021). Oxidative stress and antioxidant metabolism under adverse environmental conditions: A review. *The Botanical Review, 87*(4), 421–466. https://doi.org/10.1007/s12229-020-09231-1

Hong, Y., Yan, W., Chen, S., Sun, C.-R., & Zhang, J.-M. (2010). The role of Nrf2 signaling in the regulation of antioxidants and detoxifying enzymes after traumatic brain injury in rats and mice. *Acta Pharmacologica Sinica, 31*(11), 1421–1430. https://doi.org/10.1038/aps.2010.101

Horie, Y., Suzuki, T., Inoue, J., Iso, T., Wells, G., Moore, T. W., Mizushima, T., Dinkova-Kostova, A. T., Kasai, T., Kamei, T., Koshiba, S., & Yamamoto, M. (2021). Molecular basis for the disruption of Keap1–Nrf2 interaction via Hinge and Latch mechanism. *Communications Biology, 4*(1), 576. https://doi.org/10.1038/s42003-021-02100-6

Hu, S., Li, S., Liu, Y., Sun, K., Luo, L., & Zeng, L. (2021). Aged ripe Pu-erh tea reduced oxidative stress-mediated inflammation in dextran sulfate sodium-induced colitis mice by regulating intestinal microbes. *Journal of Agriculture and Food Chemistry, 69*(36), 10592–10605. https://doi.org/10.1021/acs.jafc.1c04032

Intra, J., & Kuo, S. M. (2007). Physiological levels of tea catechins increase cellular lipid antioxidant activity of vitamin C and vitamin E in human intestinal caco-2 cells. *Chemico-Biological Interactions, 169*(2), 91–99. https://doi.org/10.1016/j.cbi.2007.05.007

Ishii, T., Mori, T., Ichikawa, T., Kaku, M., Kusaka, K., Uekusa, Y., Akagawa, M., Aihara, Y., Furuta, T., Wakimoto, T., Kan, T., & Nakayama, T. (2010). Structural characteristics of green tea catechins for formation of protein carbonyl in human serum albumin. *Bioorganic & Medicinal Chemistry, 18*(14), 4892–4896. https://doi.org/10.1016/j.bmc.2010.06.021

Jakobek, L. (2015). Interactions of polyphenols with carbohydrates, lipids and proteins. *Food Chemistry, 175*, 556–567. https://doi.org/10.1016/j.foodchem.2014.12.013

Jomova, K., Raptova, R., Alomar, S. Y., Alwasel, S. H., Nepovimova, E., Kuca, K., & Valko, M. (2023). Reactive oxygen species, toxicity, oxidative stress, and antioxidants: Chronic diseases and aging. *Archives of Toxicology, 97*(10), 2499–2574. https://doi.org/10.1007/s00204-023-03562-9

Krajka-Kuźniak, V., Paluszczak, J., & Baer-Dubowska, W. (2017). The Nrf2-ARE signaling pathway: An update on its regulation and possible role in cancer prevention and treatment. *Pharmacological Reports, 69*(3), 393–402. https://doi.org/10.1016/j.pharep.2016.12.011

Kumar, M., & Bansal, N. (2022). Implications of phosphoinositide 3-kinase-Akt (PI3K-Akt) pathway in the pathogenesis of Alzheimer's disease. *Molecular Neurobiology, 59*(1), 354–385. https://doi.org/10.1007/s12035-021-02611-7

Kwon, O. Y., & Lee, S. H. (2020). Ameliorating activity of Ishige okamurae on the amyloid beta-induced cognitive deficits and neurotoxicity through regulating ERK, p38 MAPK, and JNK signaling in Alzheimer's disease-like mice model. *Molecular Nutrition & Food Research, 64*(12), 1901220. https://doi.org/10.1002/mnfr.201901220

Li, B., Wang, Y., Jiang, X., Du, H., Shi, Y., Xiu, M., Liu, Y., & He, J. (2023). Natural products targeting Nrf2/ARE signaling pathway in the treatment of inflammatory bowel disease. *Biomedicine & Pharmacotherapy, 164*, 114950. https://doi.org/10.1016/j.biopha.2023.114950

Li, Q., & Van de Wiele, T. (2023). Gut microbiota as a driver of the interindividual variability of cardiometabolic effects from tea polyphenols. *Critical Reviews in Food Science and Nutrition, 63*(11), 1500–1526. https://doi.org/10.1080/10408398.2021.1965536

Liu, H., Guan, H., He, F., Song, Y., Li, F., Sun-Waterhouse, D., & Li, D. (2024). Therapeutic actions of tea phenolic compounds against oxidative stress and inflammation as central mediators in the development and progression of health problems: A review focusing on microRNA regulation. *Critical Reviews in Food Science and Nutrition, 64*(23), 8414–8444. https://doi.org/10.1080/10408398.2023.2202762

Ma, Y., Ma, X., An, Y., Sun, Y., Dou, W., Li, M., Bao, H., & Zhang, C. (2022). Green tea polyphenols alleviate hydrogen peroxide-induced oxidative stress, inflammation, and apoptosis in bovine mammary epithelial cells by activating ERK1/2–NFE2L2–HMOX1 pathways. *Frontiers in Veterinary Science, 8*. https://doi.org/10.3389/fvets.2021.804241

Ma, Q.-Y., Xu, Q.-D., Chen, N., & Zeng, W.-C. (2025). Effects of catechins with different structure characteristics on the structure and properties of gluten-catechin covalent complex. *Food Research International, 201*, 115529. https://doi.org/10.1016/j.foodres.2024.115529

Muzolf-Panek, M., Gliszczyńska-Świgło, A., Szymusiak, H., & Tyrakowska, B. (2012). The influence of stereochemistry on the antioxidant properties of catechin epimers. *European Food Research and Technology, 235*(6), 1001–1009. https://doi.org/10.1007/s00217-012-1826-4

Pan, S.-Y., Nie, Q., Tai, H.-C., Song, X.-L., Tong, Y.-F., Zhang, L.-J.-F., Wu, X. W., Lin, Z. H., Zhang, Y. Y., Ye, D. Y., & Liang, C. (2022). Tea and tea drinking: China's outstanding contributions to the mankind. *Chinese Medicine, 17*(1), 27. https://doi.org/10.1186/s13020-022-00571-1

Qi, G., Wu, W., Mi, Y., Shi, R., Sun, K., Li, R., Liu, X., & Liu, X. (2018). Tea polyphenols direct Bmal1-driven ameliorating of the redox imbalance and mitochondrial dysfunction in hepatocytes. *Food and Chemical Toxicology, 122*, 181–193. https://doi.org/10.1016/j.fct.2018.10.031

Qi, G., Mi, Y., Fan, R., Li, R., Wang, Y., Li, X., Huang, S., & Liu, X. (2017a). Tea polyphenols ameliorate hydrogen peroxide- and constant darkness-triggered oxidative stress via modulating the Keap1/Nrf2 transcriptional signaling pathway in HepG2 cells and mice liver. *RSC Advances, 7*(51), 32198–32208. https://doi.org/10.1039/C7RA05000C

Qi, G., Mi, Y., Wang, Y., Li, R., Huang, S., Li, X., & Liu, X. (2017b). Neuroprotective action of tea polyphenols on oxidative stress-induced apoptosis through the activation of the TrkB/CREB/BDNF pathway and Keap1/Nrf2 signaling pathway in SH-SY5Y cells and mice brain. *Food & Function, 8*(12), 4421–4432. https://doi.org/10.1039/C7FO00991G

Rana, A., Samtiya, M., Dhewa, T., Mishra, V., & Aluko, R. E. (2022). Health benefits of polyphenols: A concise review. *Journal of Food Biochemistry, 46*(10), e14264. https://doi.org/10.1111/jfbc.14264

Sabharwal, S. S., & Schumacker, P. T. (2014). Mitochondrial ROS in cancer: Initiators, amplifiers or an Achilles' heel? *Nature Reviews Cancer, 14*(11), 709–721. https://doi.org/10.1038/nrc3803

Shanmugam, T., Selvaraj, M., & Poomalai, S. (2016). Epigallocatechin gallate potentially abrogates fluoride induced lung oxidative stress, inflammation via Nrf2/Keap1 signaling pathway in rats: An in-vivo and in-silico study. *International Immunopharmacology, 39*, 128–139. https://doi.org/10.1016/j.intimp.2016.07.022

Shi, J., Yang, G., You, Q., Sun, S., Chen, R., Lin, Z., Simal-Gandara, J., & Lv, H. (2023). Updates on the chemistry, processing characteristics, and utilization of tea flavonoids in last two decades (2001–2021). *Critical Reviews in Food Science and Nutrition, 63*(20), 4757–4784. https://doi.org/10.1080/10408398.2021.2007353

Sies, H., & Jones, D. P. (2020). Reactive oxygen species (ROS) as pleiotropic physiological signalling agents. *Nature Reviews Molecular Cell Biology, 21*(7), 363–383. https://doi.org/10.1038/s41580-020-0230-3

Suomalainen, A., & Nunnari, J. (2024). Mitochondria at the crossroads of health and disease. *Cell, 187*(11), 2601–2627. https://doi.org/10.1016/j.cell.2024.04.037

Trosko, J. E. (2021). On the potential origin and characteristics of cancer stem cells. *Carcinogenesis, 42*(7), 905–912. https://doi.org/10.1093/carcin/bgab042

Wang, Q., Huang, J., Zheng, Y., Guan, X., Lai, C., Gao, H., Ho, C. T., & Lin, B. (2022). Selenium-enriched oolong tea (Camellia sinensis) extract exerts anti-inflammatory potential via targeting NF-κB and MAPK pathways in macrophages. *Food Science and Human Wellness, 11*(3), 635–642. https://doi.org/10.1016/j.fshw.2021.12.020

Wu, Z., Yu, W., Ni, W., Teng, C., Ye, W., Yu, C., & Zeng, Y. (2023). Improvement of obesity by Liupao tea is through the IRS-1/PI3K/AKT/GLUT4 signaling pathway according to network pharmacology and experimental verification. *Phytomedicine, 110*, 154633. https://doi.org/10.1016/j.phymed.2022.154633

Xie, S., Xu, S. C., Deng, W., & Tang, Q. (2023). Metabolic landscape in cardiac aging: Insights into molecular biology and therapeutic implications. *Signal Transduction and Targeted Therapy, 8*(1), 114. https://doi.org/10.1038/s41392-023-01378-8

Xiong, W., Li, J., Zhang, E., & Huang, H. (2016). BMAL1 regulates transcription initiation and activates circadian clock gene expression in mammals. *Biochemical and Biophysical Research Communications, 473*(4), 1019–1025. https://doi.org/10.1016/j.bbrc.2016.04.009

Xu, S., Li, X., Zhang, S., Qi, C., Zhang, Z., Ma, R., Xiang, L., Chen, L., Zhu, Y., Tang, C. & Bourgonje, A. R., & Chen, M. (2023). Oxidative stress gene expression, DNA methylation, and gut microbiota interaction trigger Crohn's disease: a multi-omics Mendelian randomization study. *BMC Medicine, 21*(1), 179. https://doi.org/10.1186/s12916-023-02878-8

Xu, L., Qi, T., Xu, L., Lu, L., & Xiao, M. (2016). Recent progress in the enzymatic glycosylation of phenolic compounds. *Journal of Carbohydrate Chemistry, 35*(1), 1–23. https://doi.org/10.1080/07328303.2015.1137580

Yan, Z., Zhong, Y., Duan, Y., Chen, Q., & Li, F. (2020). Antioxidant mechanism of tea polyphenols and its impact on health benefits. *Animal Nutrition, 6*(2), 115–123. https://doi.org/10.1016/j.aninu.2020.01.001

Yang, H., Xue, X., Li, H., Apandi, S. N., Tay-Chan, S. C., Ong, S. P., & Tian, E. F. (2018). The relative antioxidant activity and steric structure of green tea catechins—A kinetic approach. *Food Chemistry, 257*, 399–405. https://doi.org/10.1016/j.foodchem.2018.03.043

Zagury, Y., Kazir, M., & Livney, Y. D. (2019). Improved antioxidant activity, bioaccessibility and bioavailability of EGCG by delivery in β-lactoglobulin particles. *Journal of Functional Foods, 52*, 121–130. https://doi.org/10.1016/j.jff.2018.10.025

Zhang, J., Cui, H., Qiu, J., Wang, X., Zhong, Y., Yao, C., Yao, L., Zheng, Q., & Xiong, C. (2023). Stability of glycosylated complexes loaded with Epigallocatechin 3-gallate (EGCG). *Food Chemistry, 410*, 135364. https://doi.org/10.1016/j.foodchem.2022.135364

Zhao, X., Shi, X., Liu, Q., & Li, X. (2022). Tea polyphenols alleviates acetochlor-induced apoptosis and necroptosis via ROS/MAPK/NF-κB signaling in Ctenopharyngodon idellus kidney cells. *Aquatic Toxicology, 246*, 106153. https://doi.org/10.1016/j.aquatox.2022.106153

Zhao, M., Meng, K., Zhao, M., Shi, H., Liu, Z., Yun, Y., Zhang, W., Duan, Z., & Xia, G. (2024). Partridge tea polyphenols alleviated STZ-induced diabetic nephropathy by regulating Keap1/Nrf2/ARE signaling pathway in C57BL/6 mice. *Journal of Functional Foods, 115*, 106127. https://doi.org/10.1016/j.jff.2024.106127

Zhao, X., Shi, X., Liu, Q., & Li, X. (2022a). Tea polyphenols alleviates acetochlor-induced apoptosis and necroptosis via ROS/MAPK/NF-κB signaling in Ctenopharyngodon idellus kidney cells. *Aquatic Toxicology, 246*, 106153. https://doi.org/10.1016/j.aquatox.2022.106153

Zhu, J., Wu, M., Zhou, H., Cheng, L., Wei, X., & Wang, Y. (2021). Liubao brick tea activates the PI3K-Akt signaling pathway to lower blood glucose, metabolic disorders and insulin resistance via altering the intestinal flora. *Food Research International, 148*, 110594. https://doi.org/10.1016/j.foodres.2021.110594

Zhuang, Y., Quan, W., Wang, X., Cheng, Y., & Jiao, Y. (2024). Comprehensive review of EGCG modification: Esterification methods and their impacts on biological activities. *Foods, 13*.

Chapter 10
Anti-Inflammatory Effect of Tea

Shanshan Hu

Abstract This chapter systematically elucidates the biological mechanisms of inflammation and its pathological consequences, with an in-depth analysis of the anti-inflammatory effects and molecular pathways of tea bioactive compounds. Inflammation, as a core defense response to injury, is triggered by multiple factors including microbial infections, physicochemical stimuli, and immune dysregulation. The process involves vascular activation, immune cell infiltration, and release of inflammatory mediators, regulated by transcription factors such as NF-κB/AP-1. Chronic inflammation contributes to tissue fibrosis and is closely associated with multiple diseases (e.g., IBD, liver disease, brain dysfunction). Studies demonstrate that tea components—notably polyphenols, theanine, polysaccharides, and theaflavins—exert synergistic anti-inflammatory effects through multiple pathways: Direct actions: Suppressing COX-2/iNOS activity to reduce PGE2/NO production and inflammatory mediator secretion. Signaling modulation: Inhibiting IKK/IκB phosphorylation to block NF-κB nuclear translocation, downregulating MAPK pathway activation, and balancing AKT/JNK signaling. Immunoregulation: Promoting macrophage polarization toward the M2 phenotype, modulating Treg/Th17 ratios, and enhancing gut barrier function. Microecological remodeling: Reshaping gut/oral microbiota, promoting SCFA and bile acid metabolism. In disease applications, tea components ameliorate colonic injury in IBD, hepatic steatosis in alcoholic liver disease, and depression-like behaviors via the gut-liver/brain axis, while inhibiting pathogenic biofilm formation and periodontal inflammation in oral disorders.

Keywords Inflammation · Tea · Anti-inflammatory mechanisms · Inflammation-related diseases · Molecular mechanisms

---

S. Hu (✉)
State Key Laboratory of Tea Plant Germplasm Innovation and Resource Utilization, School of Tea Science, Anhui Agricultural University, Hefei, Anhui, China
e-mail: Hshan1008@163.com

International Joint Laboratory on Tea Chemistry and Health Effects of Ministry of Education, Anhui Agricultural University, Hefei, Anhui, China

X. Zhai (ed.), *Tea Chemistry and Health Benefits*,
https://doi.org/10.1007/978-3-032-11828-8_10

10.1 Introduction

Inflammation is a complex biological response of the body to various types of injury or stimuli, including infections, physical trauma, chemical irritation, or immune reactions. It is also a natural defense mechanism of the organism against injury or infection, characterized by local redness, swelling, heat, pain, and loss of function.

10.2 The Biological Mechanism of Inflammation

10.2.1 Inducements of Inflammation

The causes of inflammation are diverse and include microbial infections, physical and chemical injuries, and autoimmune abnormalities (Akdis, 2021; Pisetsky, 2023).

(a) *Microbial Infections* Microbial infections are the most common cause of inflammation. Pathogens such as bacteria, viruses, fungi, and parasites activate the immune cells of the host through their surface molecules, triggering an inflammatory response.
(b) *Physical and Chemical Injuries* Physical injuries, such as trauma and burns, and chemical injuries, such as exposure to toxins, as well as ultraviolet radiation, can directly damage cells and tissues, activating an inflammatory response.
(c) *Autoimmune* Abnormalities In autoimmune abnormalities, the immune system mistakenly attacks the body's own tissues, leading to inflammation. The activation of autoreactive T and B cells and the production of autoantibodies are key factors in autoimmune inflammation. Conditions such as rheumatoid arthritis and systemic lupus erythematosus involve complex interactions between multiple autoantigens and immune cells in the inflammatory process.

In addition to the direct biological factors mentioned above, environmental and lifestyle factors also play a significant role in the development and progression of inflammation (Fang et al., 2025; Ramos-Lopez et al., 2021). Environmental pollution, particularly fine particulate matter (PM2.5) and ozone in air pollution, can enter the body through the respiratory tract and activate inflammatory responses in the lungs and throughout the body. Climate change, such as extreme temperature and humidity variations, can also disrupt physiological balance and trigger inflammatory responses. Moreover, unhealthy lifestyle factors, including smoking, alcohol consumption, high-fat diets, and lack of exercise, have been shown to be associated with chronic inflammatory states. These factors influence the production and metabolism of inflammatory mediators through various mechanisms, thereby subtly promoting the development of the inflammatory process.

10.2.2 Inflammation Types and Initiation

Inflammation can be categorized into acute and chronic types, which differ significantly in their biological mechanisms, clinical manifestations, and therapeutic strategies (Bezerra et al., 2023; Maccarrone & Brüne, 2009; Zotova et al., 2023).

(a) *Acute Inflammation* Acute inflammation is typically characterized by redness, swelling, and pain. It is marked by the dilation of local blood vessels and increased permeability, leading to the rapid movement of plasma and white blood cells to the site of injury. The goal is to quickly eliminate pathogens and damaged tissue, prevent the spread of infection, and initiate the repair process.
(b) *Chronic Inflammation* Chronic inflammation is a long-lasting process that may arise from unresolved acute inflammation or persistent exposure to irritants. It is characterized by the continuous infiltration of inflammatory cells, including lymphocytes, macrophages, and plasma cells, and gradual changes in tissue structure. Unlike acute inflammation, chronic inflammation does not have distinct features but is associated with local tissue degeneration, exudation, and proliferation.

The initiation of inflammation typically begins with local tissue changes induced by injury or infection. These changes lead to the release of damage-associated molecular patterns (DAMPs) and pathogen-associated molecular patterns (PAMPs) by cells, which in turn activate the immune system (Zindel & Kubes, 2020). One of the early hallmarks of inflammation is the activation of blood vessels. The activation of vascular endothelial cells results in vasodilation and increased vascular permeability, driven by the release of vasoactive substances such as histamine, kinins, and prostaglandins from endothelial cells (Pober & Sessa, 2007; Vanhoutte et al., 2017). Following the increase in vascular permeability, white blood cells (such as neutrophils, monocytes, and lymphocytes) extravasate from the blood vessels into the damaged tissue. This process involves multiple adhesion molecules and chemokines that guide the migration of white blood cells to the inflamed area. In the inflamed area, white blood cells and other cells release a variety of inflammatory mediators, including cytokines (such as tumor necrosis factor (TNF)-α, interleukin (IL)-1β, etc.), chemokines, growth factors, and enzymes. These mediators act in concert to amplify the inflammatory response and regulate the recruitment and activation of cells. For example, transcription factors such as NF-κB and AP-1 are activated under the influence of inflammatory stimuli, translocate from the cytoplasm to the nucleus, and upregulate the expression of multiple inflammatory genes, thereby causing an intensified inflammatory response and tissue damage (Rius-Pérez et al., 2019; Zhang et al., 2021a, 2021b). Inflammatory cells are activated by these mediators and perform their specific functions. For instance, neutrophils kill pathogens through phagocytosis and the release of oxidants, while macrophages engulf and digest cellular debris and pathogens. The inflammatory response needs to be precisely regulated to avoid excessive damage. The release of anti-inflammatory cytokines (such as IL-10 and transforming growth factor (TGF)-β) helps to suppress the inflammatory process.

Moreover, the apoptosis and clearance of inflammatory cells are also important aspects of resolving inflammation. It is noteworthy that inflammation is not only the initial response to injury but also the starting point for tissue repair. Growth factors released by inflammatory cells can promote cell proliferation and tissue reconstruction (Werner & Grose, 2003). However, chronic inflammation can lead to tissue fibrosis and functional impairment. In some cases, inflammation may become chronic, resulting in persistent tissue damage and dysfunction. Chronic inflammation is associated with the development of various diseases, including cardiovascular disease, diabetes, and cancer.

10.3 Anti-inflammatory Effect of Tea

10.3.1 Anti-inflammatory Active Ingredients in Tea

The anti-inflammatory active components in tea constitute the cornerstone of its biological activity, working synergistically through multiple mechanisms to combat inflammatory processes. Tea polyphenols, one of the principal chemical constituents of tea, account for 18% to 36% of the dry weight of tea leaves and are the primary source of various bioactivities, including anti-inflammatory and antioxidant properties (Ma et al., 2021).

(a) *Tea Polyphenols* Tea polyphenols mainly consist of catechins, flavonoids, flavonols, and anthocyanins. Among these, catechin compounds, particularly EGCG, have garnered extensive attention due to their remarkable anti-inflammatory and antioxidant characteristics. EGCG can inhibit the production of inflammatory mediators such as prostaglandin E2 (PGE2), nitric oxide (NO), and TNF-α, thereby attenuating inflammatory responses (Ansari et al., 2020).
(b) *Theaflavins* Theaflavins, a class of pigments formed during the fermentation process of black tea, exhibit significant anti-inflammatory activity. They function by inhibiting the activation of inflammation-related signaling pathways, such as NF-κB and MAPKs (Liu et al., 2022; Yang et al., 2021).
(c) *Theanine* Theanine, a non-protein amino acid, is also an important anti-inflammatory component in tea. It exerts its anti-inflammatory effects by modulating the function of immune cells and reducing the production of inflammatory mediators (Chen et al., 2023a, 2023b, 2023c).
(d) *Tea Polysaccharide* Tea polysaccharide, a category of polysaccharide compounds, can activate immune cells and inhibit the production of inflammatory mediators, thus playing a role in immune regulation and anti-inflammation (Zhang et al., 2024).
(e) *Saponins* Saponins, a class of saponin compounds, exhibit anti-inflammatory activity by suppressing the expression of inflammatory cytokines and reducing the infiltration of inflammatory cells (Wijesekara et al., 2024).

(f) *Vitamins and Minerals* Vitamins and minerals in tea, such as vitamin C, vitamin E, zinc, and selenium, also contribute to the inhibition of inflammatory responses. These micronutrients, through their antioxidant properties, help to scavenge free radicals and mitigate inflammation caused by oxidative stress.

(g) *Volatile Aromatic Components* The volatile aromatic components in tea, such as phenylethanol, benzyl alcohol, and linalool, although present in relatively low concentrations, have been shown in recent studies to possess certain anti-inflammatory activities (Łyko et al., 2024; Yin et al., 2022). They may alleviate inflammation by modulating the function of immune cells and influencing the production of inflammatory mediators.

10.3.2 Anti-inflammatory Effects of Tea and Its Functional Compounds

Theanine

L-Theanine is a unique free amino acid in tea, which is an important substance determining umami and sweet taste. It is widely used as a functional ingredient and dietary supplement with sedative, anti-inflammatory, and antioxidant properties. L-theanine can cross the blood–brain barrier and directly affect the secretion of neurotransmitters in the central nervous system of the brain. *L*-theanine has the effect of regulating emotions and calming nerves. Studies have shown that the intake of L-theanine increases participants' sleep duration (Unno et al., 2013), and significantly reduces the increase in heart rate and salivary immunoglobulin induced by acute stress (Kimura et al., 2007). The intake of theanine decreased the expression of transcription factor NPAS4 (neuronal PAS domain protein 4) and Lipocalin 2 (LCN2) in the hippocampus and prefrontal cortex, preventing brain atrophy caused by chronic stress (Unno et al., 2020). Moreover, theanine also regulates γ-aminobutyric acid (GABA), dopamine, and glutamate levels in neural areas (Suhyeon et al., 2019), which is noted to be neuroprotective. Pretreatment with L-theanine can directly block Δ-9-tetrahydrocannabinol (THC)-induced down-regulation of local GSK-3 (glycogen synthase kinase 3) and Akt signaling pathways in prefrontal cortex (PFC), Thereby blocking the emotional and cognitive abnormalities caused by THC exposure in adolescents, and effectively prevent the long-term THC-induced dysregulation of PFC and DA activity (De Felice et al., 2021). In addition, theanine plays an important role in improving cognitive impairment and memory by affecting the transport of glutamine in the brain (Kakuda, 2011).

Theanine has antioxidant and anti-inflammatory properties. Of dextran sulfate sodium (DSS) induced colitis C57BL/6J male mice fed with 0.1% (w/v) of *L*-theanine drinking water, the results showed that the *L*-theanine inhibited DSS induced colon tissue lesions, decreased the expression of proinflammatory cytokines (IL-1β, IL-6 and TNF-α), regulated intestinal flora, and increased the production of short-chain fatty acids (Wang et al., 2020a, 2020b). Similarly, *L*-theanine significantly reduced the development of inflammation by decreasing the pro-inflammatory PKC/ERK/NF-κB/ICAM-1/IL-33 signaling in the bladder of female Wistar rats. The activity of H_2O_2 and HOCl was decreased in a dose-dependent manner in vitro (Tsai et al., 2017). *L*-theanine also decreased inflammatory cell infiltration in the blood vessels of the ovalbumin (OVA)-induced asthma mouse model, reduced the production of reactive oxygen species and the activation of NF-κB and matrix metalloprotease-9 in bronchoalveolar lavage fluid (Hwang et al., 2017).

Tea Polyphenols

According to their main chemical components, tea polyphenols can be divided into catechins, flavonoids, anthocyanins, and phenolic acids. Among them, catechin content is the highest, accounting for 60–80% of tea polyphenols, is an important health care substance (Liang et al., 2017). The main catechins are EC, ECG, EGC, and EGCG. A large number of studies have shown that tea polyphenols play an important role in anti-oxidation, anti-inflammation, balance lipid metabolism, anti-aging and other aspects. Tea polyphenols are a highly active antioxidant substance that effectively scavenges free radicals. Tea polyphenols regulate the oxidoreductase system and improve the body's antioxidant capacity by increasing the levels of serum catalase (CAT), GSH-Px, and SOD (Wang et al., 2021a, 2021b). Dietary EGCG could up-regulate the expression of NRF2 and HO-1 in liver of laying hens, and improve the antioxidant capacity of eggs (Wang et al., 2019a, 2019b, 2019c). In addition, tea polyphenols can also increase the antioxidant activity of the compounds. The complex of tea polyphenols with oat β-glucan increased antioxidant capacity in vitro and in vivo, especially in liver, showing higher superoxide dismutase and glutathione peroxidase activities (Wu et al., 2011). Combined with docosahexaenoic acid (DHA) algae oil can improve its oxidation stability and 3.80-fold longer than the control sample (Shen et al., 2020).

Tea polyphenols can inhibit the expression of inflammatory pathway proteins, down-regulate inflammatory factors, and reduce the recruitment of white blood cells and B lymphocytes, thus reducing the inflammatory response. Tea polyphenols treatment significantly reduced the expression of monocyte chemotactic protein 1 (MCP-1) in liver tissue and epididymal fat in mice fed with high fat/high sucrose (Heber et al., 2014). Supplementation of tea polyphenols in rats fed a high-fructose diet reduced serum levels of pro-inflammatory cytokines (TNF-α and IL-6) and down-regulated the expression of inflammatory cytokine-related genes (TNF-α, IL1β, and IL-6 mRNA levels) in the heart. Alleviate the effect of high fructose diet on myocardial inflammation in rats (Qin et al., 2010). In the DSS colitis mouse model, 0.01 or 0.05% EGCG diet was supplemented. It can alleviate the weight loss and colon shortening caused by DSS, improve the colon injury, and reduce serum TNF-α and IFN-γ (Abboud et al., 2005). However, high doses of tea polyphenols can increase inflammation. Mice fed 0.5% EGCG developed more severe rectal bleeding than those treated with DSS (Guan et al., 2012). Similarly, 0.5 and 1% doses of tea polyphenols did not prevent DSS-induced colon shortening, but instead increased

IL-1β expression and decreased the expression of macrophage migration inhibitors, thereby increasing the occurrence of colon cancer (Kim et al., 2010). In addition, EGCG disrupted the p53-MDM2 interaction and inhibited MDM2-mediated p53 ubiquitination. It plays an anticancer role by increasing the expression of the p53 gene and inducing cell cycle arrest or apoptosis (Liu et al., 2018b; Zhao et al., 2021).

Tea polyphenols also effectively block the process of lipid peroxidation, accelerate the decomposition and metabolism of excess fat, and prevent hypertension, hyperlipidemia and hyperglycemia. Tea polyphenols restored fasting blood glucose, reduced serum triglyceride (TG) and total cholesterol (TCHO) levels in mice fed high fat diet (HFD), and had protective effects on hepatic steatosis and insulin resistance caused by HFD (Pan et al., 2017; Xia et al., 2019). Tea polyphenols was shown to improve mouse muscle insulin sensitivity and hyperlipidemia by increasing the expression of glucose transporter 4 and pSer473Akt2 in skeletal muscle and raising the expression level of lipid metabolism genes in the liver (Cheng et al., 2020). In addition, by reducing the number of *Firmicutes/Bacteroidetes* and increasing the abundance of *Lachnospiraceae family/Barnesiella*, tea polyphenols regulate the disturbance of intestinal flora caused by HFD, which may provide a beneficial effect in the metabolism of glucose and lipids (Chen et al., 2019; Ma et al., 2019).

Tea Polysaccharide

Tea polysaccharide is a kind of complex composed of sugar, pectin, protein and so on, which has the effect of regulating glucose-lipid metabolism, anti-oxidation, anti-bacteria and regulating immunity. The activity of TC, TG, aspartate aminotransferase (AST), alanine aminotransferase (ALT) was significantly reduced upon administration of Tea polysaccharide to high fructose-treated rats, and had a protective effect on hepatic steatosis and intravascular dysfunction caused by high fructose feeding (Zhai et al., 2017). Similar results were reported by Chen et al. (2010), where the administration of Tea polysaccharide was shown to inhibit spontaneous diabetes by reducing blood glucose level and down-regulating anti-glutamate decarboxylase (GAD) antibody expression in non-obese diabetic mice. In a study by Zhou et al. (2007), through testing the effects of crude tea polysaccharide and tea polysaccharide fraction on hyperglycemic diabetic mice, it was shown that soluble tea polysaccharide is the main hypoglycemic factor in tea. High-performance gel permeation chromatography (HPGPC) and composition analysis showed that the 100–120 kDa fractions contained the hypoglycemic activity, among which the arabinogalactan protein played an important role. Moreover, tea polysaccharide had significant inhibitory effects on hydroxyl group, superoxide free radical and lipid peroxidation, and increased the activity of SOD, suggesting that tea polyphenol treatment might provide resistance to oxidative stress by regulating the enzyme activity (Chen et al., 2005). Studies have shown that the antioxidant capacity of tea polysaccharides is also related to their protein and polyphenol content (Fan et al., 2020). With the increase of the content of uronic acids in polysaccharide conjugates, the scavenging effect of reactive oxygen species was significantly increased (Chen et al., 2004).

Moreover, the change of the spatial structure of tea polysaccharide can also enhance its antioxidant activity (Wang et al., 2021a, 2021b).

Tea polysaccharide showed antibacterial and immunomodulatory effects. Stimulation of *Escherichia coli* with tea polysaccharide was shown that the cell wall was blurred and the cytoplasm leaked out in a dose-dependent manner (Zhou et al., 2020). Tea polysaccharide has obvious inhibitory effect on the pathogenic-mediated hemagglutination by inhibiting the adhesion of *Helicobacter pylori*, *Propionibacterium acnes* and *Staphylococcus aureus* (Lee et al., 2006). Besides, tea polysaccharide also reduced the incidence of azomethane/sodium dextran sulfate (AOM/DSS)-induced colon tumor in mice by down-regulating the expression levels of the cell cycle protein cyclin D1, matrix metalloproteinase (MMP-2) and MMP-9 (Liu et al., 2018a). In vitro, tea polysaccharide inhibited the proliferation of HCT116 cells by activating transcription factor EB (TFEB) and inducing cytotoxic autophagy by targeting lysosomes (Zhou et al., 2021a, 2021b). Lots of evidence shows that tea polysaccharide complex has more advantages in regulating immunity. The selenium-polysaccharide complex significantly inhibited the proliferation of human osteosarcoma U-2 OS cells in a concentration-dependent manner. In vivo, tumor regression was significantly induced compared to model (Wang et al., 2013). Monobe et al. reported similar results. Compared with tea polysaccharide without polyphenols, catechin-polysaccharide complex significantly increased the phagocytic activity of macrophage-like cells, indicating the importance of catechin-polysaccharide complex in immunomodulating activity (Monobe et al., 2008).

10.4 The Role of Tea and Its Functional Components in Inflammation-Related Diseases

10.4.1 Inflammatory Bowel Disease

Inflammatory bowel disease (IBD), which includes ulcerative colitis and Crohn's disease, is characterized by chronic intestinal inflammation. The pathogenesis of IBD is multifactorial, involving complex interactions among dysregulated immune responses, disruption of the intestinal mucosal barrier, abnormal nutrient absorption, and imbalances in the symbiotic relationship with gut microbiota. Among these factors, the interplay between impaired intestinal mucosal barrier function, dysbiosis of gut microbiota, and overactivation of immune cells has garnered significant attention in the pathogenesis of IBD. Recent studies have gradually unveiled the ameliorative effects of tea and its bioactive components on IBD. In a DSS-induced mouse model, supplementation with L-theanine was found to reduce serum diamine oxidase levels in colonic tissues, promote the production of short-chain fatty acids (SCFAs), and restore mucosal barrier expression (ZO-1 and claudin-1), thereby maintaining the integrity of the intestinal epithelial barrier (Wang et al.,

2020a, 2020b). Tea polysaccharides, through alleviating oxidative stress, downregulate pro-inflammatory cytokines (IL-6, IL-1β, IFN-γ, TNF-α, and LPS) and inflammatory-related mRNA (tight junction proteins) in colonic tissues, thereby enhancing intestinal barrier function in mice (Xu et al., 2022; Zeng et al., 2022). Moreover, tea polysaccharides can act as key regulators of gut homeostasis by upregulating glucagon-like peptide-1 (GLP-1) and increasing goblet cell numbers to strengthen intestinal barrier function, balance gut immunity, and suppress inflammation (Ma et al., 2024). Additionally, theabrownin can enhance the levels of IAld and IAA, thereby activating the aryl hydrocarbon receptor (AhR) and boosting IL-22 production to repair the intestinal barrier (Yang et al., 2023).

In current intervention studies, the reshaping of gut microbiota and their metabolites by tea has been a focal point in ameliorating intestinal inflammation (Fig. 10.1). Studies have shown that supplementation with green tea, oolong tea, and Pu-erh tea can increase gut microbial diversity and the Firmicutes/Bacteroidetes ratio in mice, thereby alleviating dysbiosis (Li et al., 2022). Among these, tea polyphenols, tea polysaccharides, and theabrownin have been identified as key bioactive compounds in tea that improve gut microbiota (Wei et al., 2023). The intake of tea polyphenols increases the abundance of *Akkermansia* and *Allobaculum*, which modulate mitochondrial dysfunction by inhibiting the Cytc/Cleaved-caspase-3 signaling pathway, thereby alleviating colonic cell apoptosis. EGCG, a key component of tea polyphenols, enhances the abundance of Firmicutes and Anaerobe, downregulates the phosphorylation of PI3K/Akt/NFκB and the acetylation of p65, and reduces intestinal permeability (Chiou et al., 2012). The increase in Akkermansia and Firmicutes promotes the synthesis of SCFAs and bile acids (BAs) (Dey et al., 2020). These changes further increase the expression of extracellular signal-regulated kinase (ERK)1/2 and activate the Nrf2-HO-1 pathway, thereby alleviating intestinal inflammation. Notably, the preparation of non-extractable polyphenols (NEPs) targeting the colon can achieve selective modulation of gut microbiota composition to alleviate intestinal inflammation (Chiou et al., 2012). The increased abundance of *Dubosiella* and *Enterorhabdus*, along with the decreased abundance of *Alistipes* and *Romboutsia*, may be the key mechanisms by which NEPs mitigate intestinal inflammation (Yang et al., 2024).

It is noteworthy that recent studies have demonstrated that aged teas, compared to fresh teas, exhibit superior regulatory effects on gut microbiota and their metabolites, with aged white tea and ripe Pu-erh tea being particularly prominent. Supplementation with aged white tea increases the abundance of butyrate, isobutyrate, and isovalerate in feces, promotes the conversion of primary BAs to secondary BAs, and significantly inhibits the relative abundance of *Bacteroides* and *Escherichia-Shigella* (Lin et al., 2024). Supplementation with ripe Pu-erh tea reshapes the gut microbiota (increasing the levels of *Lachnospiraceae_NK4A136_group* and *Akkermansia*), regulates tryptophan and BA metabolism, and significantly suppresses intestinal oxidative stress-mediated inflammatory pathways (TLR4/MyD88/ROS/p38MAPK/NF-κB p65), thereby improving intestinal function and lipid metabolism disorders (Hu et al., 2021a, 2021b, 2021c). Interestingly, Eurotium cristatum in Fu brick tea promotes the production of SCFAs, activates SCFA receptors, and enriches

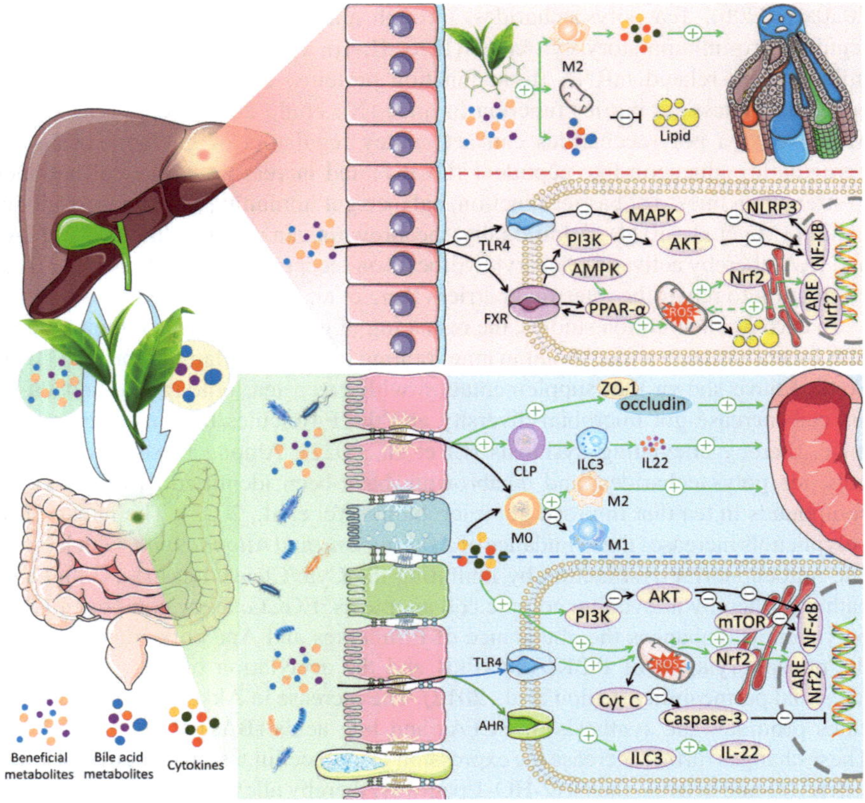

Fig. 10.1 The mechanism of tea in alleviating gut inflammation and liver lipid metabolism disorders (Fang et al., 2025)

Bacteroides, *Parabacteroides*, and *Faecalibacterium*, thereby exerting prebiotic effects (Xie et al., 2023). With increased post-fermentation time, the key fungus *Eurotium cristatum* in Fu brick tea significantly increases the levels of tryptophan metabolites in the feces of colitis mice and repairs tight junction protein expression to resist colonic injury (Zhang et al., 2022a, 2022b, 2022c).

10.4.2 Alcoholic Liver Disease

Alcoholic liver disease (ALD) is one of the leading causes of liver disease worldwide, with alcoholic fatty liver disease (AFLD) representing its early stage. The incidence of AFLD is rising in parallel with the increasing prevalence of alcohol abuse. The pathogenesis of AFLD is a complex process involving multiple intertwined factors (Cheng et al., 2024; Wu et al., 2023). Upon entering the liver, ethanol is metabolized via pathways such as alcohol dehydrogenase and cytochrome P450 2E1, generating highly toxic acetaldehyde. Acetaldehyde can form adducts with biomacromolecules and induce apoptosis. Ethanol metabolism also triggers oxidative stress, characterized by the production of large amounts of reactive oxygen species (ROS) and the suppression of antioxidant enzyme activity, leading to lipid peroxidation and disruption of cell membrane structure and function. This, in turn, activates inflammatory signaling pathways. Inflammatory cells, such as neutrophils and macrophages, are recruited to the liver by chemotactic factors, releasing pro-inflammatory cytokines like TNF-α, which promote steatosis, hepatocyte apoptosis, and the initiation of an inflammatory cascade. Alcohol consumption also disrupts the intestinal barrier, leading to gut dysbiosis, translocation of endotoxins into the bloodstream, and activation of inflammatory and oxidative stress pathways via Toll-like receptor 4 (TLR4) on Kupffer cells. Additionally, alcohol affects bile acid and lipid metabolism.

Tea and its bioactive components not only alleviate intestinal inflammation but also mitigate liver complications induced by enteritis through the gut-liver axis or by directly acting on the liver. Tea has been widely reported to possess anti-inflammatory and lipid metabolism regulatory effects on the liver. Studies have shown that tea consumption alleviates liver injury by modulating the PI3K/AKT/eNOS pathway and inhibiting the activation of the hepatic NF-κB signaling pathway and the NLRP3 inflammasome (Wang et al., 2017; Wang et al., 2019a, 2019b, 2019c). *L*-theanine, a key component, reduces hepatic quinoprotein levels and liver injury by modulating the overactivation of the hypothalamic–pituitary–adrenal (HPA) axis and inhibiting dopamine oxidation (Jia et al., 2022; Wang et al., 2018). Furthermore, by activating insulin and AMPK and their downstream signaling pathways, theanine helps regulate hepatic glucose, lipid, and protein metabolism (Lin et al., 2020). As a star component in tea, polyphenols reduce oxidative stress, pro-inflammatory cytokines, and the expression levels of 8-OHdG and AGEs in the liver by modulating the Nrf2 signaling pathway, thereby alleviating D-Gal-induced liver dysfunction in mice (Wang et al., 2022). Among them, catechins (ECG, EGC, and EGCG), as major components of tea polyphenols, can mitigate liver fibrosis by reducing the expression of α-smooth muscle actin, transforming growth factor-β (TGF-β), and downstream phosphorylation of ERK1/2 and Smad1/2 (Wang et al., 2019a, 2019b, 2019c). Additionally, post-fermented tea (e.g., dark tea) alleviates gut-derived liver injury by inhibiting the activation of hepatic inflammatory pathways such as TLR4/NF-κB and the NLRP3 inflammasome (Wang et al., 2024). Notably, tea polysaccharides in dark tea restore the expression levels of bile acid receptors (FXR, TGR5) in the gut-liver axis, reducing hepatic steatosis and oxidative stress damage, and are considered a potential mechanism for alleviating enteritis-mediated liver inflammation (Ren et al., 2015; Xiang et al., 2023).

Regarding the regulation of hepatic lipid metabolism, tea has been reported to modulate hepatic lipid metabolic disorders and promote fatty acid oxidation (Xu et al., 2023a, 2023b). Supplementation with tea water extracts activates the AMPK/PPAR-α signaling pathway, reducing hepatic fat production and enhancing fatty acid β-oxidation (Zhang et al., 2022a, 2022b, 2022c). Tea consumption or gavage with EGCG can alter bile acid (BA) metabolism in the body, characterized by increased levels of glycohyodeoxycholic acid and tauroursodeoxycholic acid, reducing intestinal BA reabsorption and thereby promoting hepatic lipid metabolism (Huang et al., 2018). In recent years, the lipid-lowering effects of dark tea have garnered significant attention. Theabrownin, as a key active component in dark tea, promote hepatic fatty acid oxidation and reduce hepatic lipid accumulation by inhibiting intestinal FXR-FGF15 signaling and modulating serotonin-related signaling pathways (HTR2A-PPARα-CYP4A14/CPT-1 and SERT-MAO-A) (Li et al., 2023). Additionally, caffeine in tea, by alleviating oxidative stress, has been proven to mitigate hepatic steatosis and fibrosis (Xin et al., 2021).

10.4.3 Brain Dysfunction

Brain dysfunction encompasses a variety of neurological disorders, including neurodegenerative diseases (such as depression, Alzheimer's disease, and Parkinson's disease) and cerebrovascular diseases. The pathogenesis of these brain disorders is a complex and multifactorial process involving the interplay of genetic susceptibility, environmental stressors, neurobiological mechanisms, and psychosocial factors. Among these disorders, depression is particularly notable as one of the most prevalent brain dysfunction diseases globally (Machado et al., 2018). Due to its complex pathogenesis, concealed manifestations, and low awareness, the cure rate for depression remains low, accompanied by a high suicide rate. In recent years, with the deepening of research in neuroscience and psychiatry, the pathogenesis of depression has become a hot topic, with many studies focusing on key aspects such as neurotransmitter disorders, dysfunction of neural circuits, and abnormalities in brain structure and functional plasticity.

Currently, selective serotonin reuptake inhibitors (SSRIs) including fluoxetine, fluvoxamine, and sertraline are the primary treatment options for depression. However, over 30% of patients experience delayed efficacy or non-response (Huang et al., 2023), and approximately 30% of patients with major depression develop resistance to SSRIs (Vadodaria et al., 2019). Additionally, about 50% of patients may experience side effects such as nausea, diarrhea, indigestion, gastrointestinal bleeding, and abdominal pain during initial treatment (Carvalho et al., 2016). Notably, the use of SSRIs during pregnancy and childbirth can lead to cardiac defects in offspring or affect the development of brain regions related to emotional regulation (Scarborough et al., 2021). Against this backdrop, natural phytochemicals have rapidly emerged as a research focus in the treatment of depression due to their potential multi-target effects, relatively mild toxicity profiles, and better tolerance in the body. These compounds may provide new ideas and strategies for breaking through the treatment bottleneck of this refractory brain dysfunction disease.

Tea is rich in various bioactive components, including *L*-theanine, tea polyphenols, tea pigments, tea polysaccharides, and GABA, which play key roles in neuroprotection (Chen et al., 2023), anxiety relief (Teng et al., 2017), antioxidant and anti-inflammatory effects (Liu et al., 2023), and regulation of metabolism and immune function. Studies have shown that tea and its phytochemicals can alter the structure of gut microbiota, reduce intestinal inflammation, and alleviate intestinal barrier dysfunction (Disease et al., 2024). Furthermore, through the immune, neural, and endocrine pathways of the gut-brain axis and communication with the central nervous system, tea can inhibit the overactivity of the hypothalamic–pituitary–adrenal (HPA) axis and glial cells, restore the monoamine system, and alleviate cognitive impairments (Shao et al., 2022).

Tea and its bioactive substances have been shown to stabilize levels of neurotransmitters in the brain (monoamines, BDNF, GABA, etc.), inhibit the overactivation of the HPA axis and adrenal hypertrophy (Fig. 10.2), thereby improving brain neurofunction (e.g., relieving anxiety and depression) and cognitive abilities (e.g., enhancing memory and attention) (Unno et al., 2022). L-theanine in tea can upregulate the levels of monoamine neurotransmitters (5-HT, NE, and DA) in brain regions related to the limbic-cortical-striatal-pallidal-thalamic (LCSPT) circuit (PFC, NAC, HIP) and increase BDNF levels (Anne, 2013; Manjun et al., 2018). This helps alleviate anxiety and depression, as well as memory decline and hippocampal long-term potentiation impairments (Zhu et al., 2018). Notably, theanine, with its structural similarity to glutamine (Gln), is considered a natural analog of glutamate (Glu) and is closely related to the 5-HT-Glu/GABA long neural circuit (Yunfeng, 2020). After tea consumption, elevated levels of *L*-theanine in the body compete with binding sites on Glu receptors (NMDAR, AMPAR), reducing the conversion efficiency of Gln to Glu and thereby alleviating glutamate metabolism imbalances that induce neuronal excitotoxicity (Shao et al., 2022). Additionally, theanine intake helps normalize brain kynurenine (Kyn) metabolism levels, increasing the proportion of the 5-HT metabolic pathway, thereby alleviating brain dysfunction such as depression (Chao et al., 2023).

Besides theanine, GABA in tea also has calming and sleep-improving effects, with the best synergistic effect when combined with theanine. Drinking GABA-rich tea can upregulate the $\alpha 1$ subunit of GABA receptors in the cerebral cortex, exerting antioxidant effects to alleviate depressive symptoms (Teng et al., 2017; Kim et al., 2019). GABAergic neurons and neurotransmitters play important roles in regulating brain circuits in the amygdala to cope with stress and anxiety responses under pathological conditions, thereby relieving brain nerve tension (Piril et al., 2020). Tea consumption increases GABA function in the medial prefrontal cortex (mPFC), reversing the effects of chronic stress exposure, similar to the mechanism of rapid-acting antidepressants like ketamine (Ronald et al., 2019). Compared to pharmacological treatments, tea and its natural active components offer advantages such as low toxicity and no side effects.

Interestingly, EGCG intake can also increase GABAergic levels in the host organism. EGCG reverses the inhibition of GABA currents by the GABA receptor negative regulator β-carboline-3-carboxylic acid methyl ester in a dose-dependent manner, maintaining neuronal activity stability and achieving anxiolytic effects (Vignes et al., 2006). EGCG's antidepressant effects are primarily focused on alleviating the overactivation of the HPA axis (Bombi et al., 2018). EGCG restores abnormal levels of plasma GCs, CRH, ACTH, and BDNF induced by depression,

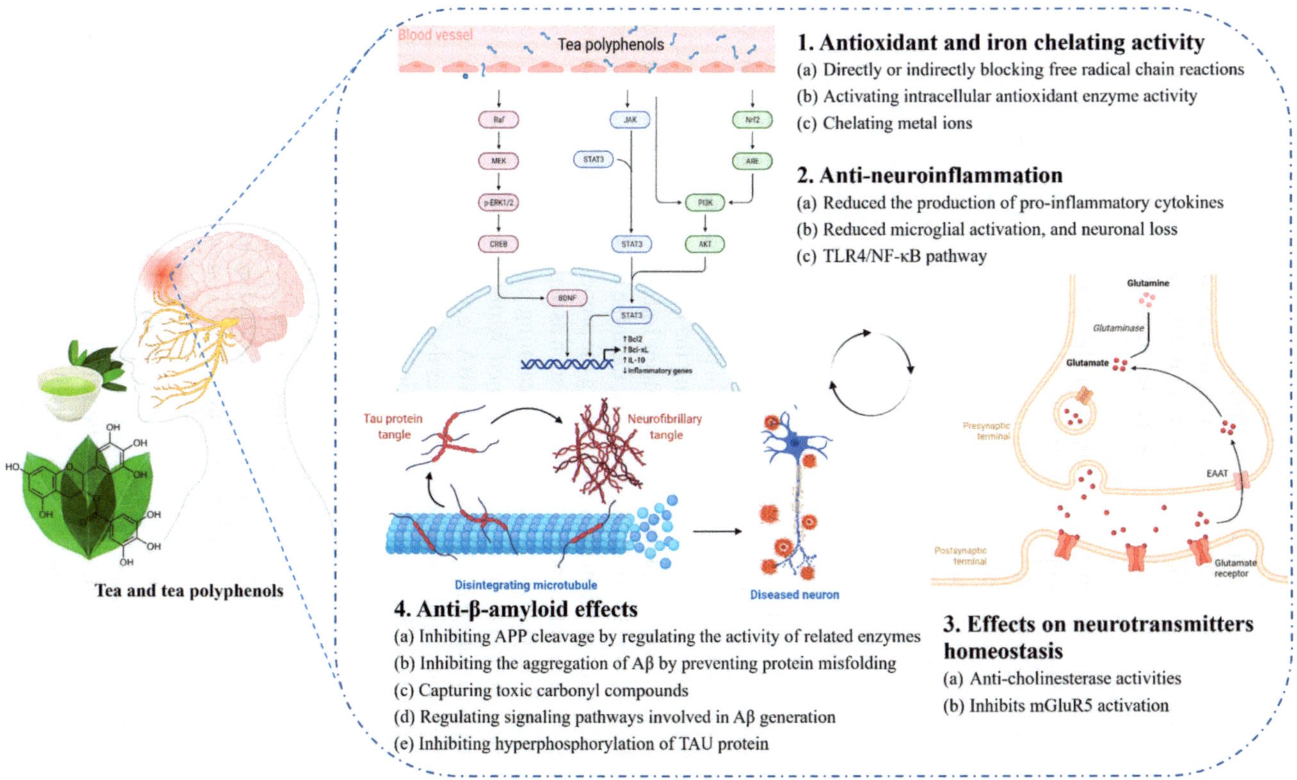

Fig. 10.2 The protective effects and mechanisms of tea on aging associated neurodegenerative cognitive disorders (Quan et al., 2024)

which may be a key factor in significantly altering the structure of tryptophan metabolism pathways (increased 5-HT metabolic pathway, decreased Kyn metabolic pathway). After EGCG intake, it acts through the PKCα and extracellular signal-regulated kinase 1/2 (ERK1/2) signaling pathways, reducing mRNA expression of IL-6, Caspase-3, -8, and -9 in serum and hippocampus, thereby alleviating neuronal apoptosis and protecting against neurodamage (Jinpeng et al., 2020; Xiaoling et al., 2017).

In addition to tea polyphenols, theaflavin formed by the oxidation and condensation of tea polyphenols also have prominent benefits in reducing oxidative stress and neuroinflammation. Theaflavin increases the conversion ratio of DA to high-melanin acid in the frontal cortex, promoting DA production, reducing microglial inflammatory responses in the brain and dendritic atrophy of pyramidal neurons in the hippocampus and prefrontal cortex, and improving memory impairments and depressive-like behaviors (Masahiro et al., 2019; Yasuhisa et al., 2019). Theaflavin also reduces the content of Kyn metabolites (Quin and Kyna) by alleviating the over-activation of astrocytes and microglia, thereby improving neurotoxicity induced by LPS. Theaflavin significantly inhibit the activation of NF-κB and IκB kinase (IKK), reducing levels of IL-1β, IL-6, and TNF-α in the hippocampus and systemic LPS coefficients, ultimately exerting anti-inflammatory, anti-anxiety, and antidepressant effects (Yasuhisa et al., 2019).

10.4.4 Oral Health

The oral cavity, as an open ecosystem, maintains the dynamic equilibrium of its microbiota (including bacteria, viruses, fungi, etc.) through the synergistic regulation of microbial invasion, colonization mechanisms, nutritional metabolic homeostasis, and host clearance (Kumar et al., 2016). Thus, oral health is a crucial part of overall health, facing multiple challenges such as microbial imbalance, inflammatory responses, and tissue repair disorders. If the mouth isn't cleaned promptly after eating, the long-term anaerobic environment will gradually degrade Gram-negative anaerobic oral bacteria. This increases the production of volatile sulfur compounds (VSCs) like hydrogen sulfide (H_2S), methyl mercaptan (CH_3SH), and dimethyl sulfide ($(CH_3)_2S$), causing oral malodor, namely halitosis (Bollen & Beikler, 2012). It's reported that over a quarter of 2,000 people in China suffer from halitosis (Apatzidou et al., 2013). Mechanical damage to the oral mucosa, lips, and peri-oral areas from dry, hard foods can lead to angular cheilitis caused by microbes such as *Staphylococcus aureus*, *Candida*, and *Streptococci* (Oza & Doshi, 2017; Silva et al., 2010). Pathogens like *Helicobacter pylori*, *Tannerella forsythia*, *Prevotella intermedia*, and *Prevotella nigrescens* can also stimulate the release of inflammatory factors like IL-6 and TNF-α, causing oral inflammation and diseases like oral ulcers and periodontitis (Hao et al., 2024). Some reports indicate that the severity of chronic periodontitis is associated with the severity and mortality rate of COVID-19 (Marouf et al., 2021). As the oral cavity is one of the entry points for microorganisms into the lower respiratory tract, aspirating excessive pathogenic bacteria like *Streptococcus pneumoniae* and *Haemophilus influenzae* can cause lower respiratory tract diseases such as bronchitis and pneumonia (Imai et al., 2021). Therefore, while oral health is challenged, it also indirectly affects the incidence of systemic diseases. However, drugs such as antibiotics and surgical treatments for oral diseases have limited effects. For example, they cannot act on residual bacteria in deep periodontal pockets and dental calculus, have a limited role in regulating inflammation, and offer limited regeneration of periodontal tissues (Zhang et al., 2022a, 2022b, 2022c). Against this background, functional foods, which provide basic nutrition and are rich in bioactive components closely related to health, have become a research focus in various fields. Tea, one of the world's three major non-alcoholic beverages, contains bioactive substances such as tea polyphenols, theanine, tea polysaccharides, and tea pigments. It may target the core pathological links of oral diseases through multiple approaches such as reshaping the microbiota, anti-inflammation, anti-oxidation, immune modulation, and promoting tissue repair, thus playing a role in the prevention and treatment of oral diseases (Zhou et al., 2021a; Liu et al., 2023).

Firstly, tea and its bioactive components can boost the abundance of oral probiotics and curb the concentration of opportunistic pathogens (Hu et al., 2021a, 2021b, 2021c). Tea polysaccharides enhance the abundance of Bifidobacteria (Yang et al., 2021). Tea polyphenols show remarkable antibacterial properties. The phenolic hydroxyl groups in their molecular structure can bind to bacterial cell membrane lipids, damaging membrane integrity and causing intracellular material leakage (Flemming et al., 2021). Studies indicate that oral epigallocatechin gallate (EGCG) inhibits the upregulation of IL-1β, IL-6, IL-17, and TNF-α in mice infected with *P. gingivalis,* thereby alleviating periodontitis (Lombardo Bedran et al., 2015a, 2015b). Theaflavins, major black tea polyphenols, exhibit antibacterial effects on planktonic cells and *P. gingivalis* biofilms by suppressing the protease activity of gingipains and the secretion of MMP-1 and MMP-2 at the transcriptional level (Kong et al., 2015). Moreover, theaflavins concentration dependently inhibit *Candida albicans* hyphae formation, a pathogen causing angular cheilitis, by interfering with cell wall β-glucan synthase activity. When used in combination with epicatechin, they show enhanced antibacterial effects (Betts et al., 2013). In terms of inflammation regulation, green tea consumption lowers IL-1β levels in the saliva of chronic periodontitis patients (Rezvani et al., 2022). Further studies reveal that green tea extract inhibits attachment loss, alveolar bone resorption, inflammatory cell infiltration, and RANKL expression in LPS-induced periodontitis rat models (Yoshinaga et al., 2014). Theaflavin supplementation reduces IL-8 secretion in oral epithelial cells stimulated by LPS (Lombardo Bedran et al., 2015a, 2015b), and decreases IL-6, CXCL-8 (IL-8), and MMP-9 secretion in macrophages stimulated by *Porphyromonas gingivalis*, thus alleviating periodontal tissue inflammation (Wu et al., 2018). When used as a dressing for alveolar sockets in periapical or periapical periodontal lesions, EGCG not only reduces local inflammation but also significantly promotes periapical tissue fibrosis and bone regeneration (Li et al., 2024). In addition, tea has significant therapeutic effects on antioxidant stress and tissue repair in the oral cavity. Tea consumption can mitigate excessive ROS generation in the oral cavity caused by smoking, chemotherapy, etc., which leads to DNA damage and cell apoptosis. Black tea consumption exerts antigonotic effects by downregulating IκB, NF-κB, p53, and MLH1 expression, protecting the oral mucosa (Pal et al., 2012). EGCG can inhibit the upregulation of IL-16 and TNF-α levels in mice induced by acetic acid and reduce histological damage in the oral cheeks, tongue, and throat, such as mucosal lesions, submucosal edema, and muscle diseases (Pan et al., 2023). Furthermore, EGCG can non-competitively inhibit α-amylase activity, effectively suppressing the formation of fermentable carbohydrates involved in caries development (Hara et al., 2012). Overall, tea and its bioactive components, particularly tea polyphenols, can reshape the oral microbiota, reduce inflammation, and enhance mucosal barriers and immune function, showing great potential in regulating oral disease and health.

Owing to the natural safety and multi-target regulation of tea and its bioactive components, they are becoming central to oral health product development. Tea-based functional products, by targeting oral microbial imbalances and immune inflammatory responses, show significant potential in preventing and treating dental caries, periodontal disease, oral ulcers, etc. For instance, 5% tea tree oil gel reduces bleeding in moderate-stage periodontitis (Taalab et al., 2021), green tea-enriched periodontal health products inhibit plaque accumulation and improve gingival health (Laleman & Teughels, 2020), and tea-polyphenol-containing mouthwashes alleviate radiation-induced oral mucosal damage (Zhu et al., 2020). Despite their potential, practical applications of tea's active components face challenges such as stability, target specificity, and clinical evidence.

Oral disease prevention and treatment involve complex microbe-host-environment interactions. Tea and its components, through multi-dimensional regulation, offer natural advantages in antibacterial, anti-inflammatory, repair, and micro-ecological adjustment. From traditional tea drinks to modern functional preparations, tea is shifting from a dietary culture carrier to a key oral health intervention resource. With in-depth basic research and industrial innovation, tea-based natural oral care solutions are expected to be a vital part of precision medicine, providing safe and efficient global oral health maintenance strategies.

10.5 Molecular Mechanism of Anti-inflammatory Effect of Tea

10.5.1 Regulating Inflammatory Signaling Pathways and Inhabiting Inflammatory Mediators Production

Tea polyphenols, particularly EGCG, play a crucial role in regulating inflammation. EGCG can reduce ROS levels, thereby decreasing the stimulation of NF-κB and inhibiting its activation. In multiple studies, EGCG has been shown to significantly attenuate Hla-induced excessive ROS production and the activation of the MAPK signaling pathway, thereby reducing the expression of NLRP3 inflammasome-related proteins in THP-1 cells (Liu et al., 2021). It also dose-dependently inhibits the expression of inflammasome proteins NLRP3, ASC, and caspase-1, and reduces the production of IL-1β and IL-18 in damaged mouse liver tissues, effectively preventing the occurrence of hepatitis (Jena et al., 2022; Zhang et al., 2021a, 2021b). At the transcriptional and translational levels, tea polyphenols can inhibit the activity of IκB kinase (IKK), preventing the phosphorylation and degradation of IκB (Kim et al., 2022; Xu et al., 2023a, 2023b). This keeps NF-κB sequestered in the cytoplasm, preventing it from entering the nucleus to initiate the transcription of pro-inflammatory cytokine genes (such as TNF-α, IL-6, and IL-1β), thereby fundamentally reducing the production of these pro-inflammatory cytokines.

In addition to these effects, EGCG can inhibit neutrophil infiltration, disrupt the expression of neutrophil cytosolic factor 1 and NLRP3 protein, and limit the secretion of pro-inflammatory mediators, thereby reducing the likelihood of monosodium urate-induced gouty arthritis (Chen et al., 2023a, 2023b, 2023c; Wu et al., 2022). Green tea polyphenols can also promote the phosphorylation of STAT3 and NF-κB in rat retinas, reducing the expression of pro-inflammatory mediators (including IL-1β, TNF-α, and IL-6) in the retina and vitreous humor, thereby alleviating retinal inflammation and protecting ocular health (Ren et al., 2018). In terms of chemokines, tea polyphenols can inhibit the production of monocyte chemoattractant protein-1 (MCP-1), reducing the migration of monocytes to inflammatory sites and thereby lowering the intensity of the inflammatory response (Pan et al., 2019; Rahman et al., 2020). Additionally, tea polyphenol supplementation can inhibit the activity of inducible nitric oxide synthase (iNOS) and cyclooxygenase-2 (COX-2), reducing the production of NO and PGE2, thereby alleviating inflammatory responses.

The oxidation products of tea polyphenols, such as theaflavin, also possess the ability to inhibit the NF-κB signaling pathway (Chen et al., 2023a, 2023b, 2023c; Sen et al., 2025; Shan et al., 2021). For example, TFDG can inhibit the expression of iNOS, block the activation of NF-κB, and reduce the nuclear accumulation of NF-κB P65 and P50 subunits induced by IκB and LPS, thereby preventing the activation of NF-κB and effectively preventing inflammation in macrophages. Theaflavin can also reduce the expression levels of key factors such as LPS-induced IL-6, inhibit the degradation of intracellular IκBα protein, and block the nuclear translocation of RelA (NF-κB p65), further interrupting the NF-κB signaling pathway and exerting anti-inflammatory effects.

10.5.2 Improve the Function of Immune Cells to Regulate Cell Apoptosis and Proliferation

Tea bioactive components, particularly EGCG, play a significant role in modulating T cell function. Under inflammatory conditions, the abnormal activation of T cells can lead to excessive production of pro-inflammatory cytokines (Ferreira et al., 2024;

Ge et al., 2024; Wei et al., 2024; Wu et al., 2021). EGCG can regulate the inflammatory environment in the gut by increasing the content of SCFAs, promoting the growth of beneficial gut bacteria, and reducing harmful bacteria, thereby alleviating intestinal inflammation. Additionally, EGCG can modulate the Treg/Th17 ratio in the spleen, increase plasma levels of IL-10 and TGF-β, and reduce the expression of hypoxia-inducible factor 1α (HIF-1α) and STAT3 in the colon, thus improving intestinal inflammation. Notably, tea components can influence the polarization state of macrophages, promoting their shift toward the anti-inflammatory M2 phenotype and inhibiting the pro-inflammatory M1 phenotype (Martins-Gomes et al., 2024; Shakoor et al., 2021). In mice, oral administration of green tea polyphenols (250 mg/kg/day) or EGCG (25 mg/kg/day) for 7 days significantly reduced LPS-induced CD80 expression in T lymphocytes, increased CD163 expression, decreased M1 macrophages, and increased Tregs in the bone marrow, as well as M2 macrophages, N2 neutrophils, and Tregs in the spleen and blood, preventing LPS-induced inflammatory damage and helping restore immune homeostasis. Similarly, aged Pu-erh tea administration in DSS-treated mice significantly promoted M2 macrophage polarization and upregulated the expression of tight junction proteins (MUC-2, ZO-1, occludin) in the gut, thereby improving gut immunity and reducing inflammation (Hu et al., 2021a, 2021b, 2021c).

Tea extracts can also inhibit the secretion of pro-inflammatory cytokines (e.g., IL-1β, IL-6) in macrophages, reducing the production of inflammatory mediators and providing a protective effect against inflammation. In inflammatory-related diseases, excessive apoptosis can exacerbate tissue damage. Tea components, such as theaflavin, have been shown to inhibit apoptosis by modulating the Keap1/Nrf2/HO-1 pathway, protecting chondrocytes from apoptosis and aging, and improving osteoarthritis in mice (Xu et al., 2021). EGCG can also prevent pulp stem cell apoptosis induced by hypoxia, effectively mitigating periodontitis (Song et al., 2024). Tea components can participate in the regulation of cell proliferation and promote tissue repair. For example, EGCG has been shown to synergize with mechanical stimulation to enhance osteogenic differentiation in human bone marrow mesenchymal stem cells (hMSCs) (Shin et al., 2017). When EGCG and 3% stretching were applied simultaneously, significant increases in runt-related transcription factor 2 (RUNX2) expression, decreases in myogenin expression, and enhanced osteogenic marker expression were observed, along with increased mitochondrial biogenesis and antioxidant capacity. In skin injury, long-term tea consumption has been shown to mitigate follicle cell apoptosis induced by circadian rhythm disruption. Long-term consumption of Pu-erh tea significantly increased the expression of Ku70, OGG1, and Rad51 genes, while reducing the activation of the Cyto c/Cas 3 pathway (Hu et al., 2024). The mechanism underlying hair loss improvement by Pu-erh tea may involve promoting the regular expression of genes related to the Wnt signaling pathway, Hippo signaling pathway, protein digestion and absorption, and nitrogen metabolism, thereby restoring follicle DNA repair function. However, the specific molecular mechanisms remain to be fully elucidated. Future research is necessary to explore how tea bioactive components affect the expression or activity of cell cycle-related proteins to maintain normal tissue structure and function and mitigate inflammation-induced tissue damage.

References

Abboud, P. A., Hake, P. W., Burroughs, T. J., Odoms, K., O'Connor, M., Wong, H. R., & Zingarelli, B. (2005). Therapeutic effect of epigallocatechin gallate in a mouse model of colitis. *Critical Care Medicine, 33.*

Akdis, C. A. (2021). Does the epithelial barrier hypothesis explain the increase in allergy, autoimmunity and other chronic conditions? *Nature Reviews Immunology, 21*(11), 739–751. https://doi.org/10.1038/s41577-021-00538-7

Anne, L. L. (2013). Neurobiological effects of the green tea constituent theanine and its potential role in the treatment of psychiatric and neurodegenerative disorders. *Nutritional Neuroscience.* https://doi.org/10.1179/1476830513y.0000000079

Ansari, M. Y., Ahmad, N., & Haqqi, T. M. (2020). Oxidative stress and inflammation in osteoarthritis pathogenesis: Role of polyphenols. *Biomedicine & Pharmacotherapy, 129*, 110452. https://doi.org/10.1016/j.biopha.2020.110452

Apatzidou, A. D., Bakirtzoglou, E., Vouros, I., Karagiannis, V., Papa, A., & Konstantinidis, A. (2013). Association between oral malodour and periodontal disease-related parameters in the general population. *Acta Odontologica Scandinavica, 71*(1), 189–195. https://doi.org/10.3109/00016357.2011.654259

Betts, J. W., Wareham, D. W., Haswell, S. J., & Kelly, S. M. (2013). Antifungal synergy of theaflavin and epicatechin combinations against Candida albicans. *Journal of Microbiology and Biotechnology, 23*(9), 1322–1326. https://doi.org/10.4014/jmb.1303.03010

Bezerra, F. S., Lanzetti, M., Nesi, R. T., Nagato, A. C., Silva, C. P., Kennedy-Feitosa, E., Melo, A. C., Cattani-Cavalieri, I., Porto, L. C., & Valenca, S. S. (2023). Oxidative stress and inflammation in acute and chronic lung injuries. *Antioxidants, 12.*

Bollen, C. M. L., & Beikler, T. (2012). Halitosis: The multidisciplinary approach. *International Journal of Oral Science, 4*(2), 55–63. https://doi.org/10.1038/ijos.2012.39

Bombi, L., Insop, S., Hyejung, L., & Dae-Hyun, H. (2018). Effects of epigallocatechin gallate on behavioral and cognitive impairments, hypothalamic–pituitary–adrenal axis dysfunction, and alternations in hippocampal BDNF expression under single prolonged stress. *Journal of Medicinal Food.* https://doi.org/10.1089/jmf.2017.4161

Carvalho, A. F., Sharma, M. S., Brunoni, A. R., Vieta, E., & Fava, G. A. (2016). The safety, tolerability and risks associated with the use of newer generation antidepressant drugs: A critical review of the literature. *Psychotherapy and Psychosomatics, 85*(5), 270–288. https://doi.org/10.1159/000447034

Chao, X., Ganglei, L., Qiuxian, Z., Xinyu, G., Qingmiao, S., Yuanshuai, S., Chu, Q., Yuan, X., Bao, Z., Lu, J., & Lanjuan, L. (2023). Tryptophan metabolism in health and disease. *Cell Metabolism.* https://doi.org/10.1016/j.cmet.2023.06.004

Chen, H., Zhang, M., & Xie, B. (2005). Components and antioxidant activity of polysaccharide conjugate from green tea. *Food Chemistry, 90*(1), 17–21. https://doi.org/10.1016/j.foodchem.2004.03.001

Chen, X., Lin, Z., Ye, Y., Zhang, R., Yin, J., Jiang, Y., & Wan, H. (2010). Suppression of diabetes in non-obese diabetic (NOD) mice by oral administration of water-soluble and alkali-soluble polysaccharide conjugates prepared from green tea. *Carbohydrate Polymers, 82*(1), 28–33. https://doi.org/10.1016/j.carbpol.2010.04.017

Chen, T., Liu, A. B., Sun, S., Ajami, N. J., Ross, M. C., Wang, H., Zhang, L., Reuhl, K., Kobayashi, K., Onishi, J. C., & Yang, C. S. (2019). Green tea polyphenols modify the gut microbiome in db/db mice as co-abundance groups correlating with the blood glucose lowering effect. *Molecular Nutrition & Food Research, 63*(8), 1801064. https://doi.org/10.1002/mnfr.201801064

Chen, X., Luo, D., Jia, G., Zhao, H., Liu, G., & Huang, Z. (2023). L-Theanine attenuates porcine intestinal tight junction damage induced by LPS via p38 MAPK/NLRP3 signaling in IPEC-J2 cells. *Food and Chemical Toxicology, 178*, 113870. https://doi.org/10.1016/j.fct.2023.113870

Chen, Y., Liu, Z., & Gong, Y. (2023). Neuron-immunity communication: Mechanism of neuroprotective effects in EGCG. *Critical Reviews in Food Science and Nutrition*, 1–20. https://doi.org/10.1080/10408398.2023.2212069

Chen, H., Zhang, M., & Xie, B. (2004). Quantification of uronic acids in tea polysaccharide conjugates and their antioxidant properties. *Journal of Agricultural and Food Chemistry, 52*(11), 3333–3336. https://doi.org/10.1021/jf0349679

Chen, S.-Y., Li, Y.-P., You, Y.-P., Zhang, H.-R., Shi, Z.-J., Liang, Q.-Q., Yuan, T., Xu, R., Xu, L. H., Zha, Q. B., & He, X.-H. (2023a). Theaflavin mitigates acute gouty peritonitis and septic organ injury in mice by suppressing NLRP3 inflammasome assembly. *Acta Pharmacologica Sinica, 44*(10), 2019–2036. https://doi.org/10.1038/s41401-023-01105-7

Chen, Y., Luo, L., Hu, S., Gan, R., & Zeng, L. (2023c). The chemistry, processing, and preclinical anti-hyperuricemia potential of tea: A comprehensive review. *Critical Reviews in Food Science and Nutrition, 63*(24), 7065–7090. https://doi.org/10.1080/10408398.2022.2040417

Cheng, Z., Yang, L., & Chu, H. (2024). The role of gut microbiota, exosomes, and their interaction in the pathogenesis of ALD. *Journal of Advanced Research*. https://doi.org/10.1016/j.jare.2024.07.002

Cheng, J., Tan, Y., Zhou, J., Xiao, L., Johnson, M., & Qu, X. (2020). Green tea polyphenols ameliorate metabolic abnormalities and insulin resistance by enhancing insulin signalling in skeletal muscle of Zucker fatty rats. *Clinical Science, 134*(10), 1167–1180. https://doi.org/10.1042/CS20200107

Chiou, Y. S., Ma, N. J., Sang, S., Ho, C. T., Wang, Y. J., & Pan, M. H. (2012). Peracetylated (-)-epigallocatechin-3-gallate (AcEGCG) potently suppresses dextran sulfate sodium-induced colitis and colon tumorigenesis in mice. *Journal of Agriculture and Food Chemistry, 60*(13), 3441–3451. https://doi.org/10.1021/jf300441p

De Felice, M., Renard, J., Hudson, R., Szkudlarek, H. J., Pereira, B. J., Schmid, S., Rushlow, W. J., & Laviolette, S. R. (2021). L-Theanine prevents long-term affective and cognitive side effects of adolescent Δ-9-tetrahydrocannabinol exposure and blocks associated molecular and neuronal abnormalities in the mesocorticolimbic circuitry. *The Journal of Neuroscience, 41*(4), 739. https://doi.org/10.1523/JNEUROSCI.1050-20.2020

Dey, P., Olmstead, B. D., Sasaki, G. Y., Vodovotz, Y., Yu, Z., & Bruno, R. S. (2020). Epigallocatechin gallate but not catechin prevents nonalcoholic steatohepatitis in mice similar to green tea extract while differentially affecting the gut microbiota. *Journal of Nutritional Biochemistry, 84*, 108455. https://doi.org/10.1016/j.jnutbio.2020.108455

Zuo, G., Chen, M., Zuo, Y., Liu, F., Yang, Y., Li, J., Zhou, X., Li, M., Huang, J.-A., Liu, Z., & Lin, Y. (2024). Tea polyphenol epigallocatechin gallate protects against nonalcoholic fatty liver disease and associated endotoxemia in rats via modulating gut microbiota dysbiosis and alleviating intestinal barrier dysfunction and related inflammation. https://doi.org/10.1021/acs.jafc.3c04832

Fan, M., Zhu, J., Qian, Y., Yue, W., Xu, Y., Zhang, D., Yang, Y., Gao, X., He, H., & Wang, D. (2020). Effect of purity of tea polysaccharides on its antioxidant and hypoglycemic activities. *Journal of Food Biochemistry, 44*(8), e13277. https://doi.org/10.1111/jfbc.13277

Fang, Q., Qin, C., Peng, Y., Zhang, L., Li, D., & Hu, S. (2025). Navigating the gut-liver axis: The potential of tea to alleviate inflammatory bowel disease and associated metabolic syndromes. *Trends in Food Science & Technology, 157*, 104893. https://doi.org/10.1016/j.tifs.2025.104893

Ferreira, C., Vieira, P., Sá, H., Malva, J., Castelo-Branco, M., Reis, F., & Viana, S. (2024). Polyphenols: Immunonutrients tipping the balance of immunometabolism in chronic diseases. *Frontiers in Immunology, 15*. https://doi.org/10.3389/fimmu.2024.1360065

Flemming, J., Meyer-Probst, C. T., Speer, K., Kölling-Speer, I., Hannig, C., & Hannig, M. (2021). Preventive applications of polyphenols in dentistry—A review. *International Journal of Molecular Sciences, 22*.

Ge, J., Li, M., Yao, J., Guo, J., Li, X., Li, G., Han, X., Li, Z., Liu, M., & Zhao, J. (2024). The potential of EGCG in modulating the oral-gut axis microbiota for treating inflammatory bowel disease. *Phytomedicine, 130*, 155643. https://doi.org/10.1016/j.phymed.2024.155643

Guan, F., Liu, A., Li, G., Sun, Y., Yang, C., & Ju, J. (2012). Deleterious effects of high concentrations of (-)-epigallocatechin-3-gallate and atorvastatin in mice with colon inflammation. *Nutrition and Cancer, 64*, 847–855. https://doi.org/10.1080/01635581.2012.695424

Hao, Y., Zhao, B., Wu, D., Ge, X., & Han, J. (2024). Recombinant humanized collagen type XVII promotes oral ulcer healing via anti-inflammation and accelerate tissue healing. *Journal of Inflammation Research, 17*, 4993–5004. https://doi.org/10.2147/jir.S470649

Hara, K., Ohara, M., Hayashi, I., Hino, T., Nishimura, R., Iwasaki, Y., Ogawa, T., Ohyama, Y., Sugiyama, M., & Amano, H. (2012). The green tea polyphenol (−)-epigallocatechin gallate precipitates salivary proteins including alpha-amylase: Biochemical implications for oral health. *European Journal of Oral Sciences, 120*(2), 132–139. https://doi.org/10.1111/j.1600-0722.2012.00947.x

Heber, D., Zhang, Y., Yang, J., Ma, J. E., Henning, S. M., & Li, Z. (2014). Green tea, black tea, and oolong tea polyphenols reduce visceral fat and inflammation in mice fed high-fat, high-sucrose obesogenic diets. *The Journal of Nutrition, 144*(9), 1385–1393. https://doi.org/10.3945/jn.114.191007

Hu, D., Zhong, T., & Dai, Q. (2021a). Clinical efficacy of probiotics as an adjunctive therapy to scaling and root planning in the management of periodontitis: A systematic review and meta-analysis of randomized controlled trails. *Journal of Evidence Based Dental Practice, 21*(2), 101547. https://doi.org/10.1016/j.jebdp.2021.101547

Hu, S., Wang, J., Lin, Z., Zhang, B., Luo, L., & Zeng, L. (2024). Pu-erh tea reduces the transmission of CRD-mediated alopecia risk to offspring. *Food Frontiers, 5*(4), 1673–1686. https://doi.org/10.1002/fft2.409

Hu, S., Li, S., Liu, Y., Sun, K., Luo, L., & Zeng, L. (2021b). Aged ripe Pu-erh tea reduced oxidative stress-mediated inflammation in dextran sulfate sodium-induced colitis mice by regulating intestinal microbes. *Journal of Agricultural and Food Chemistry, 69*(36), 10592–10605. https://doi.org/10.1021/acs.jafc.1c04032

Hu, S., Li, S., Liu, Y., Sun, K., Luo, L., & Zeng, L. (2021c). Aged ripe Pu-erh tea reduced oxidative stress-mediated inflammation in dextran sulfate sodium-induced colitis mice by regulating intestinal microbes. *Journal of Agriculture and Food Chemistry, 69*(36), 10592–10605. https://doi.org/10.1021/acs.jafc.1c04032

Huang, J., Feng, S., Liu, A., Dai, Z., Wang, H., Reuhl, K., Lu, W., & Yang, C. S. (2018). Green tea polyphenol EGCG alleviates metabolic abnormality and fatty liver by decreasing bile acid and lipid absorption in mice. *Molecular Nutrition & Food Research, 62*(4). https://doi.org/10.1002/mnfr.201700696

Huang, J., Huang, W., Yi, J., Deng, Y., Li, R., Chen, J., Shi, J., Qiu, Y., Wang, T., Chen, X., & Zhang, X. (2023). Mesenchymal stromal cells alleviate depressive and anxiety-like behaviors via a lung vagal-to-brain axis in male mice. *Nature Communications, 14*(1), 7406. https://doi.org/10.1038/s41467-023-43150-0

Hwang, Y. P., Jin, S. W., Choi, J. H., Choi, C. Y., Kim, H. G., Kim, S. J., Kim, Y., Lee, K. J., Chung, Y. C., & Jeong, H. G. (2017). Inhibitory effects of l-theanine on airway inflammation in ovalbumin-induced allergic asthma. *Food and Chemical Toxicology, 99*, 162–169. https://doi.org/10.1016/j.fct.2016.11.032

Imai, K., Iinuma, T., & Sato, S. (2021). Relationship between the oral cavity and respiratory diseases: Aspiration of oral bacteria possibly contributes to the progression of lower airway inflammation. *Japanese Dental Science Review, 57*, 224–230. https://doi.org/10.1016/j.jdsr.2021.10.003

Jena, A. B., Dash, U. C., & Duttaroy, A. K. (2022). An in silico investigation on the interactions of curcumin and epigallocatechin-3-gallate with NLRP3 Inflammasome complex. *Biomedicine & Pharmacotherapy, 156*, 113890. https://doi.org/10.1016/j.biopha.2022.113890.

Jia, L., Wang, F., Zhang, K., Wang, D., Wang, X., Li, X., & Zhang, J. (2022). L-Theanine inhibits (-)-epigallocatechin-3-gallate oxidation via chelating copper. *Journal of Agriculture and Food Chemistry, 70*(25), 7751–7761. https://doi.org/10.1021/acs.jafc.2c01379

Jinpeng, W., Ping, L., Qing, T., Dongjie, S., Xin, Z., & Beilin, Z. (2020). Protective effect of epigallocatechin-3-gallate against neuroinflammation and anxiety-like behavior in a rat model of myocardial infarction. *Brain and Behavior*. https://doi.org/10.1002/brb3.1633

Kakuda, T. (2011). Neuroprotective effects of theanine and its preventive effects on cognitive dysfunction. *Pharmacological Research, 64*(2), 162–168. https://doi.org/10.1016/j.phrs.2011.03.010

Kim, M., Murakami, A., Miyamoto, S., Tanaka, T., & Ohigashi, H. (2010). The modifying effects of green tea polyphenols on acute colitis and inflammation-associated colon carcinogenesis in male ICR mice. *BioFactors, 36*(1), 43–51. https://doi.org/10.1002/biof.69

Kim, S.-R., Seong, K.-J., Kim, W.-J., & Jung, J.-Y. (2022). Epigallocatechin gallate protects against hypoxia-induced inflammation in microglia via NF-κB suppression and Nrf-2/HO-1 activation. *International Journal of Molecular Sciences, 23*.

Kim, S., Jo, K., Hong, K. B., Han, S. H., & Suh, H. J. (2019). GABA and l-theanine mixture decreases sleep latency and improves NREM sleep. *Pharmaceutical Biology, 57*(1), 65–73. https://doi.org/10.1080/13880209.2018.1557698

Kimura, K., Ozeki, M., Juneja, L. R., & Ohira, H. (2007). l-Theanine reduces psychological and physiological stress responses. *Biological Psychology, 74*(1), 39–45. https://doi.org/10.1016/j.biopsycho.2006.06.006

Kong, L., Qi, X., Huang, S., Chen, S., Wu, Y., & Zhao, L. (2015). Theaflavins inhibit pathogenic properties of P. gingivalis and MMPs production in P. gingivalis-stimulated human gingival fibroblasts. *Archives of Oral Biology, 60*(1), 12–22. https://doi.org/10.1016/j.archoralbio.2014.08.019

Kumar, T., Arora, N., Puri, G., Aravinda, K., Dixit, A., & Jatti, D. (2016). Efficacy of ozonized olive oil in the management of oral lesions and conditions: A clinical trial. *Contemp Clin Dent, 7*(1), 51–54. https://doi.org/10.4103/0976-237x.177097

Laleman, I., & Teughels, W. (2020). Novel natural product-based oral topical rinses and toothpastes to prevent periodontal diseases. *Periodontology 2000, 84*(1), 102–123. https://doi.org/10.1111/prd.12339

Lee, J.-H., Shim, J. S., Lee, J. S., Kim, J. K., Yang, I. S., Chung, M.-S., & Kim, K. H. (2006). Inhibition of pathogenic bacterial adhesion by acidic polysaccharide from green tea (Camellia sinensis). *Journal of Agricultural and Food Chemistry, 54*(23), 8717–8723. https://doi.org/10.1021/jf061603i

Li, Y., Cheng, L., & Li, M. (2024). Effects of green tea extract epigallocatechin-3-gallate on oral diseases: a narrative review. *Pathogens, 13*.

Li, A., Wang, J., Kou, R., Chen, M., Zhang, B., Zhang, Y., Liu, J., Xing, X., Peng, B., & Wang, S. (2022). Polyphenol-rich oolong tea alleviates obesity and modulates gut microbiota in high-fat diet-fed mice. *Frontiers in Nutrition, 9*, 937279. https://doi.org/10.3389/fnut.2022.937279

Li, H. Y., Huang, S. Y., Zhou, D. D., Xiong, R. G., Luo, M., Saimaiti, A., Han, M. K., Gan, R. Y., Zhu, H. L., & Li, H. B. (2023). Theabrownin inhibits obesity and non-alcoholic fatty liver disease in mice via serotonin-related signaling pathways and gut-liver axis. *Journal of Advanced Research, 52*, 59–72. https://doi.org/10.1016/j.jare.2023.01.008

Liang, J., Yan, H., Puligundla, P., Gao, X., Zhou, Y., & Wan, X. (2017). Applications of chitosan nanoparticles to enhance absorption and bioavailability of tea polyphenols: A review. *Food Hydrocolloids, 69*, 286–292. https://doi.org/10.1016/j.foodhyd.2017.01.041

Lin, L., Zeng, L., Liu, A., Peng, Y., Yuan, D., Zhang, S., Li, Y., Chen, J., Xiao, W., & Gong, Z. (2020). L-Theanine regulates glucose, lipid, and protein metabolism via insulin and AMP-activated protein kinase signaling pathways. *Food & Function, 11*(2), 1798–1809. https://doi.org/10.1039/c9fo02451d

Lin, Z., Dai, W., Hu, S., Chen, D., Yan, H., Zeng, L., & Lin, Z. (2024). Stored white tea ameliorates DSS-induced ulcerative colitis in mice by modulating the composition of the gut microbiota and intestinal metabolites. *Food & Function, 15*(8), 4262–4275. https://doi.org/10.1039/d3fo05176e

Liu, L.-Q., Li, H.-S., Nie, S.-P., Shen, M.-Y., Hu, J.-L., & Xie, M.-Y. (2018a). Tea polysaccharide prevents colitis-associated carcinogenesis in mice by inhibiting the proliferation and invasion of tumor cells. *International Journal of Molecular Sciences, 19*(2). https://doi.org/10.3390/ijms19020506

Liu, L.-Q., Li, H.-S., Nie, S.-P., Shen, M.-Y., Hu, J.-L., & Xie, M.-Y. (2018b). Tea polysaccharide prevents colitis-associated carcinogenesis in mice by inhibiting the proliferation and invasion of tumor cells. *International Journal of Molecular Sciences, 19*(2), 506. https://www.mdpi.com/1422-0067/19/2/506

Liu, C., Hao, K., Liu, Z., Liu, Z., & Guo, N. (2021). Epigallocatechin gallate (EGCG) attenuates staphylococcal alpha-hemolysin (Hla)-induced NLRP3 inflammasome activation via ROS-MAPK pathways and EGCG-Hla interactions. *International Immunopharmacology, 100*, 108170. https://doi.org/10.1016/j.intimp.2021.108170

Liu, H., Guan, H., He, F., Song, Y., Li, F., Sun-Waterhouse, D., & Li, D. (2023). Therapeutic actions of tea phenolic compounds against oxidative stress and inflammation as central mediators in the development and progression of health problems: A review focusing on microRNA regulation. *Critical Reviews in Food Science and Nutrition*, 1–31. https://doi.org/10.1080/10408398.2023.2202762

Liu, C., Liu, A., Zhou, J., Zhang, Y., Zhou, F., Chen, H., Liu, Q., Zhang, S., Huang, J., & Liu, Z. (2022). Role and mechanism of theaflavins in regulating skeletal muscle inflammation. *Journal of Agricultural and Food Chemistry, 70*(41), 13233–13250. https://doi.org/10.1021/acs.jafc.2c04063

Lombardo Bedran, T. B., Palomari Spolidorio, D., & Grenier, D. (2015b). Green tea polyphenol epigallocatechin-3-gallate and cranberry proanthocyanidins act in synergy with cathelicidin (LL-37) to reduce the LPS-induced inflammatory response in a three-dimensional co-culture model of gingival epithelial cells and fibroblasts. *Archives of Oral Biology, 60*(6), 845–853. https://doi.org/10.1016/j.archoralbio.2015.02.021

Lombardo Bedran, T. B., Morin, M. P., Palomari Spolidorio, D., & Grenier, D. (2015a). Black tea extract and its theaflavin derivatives inhibit the growth of periodontopathogens and modulate interleukin-8 and β-defensin secretion in oral epithelial cells. *PLoS ONE, 10*(11), e0143158. https://doi.org/10.1371/journal.pone.0143158

Łyko, L., Olech, M., Gawlik, U., Krajewska, A., Kalemba, D., Tyśkiewicz, K., Piórecki, N., Prokopiv, A., & Nowak, R. (2024). Rhododendron luteum sweet flower supercritical CO_2 extracts: Terpenes composition, pro-inflammatory enzymes inhibition and antioxidant activity. *International Journal of Molecular Sciences, 25*.

Ma, C., Li, X., Zheng, C., Zhou, B., Xu, C., & Xia, T. (2021). Comparison of characteristic components in tea-leaves fermented by Aspergillus pallidofulvus PT-3, Aspergillus sesamicola PT-4 and Penicillium manginii PT-5 using LC-MS metabolomics and HPLC analysis. *Food Chemistry, 350*, 129228. https://doi.org/10.1016/j.foodchem.2021.129228

Ma, H., Zhang, B., Hu, Y., Wang, J., Liu, J., Qin, R., Lv, S., & Wang, S. (2019). Correlation analysis of intestinal redox state with the gut microbiota reveals the positive intervention of tea polyphenols on hyperlipidemia in high fat diet fed mice. *Journal of Agricultural and Food Chemistry, 67*(26), 7325–7335. https://doi.org/10.1021/acs.jafc.9b02211

Ma, Y., Xie, H., Xu, N., Li, M., Wang, L., Ge, H., Xie, Z., Li, D., & Wang, H. (2024). Large yellow tea polysaccharide alleviates HFD-induced intestinal homeostasis dysbiosis via modulating gut barrier integrity, immune responses, and the gut microbiome. *Journal of Agriculture and Food Chemistry, 72*(13), 7230–7243. https://doi.org/10.1021/acs.jafc.4c00616

Maccarrone, M., & Brüne, B. (2009). Redox regulation in acute and chronic inflammation. *Cell Death & Differentiation, 16*(8), 1184–1186. https://doi.org/10.1038/cdd.2009.65

Machado, M. O., Veronese, N., Sanches, M., Stubbs, B., Koyanagi, A., Thompson, T., Tzoulaki, I., Solmi, M., Vancampfort, D., Schuch, F. B., & Maes, M. (2018). The association of depression and all-cause and cause-specific mortality: An umbrella review of systematic reviews and meta-analyses. *BMC Medicine, 16*(1), 112. https://doi.org/10.1186/s12916-018-1101-z

Manjun, S., Yi, Y., Wantao, Y., Beibei, Z., Haishan, W., Lu, W., Tang Hui, T. H., & Jindong, C. (2018). L-Theanine ameliorate depressive-like behavior in a chronic unpredictable mild stress rat model via modulating the monoamine levels in limbic–cortical–striatal–pallidal–thalamic-circuit related brain regions. *Phytotherapy Research*. https://doi.org/10.1002/ptr.6237

Marouf, N., Cai, W., Said, K. N., Daas, H., Diab, H., Chinta, V. R., Hssain, A. A., Nicolau, B., Sanz, M., & Tamimi, F. (2021). Association between periodontitis and severity of COVID-19 infection: A case-control study. *Journal of Clinical Periodontology, 48*(4), 483–491. https://doi.org/10.1111/jcpe.13435

Martins-Gomes, C., Nunes, F. M., & Silva, A. M. (2024). Natural products as dietary agents for the prevention and mitigation of oxidative damage and inflammation in the intestinal barrier. *Antioxidants, 13*.

Masahiro, K., Shunya, U., Koji, Y., & Yasuhisa, A. (2019). Anxiolytic effects of theaflavins via dopaminergic activation in the frontal cortex. *Bioscience, Biotechnology, and Biochemistry*. https://doi.org/10.1080/09168451.2019.1584523

Monobe, M., Ema, K., Kato, F., & Maeda-Yamamoto, M. (2008). Immunostimulating activity of a crude polysaccharide derived from green tea (Camellia sinensis) extract. *Journal of Agricultural and Food Chemistry, 56*(4), 1423–1427. https://doi.org/10.1021/jf073127h

Oza, N., & Doshi, J. J. (2017). Angular cheilitis: A clinical and microbial study. *Indian Journal of Dental Research, 28*(6), 661–665. https://doi.org/10.4103/ijdr.IJDR_668_16

Pal, D., Sur, S., Mandal, S., Das, S., & Panda, C. K. (2012). Regular black tea habit could reduce tobacco associated ROS generation and DNA damage in oral mucosa of normal population. *Food and Chemical Toxicology, 50*(9), 2996–3003. https://doi.org/10.1016/j.fct.2012.06.005

Pan, M.-H., Yang, G., Li, S., Li, M.-Y., Tsai, M.-L., Wu, J.-C., Badmaev, V., Ho, C. T., & Lai, C.-S. (2017). Combination of citrus polymethoxyflavones, green tea polyphenols, and lychee extracts suppresses obesity and hepatic steatosis in high-fat diet induced obese mice. *Molecular Nutrition & Food Research, 61*(11), 1601104. https://doi.org/10.1002/mnfr.201601104

Pan, M.-H., Li, M.-Y., Tsai, M.-L., Pan, C.-Y., Badmaev, V., Ho, C.-T., & Lai, C.-S. (2019). A mixture of citrus polymethoxyflavones, green tea polyphenols and lychee extracts attenuates adipogenesis in 3T3-L1 adipocytes and obesity-induced adipose inflammation in mice. *Food & Function, 10*(12), 7667–7677. https://doi.org/10.1039/C9FO02235J

Pan, Y., Lv, H., Feng, X., Zhou, S., Hu, H., Chen, S., Cheng, Y., Fan, F., Gong, S., Chen, P., & Chu, Q. (2023). Epigallocatechin gallate (EGCG) alleviates the inflammatory response and recovers oral microbiota in acetic acid-induced oral inflammation mice. *Food & Function, 14*(22), 10069–10082. https://doi.org/10.1039/d3fo03107a

Piril, H., John, A. G., Jun, N., & Andrew, S. (2020). Effects of oral gamma-aminobutyric acid (GABA) administration on stress and sleep in humans: A systematic review. *Frontiers in Neuroscience*. https://doi.org/10.3389/fnins.2020.00923

Pisetsky, D. S. (2023). Pathogenesis of autoimmune disease. *Nature Reviews Nephrology, 19*(8), 509–524. https://doi.org/10.1038/s41581-023-00720-1

Pober, J. S., & Sessa, W. C. (2007). Evolving functions of endothelial cells in inflammation. *Nature Reviews Immunology, 7*(10), 803–815. https://doi.org/10.1038/nri2171

Qin, B., Polansky, M. M., Harry, D., & Anderson, R. A. (2010). Green tea polyphenols improve cardiac muscle mRNA and protein levels of signal pathways related to insulin and lipid metabolism and inflammation in insulin-resistant rats. *Molecular Nutrition & Food Research, 54*(S1), S14–S23. https://doi.org/10.1002/mnfr.200900306

Quan, W., Lin, Y., Zou, H., Li, M., Luo, J., He, Z., Chen, J., & Liu, Z. (2024). Can habitual tea drinking be an effective approach against age-related neurodegenerative cognitive disorders: A systematic review and meta-analysis of epidemiological evidence. *Critical Reviews in Food Science and Nutrition, 64*(17), 5835–5851. https://doi.org/10.1080/10408398.2022.2158780

Rahman, S. U., Huang, Y., Zhu, L., Chu, X., Junejo, S. A., Zhang, Y., Khan, I. M., Li, Y., Feng, S., Wu, J., & Wang, X. (2020). Tea polyphenols attenuate liver inflammation by modulating obesity-related genes and down-regulating COX-2 and iNOS expression in high fat-fed dogs. *BMC Veterinary Research, 16*(1), 234. https://doi.org/10.1186/s12917-020-02448-7

Ramos-Lopez, O., Milagro, F. I., Riezu-Boj, J. I., & Martinez, J. A. (2021). Epigenetic signatures underlying inflammation: An interplay of nutrition, physical activity, metabolic diseases, and environmental factors for personalized nutrition. *Inflammation Research, 70*(1), 29–49. https://doi.org/10.1007/s00011-020-01425-y

Ren, D., Hu, Y., Luo, Y., & Yang, X. (2015). Selenium-containing polysaccharides from Ziyang green tea ameliorate high-fructose diet induced insulin resistance and hepatic oxidative stress in mice. *Food & Function, 6*(10), 3342–3350. https://doi.org/10.1039/c5fo00557d

Ren, J. L., Yu, Q. X., Liang, W. C., Leung, P. Y., Ng, T. K., Chu, W. K., Pang, C. P., & Chan, S. O. (2018). Green tea extract attenuates LPS-induced retinal inflammation in rats. *Scientific Reports, 8*(1), 429. https://doi.org/10.1038/s41598-017-18888-5

Rezvani, G., Taleghani, F., & Valizadeh, M. (2022). Effect of green tea on the level of salivary interleukin-1 beta in patients with chronic periodontitis: A randomized clinical trial. *International Journal of Dentistry, 2022*, 8992313. https://doi.org/10.1155/2022/8992313

Rius-Pérez, S., Pérez, S., Martí-Andrés, P., Monsalve, M., & Sastre, J. (2019). Nuclear factor kappa B signaling complexes in acute inflammation. *Antioxidants & Redox Signaling, 33*(3), 145–165. https://doi.org/10.1089/ars.2019.7975

Ronald, S. D., Gerard, S., & John, H. K. (2019). Altered connectivity in depression: GABA and glutamate neurotransmitter deficits and reversal by novel treatments. *Neuron*. https://doi.org/10.1016/j.neuron.2019.03.013

Scarborough, J., Mueller, F. S., Weber-Stadlbauer, U., Mattei, D., Opitz, L., Cattaneo, A., & Richetto, J. (2021). A novel murine model to study the impact of maternal depression and antidepressant treatment on biobehavioral functions in the offspring. *Molecular Psychiatry, 26*(11), 6756–6772. https://doi.org/10.1038/s41380-021-01145-7

Sen, A. K., Maheshwari, R. A., Alam, F., Dumpala, R. L., Shah, U. H., & Sen, D. B. (2025). Chapter 54—Theaflavins in tea: Features and effects. In V. R. Preedy & V. B. Patel (Eds.), *Tea in health and disease prevention* (2nd ed., pp. 635–647). Academic Press.

Shakoor, H., Feehan, J., Apostolopoulos, V., Platat, C., Al Dhaheri, A. S., Ali, H. I., Ismail, L. C., Bosevski, M., & Stojanovska, L. (2021). Immunomodulatory effects of dietary polyphenols. *Nutrients, 13*.

Shan, Z., Nisar, M. F., Li, M., Zhang, C., & Wan, C. (2021). [Retracted] Theaflavin chemistry and its health benefits. *Oxidative Medicine and Cellular Longevity, 2021*(1), 6256618. https://doi.org/10.1155/2021/6256618

Shao, J., Wei, Y., & Wei, X. (2022). A comprehensive review on bioavailability, safety and antidepressant potential of natural bioactive components from tea. *Food Research International, 158*, 111540. https://doi.org/10.1016/j.foodres.2022.111540

Shen, Y., Lu, T., Liu, X.-Y., Zhao, M.-T., Yin, F.-W., Rakariyatham, K., & Zhou, D.-Y. (2020). Improving the oxidative stability and lengthening the shelf life of DHA algae oil with composite antioxidants. *Food Chemistry, 313*, 126139. https://doi.org/10.1016/j.foodchem.2019.126139

Shin, J. W., Wu, Y., Kang, Y. G., Kim, J. K., Choi, H. J., & Shin, J.-W. (2017). The effects of epigallocatechin-3-gallate and mechanical stimulation on osteogenic differentiation of human mesenchymal stem cells: Individual or synergistic effects. *Tissue Engineering and Regenerative Medicine, 14*(3), 307–315. https://doi.org/10.1007/s13770-017-0040-7

Silva, D. G., Stevens, R. H., Macedo, J. M., Albano, R. M., Falabella, M. E., Fischer, R. G., Veerman, E. C., & Tinoco, E. M. (2010). Presence of Helicobacter pylori in supragingival dental plaque of individuals with periodontal disease and upper gastric diseases. *Archives of Oral Biology, 55*(11), 896–901. https://doi.org/10.1016/j.archoralbio.2010.06.018

Song, P., Hao, Y., Lin, D., Jin, Y., & Lin, J. (2024). Evaluation of the antibacterial effect of epigallocatechin gallate on the major pathogens of canine periodontal disease and therapeutic effects on periodontal disease mice. *Frontiers in Microbiology, 14*. https://doi.org/10.3389/fmicb.2023.1329772

Suhyeon, K., Kyungae, J., Ki-Bae, H., Han, S. H., & Hyung. (2019). GABA and l-theanine mixture decreases sleep latency and improves NREM sleep. *Pharmaceutical Biology*.

Taalab, M. R., Mahmoud, S. A., Moslemany, R. M. E., & Abdelaziz, D. M. (2021). Intrapocket application of tea tree oil gel in the treatment of stage 2 periodontitis. *BMC Oral Health, 21*(1), 239. https://doi.org/10.1186/s12903-021-01588-y

Teng, J., Zhou, W., Zeng, Z., & Zhao, W. (2017). Quality components and antidepressant-like effects of GABA green tea. *Food & Function, 8*(9), 3311–3318. https://doi.org/10.1039/c7fo01045a

Tsai, W.-H., Wu, C.-H., Yu, H.-J., & Chien, C.-T. (2017). l-Theanine inhibits proinflammatory PKC/ERK/ICAM-1/IL-33 signaling, apoptosis, and autophagy formation in substance P-induced hyperactive bladder in rats. *Neurourology and Urodynamics, 36*(2), 297–307. https://doi.org/10.1002/nau.22965

Unno, K., Tanida, N., Ishii, N., Yamamoto, H., Iguchi, K., Hoshino, M., Takeda, A., Ozawa, H., Ohkubo, T., Juneja, L. R., & Juneja, L. R. (2013). Anti-stress effect of theanine on students during pharmacy practice: Positive correlation among salivary α-amylase activity, trait anxiety and subjective stress. *Pharmacology Biochemistry & Behavior, 111*, 128–135.

Unno, K., Sumiyoshi, A., Konishi, T., Hayashi, M., Taguchi, K., Muguruma, Y., Inoue, K., Iguchi, K., Nonaka, H., Kawashima, R., & Nakamura, Y. (2020). Theanine, the main amino acid in tea, prevents stress-induced brain atrophy by modifying early stress responses. *Nutrients, 12*(1), 174. https://www.mdpi.com/2072-6643/12/1/174

Unno, K., Furushima, D., Tanaka, Y., Tominaga, T., Nakamura, H., Yamada, H., Taguchi, K., Goda, T., & Nakamura, Y. (2022). Improvement of depressed mood with green tea intake. *Nutrients, 14*(14). https://doi.org/10.3390/nu14142949

Vadodaria, K. C., Ji, Y., Skime, M., Paquola, A., Nelson, T., Hall-Flavin, D., Fredlender, C., Heard, K. J., Deng, Y., Le, A. T., & Dave, S. (2019). Serotonin-induced hyperactivity in SSRI-resistant major depressive disorder patient-derived neurons. *Molecular Psychiatry, 24*(6), 795–807. https://doi.org/10.1038/s41380-019-0363-y

Vanhoutte, P. M., Shimokawa, H., Feletou, M., & Tang, E. H. C. (2017). Endothelial dysfunction and vascular disease—A 30th anniversary update. *Acta Physiologica, 219*(1), 22–96. https://doi.org/10.1111/apha.12646

Vignes, M., Maurice, T., Lanté, F., Nedjar, M., Thethi, K., Guiramand, J., & Récasens, M. (2006). Anxiolytic properties of green tea polyphenol (−)-epigallocatechin gallate (EGCG). *Brain Research, 1110*(1), 102–115. https://doi.org/10.1016/j.brainres.2006.06.062

Wang, Y., Chen, J., Zhang, D., Zhang, Y., Wen, Y., Li, L., & Zheng, L. (2013). Tumoricidal effects of a selenium (Se)-polysaccharide from Ziyang green tea on human osteosarcoma U-2 OS cells. *Carbohydrate Polymers, 98*(1), 1186–1190. https://doi.org/10.1016/j.carbpol.2013.07.022

Wang, D., Gao, Q., Zhao, G., Kan, Z., Wang, X., Wang, H., Huang, J., Wang, T., Qian, F., Ho, C. T., & Wang, Y. (2018). Protective effect and mechanism of theanine on lipopolysaccharide-induced inflammation and acute liver injury in mice. *Journal of Agricultural and Food Chemistry, 66*(29), 7674–7683. https://doi.org/10.1021/acs.jafc.8b02293

Wang, J., Jia, R., Celi, P., Ding, X., Bai, S., Zeng, Q., Mao, X., Xu, S., & Zhang, K. (2019b). Green tea polyphenol epigallocatechin-3-gallate improves the antioxidant capacity of eggs. *Food & Function, 11*. https://doi.org/10.1039/C9FO02157D

Wang, D., Cai, M., Wang, T., Liu, T., Huang, J., Wang, Y., & Granato, D. (2020a). Ameliorative effects of L-theanine on dextran sulfate sodium induced colitis in C57BL/6J mice are associated with the inhibition of inflammatory responses and attenuation of intestinal barrier disruption. *Food Research International, 137*, 109409. https://doi.org/10.1016/j.foodres.2020.109409

Wang, R., Sun, F., Ren, C., Zhai, L., Xiong, R., Yang, Y., Yang, W., Yi, R., Li, C., & Zhao, X. (2021). Hunan insect tea polyphenols provide protection against gastric injury induced by HCl/ethanol through an antioxidant mechanism in mice. *Food & Function, 12*. https://doi.org/10.1039/D0FO02677H

Wang, H., Chen, J., Ren, P., Zhang, Y., & Omondi Onyango, S. (2021). Ultrasound irradiation alters the spatial structure and improves the antioxidant activity of the yellow tea polysaccharide. *Ultrasonics Sonochemistry, 70*, 105355. https://doi.org/10.1016/j.ultsonch.2020.105355

Wang, X., Liu, J., Wei, J., Zhang, Y., Xu, Y., Yue, T., & Yuan, Y. (2024). Protective mechanism of Eurotium amstelodami from Fuzhuan brick tea against colitis and gut-derived liver injury

induced by dextran sulfate sodium in C57BL/6 mice. *Nutrients, 16*(8). https://doi.org/10.3390/nu16081178
Wang, D., Cai, M., Wang, T., Liu, T., Huang, J., Wang, Y., & Granato, D. (2020b). Ameliorative effects of L-theanine on dextran sulfate sodium induced colitis in C57BL/6J mice are associated with the inhibition of inflammatory responses and attenuation of intestinal barrier disruption. *Food Research International, 137*, 109409. https://doi.org/10.1016/j.foodres.2020.109409
Wang, D., Gao, Q., Wang, T., Zhao, G., Qian, F., Huang, J., Wang, H., Zhang, X., & Wang, Y. (2017). Green tea infusion protects against alcoholic liver injury by attenuating inflammation and regulating the PI3K/Akt/eNOS pathway in C57BL/6 mice. *Food & Function, 8*(9), 3165–3177. https://doi.org/10.1039/c7fo00791d
Wang, D., Wang, T., Li, Z., Guo, Y., & Granato, D. (2022). Green tea polyphenols upregulate the Nrf2 signaling pathway and suppress oxidative stress and inflammation markers in D-galactose-induced liver aging in mice. *Frontiers in Nutrition, 9*, 836112. https://doi.org/10.3389/fnut.2022.836112
Wang, D., Zhang, M., Wang, T., Cai, M., Qian, F., Sun, Y., & Wang, Y. (2019a). Green tea polyphenols prevent lipopolysaccharide-induced inflammatory liver injury in mice by inhibiting NLRP3 inflammasome activation. *Food & Function, 10*(7), 3898–3908. https://doi.org/10.1039/c9fo00572b
Wang, L., Yang, G., Yuan, L., Yang, Y., Zhao, H., Ho, C. T., & Li, S. (2019c). Green tea catechins effectively altered hepatic fibrogenesis in rats by inhibiting ERK and Smad1/2 phosphorylation. *Journal of Agriculture and Food Chemistry, 67*(19), 5437–5445. https://doi.org/10.1021/acs.jafc.8b05179
Wei, F., Li, D., Chen, X., Li, Y., Zeng, Y., Cai, Y., Zeng, Y., Chen, Y., Ma, X., & Zeng, J. (2024). Therapeutic effects of epigallocatechin-3-gallate for inflammatory bowel disease: A preclinical meta-analysis. *Phytomedicine, 128*, 155408. https://doi.org/10.1016/j.phymed.2024.155408
Wei, K., Wei, Y., Wang, Y., & Wei, X. (2023). Amelioration effects and regulatory mechanisms of different tea active ingredients on DSS-induced colitis. *Journal of Agriculture and Food Chemistry, 71*(44), 16604–16617. https://doi.org/10.1021/acs.jafc.3c04524
Werner, S., & Grose, R. (2003). Regulation of wound healing by growth factors and cytokines. *Physiological Reviews, 83*(3), 835–870. https://doi.org/10.1152/physrev.2003.83.3.835
Wijesekara, T., Luo, J., & Xu, B. (2024). Critical review on anti-inflammation effects of saponins and their molecular mechanisms. *Phytotherapy Research, 38*(4), 2007–2022. https://doi.org/10.1002/ptr.8164
Wu, Z., Huang, S., Li, T., Li, N., Han, D., Zhang, B., Xu, Z. Z., Zhang, S., Pang, J., Wang, S., & Wang, J. (2021). Gut microbiota from green tea polyphenol-dosed mice improves intestinal epithelial homeostasis and ameliorates experimental colitis. *Microbiome, 9*(1), 184. https://doi.org/10.1186/s40168-021-01115-9
Wu, C., Li, F., Zhang, X., Xu, W., Wang, Y., Yao, Y., Han, Z., & Xia, D. (2022). (−)-Epicatechin ameliorates monosodium urate-induced acute gouty arthritis through inhibiting NLRP3 inflammasome and the NF-κB signaling pathway. *Frontiers in Pharmacology, 13*. https://doi.org/10.3389/fphar.2022.799552
Wu, X., Fan, X., Miyata, T., Kim, A., Cajigas-Du Ross, C. K., Ray, S., Huang, E., Taiwo, M., Arya, R., Wu, J., & Nagy, L. E. (2023). Recent advances in understanding of pathogenesis of alcohol-associated liver disease. *Annual Review of Pathology: Mechanisms of Disease, 18*(18), 411–438. https://doi.org/10.1146/annurev-pathmechdis-031521-030435
Wu, Y. H., Kuraji, R., Taya, Y., Ito, H., & Numabe, Y. (2018). Effects of theaflavins on tissue inflammation and bone resorption on experimental periodontitis in rats. *Journal of Periodontal Research, 53*(6), 1009–1019. https://doi.org/10.1111/jre.12600
Wu, Z., Ming, J., Gao, R., Wang, Y., Liang, Q., Yu, H., & Zhao, G. (2011). Characterization and antioxidant activity of the complex of tea polyphenols and oat β-glucan. *Journal of Agricultural and Food Chemistry, 59*(19), 10737–10746. https://doi.org/10.1021/jf202722w
Xia, H. M., Wang, J., Xie, X. J., Xu, L. J., & Tang, S. Q. (2019). Green tea polyphenols attenuate hepatic steatosis, and reduce insulin resistance and inflammation in high-fat diet-induced rats.

International Journal of Molecular Medicine, 44(4), 1523–1530. https://doi.org/10.3892/ijmm.2019.4285

Xiang, G., Sun, H., Chen, Y., Guo, H., Liu, Y., Li, Y., Lu, C., & Wang, X. (2023). Antioxidant and hypoglycemic activity of tea polysaccharides with different degrees of fermentation. *International Journal of Biological Macromolecules, 228*, 224–233. https://doi.org/10.1016/j.ijbiomac.2022.12.114

Xiaoling, Z., Fengqin, L., Haimin, J., Renjia, L., Yonghui, W., Wanqi, Z., Wang, H., & Weiqiang, C. (2017). Involvement of PKCα and ERK1/2 signaling pathways in EGCG's protection against stress-induced neural injuries in Wistar rats. *Neuroscience*. https://doi.org/10.1016/j.neuroscience.2017.01.025

Xie, Z., Zeng, Z., Chen, G., Dong, W., Peng, Y., Xu, W., Sun, Y., Zeng, X., & Liu, Z. (2023). Intracellular polysaccharides of Aspergillus cristatus from Fuzhuan brick tea leverage the gut microbiota and repair the intestinal barrier to ameliorate DSS-induced colitis in mice. *Journal of Agriculture and Food Chemistry, 71*(21), 8023–8037. https://doi.org/10.1021/acs.jafc.3c00611

Xin, X., Cheng, C., Bei-Yu, C., Hong-Shan, L., Hua-Jie, T., Xin, W., Zi-Ming, A., Qin-Mei, S., Yi-Yang, H., & Qin, F. (2021). Caffeine and EGCG alleviate high-trans fatty acid and high-carbohydrate diet-induced NASH in mice: Commonality and specificity. *Frontiers in Nutrition, 8*, 784354. https://doi.org/10.3389/fnut.2021.784354

Xu, H., Gan, C., Xiang, Z., Xiang, T., Li, J., Huang, X., Qin, X., Liu, T., Sheng, J., & Wang, X. (2023). Targeting the TNF-α–TNFR interaction with EGCG to block NF-κB signaling in human synovial fibroblasts. *Biomedicine & Pharmacotherapy, 161*, 114575. https://doi.org/10.1016/j.biopha.2023.114575

Xu, A., Zhao, Y., Shi, Y., Zuo, X., Yang, Y., Wang, Y., & Xu, P. (2022). Effects of oxidation-based tea processing on the characteristics of the derived polysaccharide conjugates and their regulation of intestinal homeostasis in DSS-induced colitis mice. *International Journal of Biological Macromolecules, 214*, 402–413. https://doi.org/10.1016/j.ijbiomac.2022.06.115

Xu, J., Wei, Y., Huang, Y., & Wei, X. (2023b). Regulatory effects and molecular mechanisms of tea and its active compounds on nonalcoholic fatty liver disease. *Journal of Agriculture and Food Chemistry*. https://doi.org/10.1021/acs.jafc.2c07702

Xu, X.-X., Zheng, G., Tang, S.-K., Liu, H.-X., Hu, Y.-Z., & Shang, P. (2021). Theaflavin protects chondrocytes against apoptosis and senescence via regulating Nrf2 and ameliorates murine osteoarthritis. *Food & Function, 12*(4), 1590–1602. https://doi.org/10.1039/D0FO02038A

Yang, W., Ren, D., Zhao, Y., Liu, L., & Yang, X. (2021). Fuzhuan brick tea polysaccharide improved ulcerative colitis in association with gut microbiota-derived tryptophan metabolism. *Journal of Agricultural and Food Chemistry, 69*(30), 8448–8459. https://doi.org/10.1021/acs.jafc.1c02774

Yang, W., Ren, D., Shao, H., Zhang, X., Li, T., Zhang, L., Liu, L., Zhao, Y., Niu, P., & Yang, X. (2023). Theabrownin from Fu brick tea improves ulcerative colitis by shaping the gut microbiota and modulating the tryptophan metabolism. *Journal of Agricultural and Food Chemistry, 71*(6), 2898–2913. https://doi.org/10.1021/acs.jafc.2c06821

Yang, C., Yang, W., Wang, Y., Du, Y., Zhao, T., Shao, H., Ren, D., & Yang, X. (2024). Nonextractable polyphenols from Fu brick tea alleviates ulcerative colitis by controlling colon microbiota-targeted release to inhibit intestinal inflammation in mice. *Journal of Agriculture and Food Chemistry, 72*(13), 7397–7410. https://doi.org/10.1021/acs.jafc.3c06883

Yang, Y., Han, X., Chen, Y., Wu, J., Li, M., Yang, H., Xu, W., & Wei, L. (2021). EGCG induces pro-inflammatory response in macrophages to prevent bacterial infection through the 67LR/p38/JNK signaling pathway. *Journal of Agricultural and Food Chemistry, 69*(20), 5638–5651. https://doi.org/10.1021/acs.jafc.1c01353

Yasuhisa, A., Rena, O., Masahiro, K., Yoshimasa, T., & Keiji, K. (2019). Theaflavins improve memory impairment and depression-like behavior by regulating microglial activation. *Molecules*. https://doi.org/10.3390/molecules24030467

Yin, X., Huang, J. a., Huang, J., Wu, W., Tong, T., Liu, S., Zhou, L., Liu, Z., & Zhang, S. (2022). Identification of volatile and odor-active compounds in Hunan black tea by SPME/GC-MS and multivariate analysis. *LWT, 164*, 113656. https://doi.org/10.1016/j.lwt.2022.113656

Yoshinaga, Y., Ukai, T., Nakatsu, S., Kuramoto, A., Nagano, F., Yoshinaga, M., Montenegro, J. L., Shiraishi, C., & Hara, Y. (2014). Green tea extract inhibits the onset of periodontal destruction in rat experimental periodontitis. *Journal of Periodontal Research, 49*(5), 652–659. https://doi.org/10.1111/jre.12147

Yunfeng, L. (2020). A hypothesis of monoamine (5-HT)—Glutamate/GABA long neural circuit: Aiming for fast-onset antidepressant discovery. *Pharmacology & Therapeutics*. https://doi.org/10.1016/j.pharmthera.2020.107494

Zeng, Z., Xie, Z., Chen, G., Sun, Y., Zeng, X., & Liu, Z. (2022). Anti-inflammatory and gut microbiota modulatory effects of polysaccharides from Fuzhuan brick tea on colitis in mice induced by dextran sulfate sodium. *Food & Function, 13*(2), 649–663. https://doi.org/10.1039/d1fo02702f

Zhai, X., Ren, D., Luo, Y., Hu, Y., & Yang, X. (2017). Chemical characteristics of an Ilex Kuding tea polysaccharide and its protective effects against high fructose-induced liver injury and vascular endothelial dysfunction in mice. *Food & Function*, 2536.

Zhang, X., Kang, Y., Li, X., Huang, Y., Qi, R., Han, Y., Cai, R., Gao, Y., & Qi, Y. (2021b). Potentilla discolor ameliorates LPS-induced inflammatory responses through suppressing NF-κB and AP-1 pathways. *Biomedicine & Pharmacotherapy, 144*, 112345. https://doi.org/10.1016/j.biopha.2021.112345

Zhang, C., Li, X., Hu, X., Xu, Q., Zhang, Y., Liu, H., Diao, Y., Zhang, X., Li, L., Yu, J., & Peng, J. (2021a). Epigallocatechin-3-gallate prevents inflammation and diabetes—Induced glucose tolerance through inhibition of NLRP3 inflammasome activation. *International Immunopharmacology, 93*, 107412. https://doi.org/10.1016/j.intimp.2021.107412

Zhang, Z., Yu, Y., Zhu, G., Zeng, L., Xu, S., Cheng, H., Ouyang, Z., Chen, J., Pathak, J. L., Wu, L., & Yu, L. (2022c). The emerging role of plant-derived exosomes-like nanoparticles in immune regulation and periodontitis treatment. *Frontiers in Immunology, 13*, 896745. https://doi.org/10.3389/fimmu.2022.896745

Zhang, M., Qin, H., Xiang, L., An, L., Zhang, X., Li, K., Wu, K., Fei, X., Fan, W., Xu, X., & Mu, D. (2024). Camellia sinensis polysaccharide attenuates inflammatory responses via the ROS-mediated pathway by endocytosis. *International Journal of Biological Macromolecules, 267*, 131674. https://doi.org/10.1016/j.ijbiomac.2024.131674

Zhang, B., Ren, D., Zhao, A., Shao, H., Li, T., Niu, P., Zhao, Y., & Yang, X. (2022a). Eurotium cristatum exhibited anti-colitis effects via modulating gut microbiota-dependent tryptophan metabolism. *Journal of Agriculture and Food Chemistry, 70*(51), 16164–16175. https://doi.org/10.1021/acs.jafc.2c05464

Zhang, J. K., Zhou, X. L., Wang, X. Q., Zhang, J. X., Yang, M. L., Liu, Y. P., Cao, J. X., & Cheng, G. G. (2022b). Que Zui tea ameliorates hepatic lipid accumulation and oxidative stress in high fat diet induced nonalcoholic fatty liver disease. *Food Research International, 156*, 111196. https://doi.org/10.1016/j.foodres.2022.111196

Zhao, J., Blayney, A., Liu, X., Gandy, L., Jin, W., Yan, L., Ha, J. H., Canning, A. J., Connelly, M., Yang, C., & Wang, C. (2021). EGCG binds intrinsically disordered N-terminal domain of p53 and disrupts p53-MDM2 interaction. *Nature Communications, 12*(1), 986. https://doi.org/10.1038/s41467-021-21258-5

Zhou, Y., Yao, Q., Zhang, T., Chen, X., Wu, Z., Zhang, N., Shao, Y., & Cheng, Y. (2020). Antibacterial activity and mechanism of green tea polysaccharide conjugates against Escherichia coli. *Industrial Crops and Products, 152*, 112464. https://doi.org/10.1016/j.indcrop.2020.112464

Zhou, F., Li, Y. L., Zhang, X., Wang, K. B., Huang, J. A., Liu, Z. H., & Zhu, M. Z. (2021a). Polyphenols from Fu brick tea reduce obesity via modulation of gut microbiota and gut microbiota-related intestinal oxidative stress and barrier function. *Journal of Agricultural and Food Chemistry, 69*(48), 14530–14543. https://doi.org/10.1021/acs.jafc.1c04553

Zhou, X., Wang, D., Sun, P., Bucheli, P., Li, L., Hou, Y., & Wang, J. (2007). Effects of soluble tea polysaccharides on hyperglycemia in alloxan-diabetic mice. *Journal of Agricultural and Food Chemistry, 55*(14), 5523–5528. https://doi.org/10.1021/jf070699t

Zhou, Y., Zhou, X., Huang, X., Hong, T., Zhang, K., Qi, W., Guo, M., & Nie, S. (2021b). Lysosome-mediated cytotoxic autophagy contributes to tea polysaccharide-induced colon cancer cell death via mTOR-TFEB signaling. *Journal of Agricultural and Food Chemistry, 69*(2), 686–697. https://doi.org/10.1021/acs.jafc.0c07166

Zhu, G., Yang, S., Xie, Z., & Wan, X. (2018). Synaptic modification by L-theanine, a natural constituent in green tea, rescues the impairment of hippocampal long-term potentiation and memory in AD mice. *Neuropharmacology, 138*, 331–340. https://doi.org/10.1016/j.neuropharm.2018.06.030

Zhu, W., Mei, H., Jia, L., Zhao, H., Li, X., Meng, X., Zhao, X., Xing, L., & Yu, J. (2020). Epigallocatechin-3-gallate mouthwash protects mucosa from radiation-induced mucositis in head and neck cancer patients: A prospective, non-randomised, phase 1 trial. *Investigational New Drugs, 38*(4), 1129–1136. https://doi.org/10.1007/s10637-019-00871-8

Zindel, J., & Kubes, P. (2020). DAMPs, PAMPs, and LAMPs in immunity and sterile inflammation. *Annual Review of Pathology: Mechanisms of Disease, 15*, 493–518. https://doi.org/10.1146/annurev-pathmechdis-012419-032847

Zotova, N., Zhuravleva, Y., Chereshnev, V., & Gusev, E. (2023). Acute and chronic systemic inflammation: Features and differences in the pathogenesis, and integral criteria for verification and differentiation. *International Journal of Molecular Sciences, 24*.

Chapter 11
Regulation of Gut Microflora by Tea

Shanshan Hu

Abstract This chapter comprehensively elucidates the pivotal role of gut microbiota in human health and disease prediction, highlighting that compositional and functional shifts serve as early biomarkers for obesity, metabolic syndrome, and neurodegenerative disorders. Critical tea bioactive compounds—including polyphenols, polysaccharides, and theanine—orchestrate gut microecology remodeling through multi-tiered mechanisms: structurally, they selectively enrich probiotics and SCFA-producing bacteria while suppressing conditional pathogens; metabolically, they enhance short-chain fatty acid production to fortify intestinal barrier integrity, precisely modulate bile acid metabolism, and rebalance tryptophan metabolism; systemically, they improve bile acid homeostasis and hepatic lipid metabolism via the gut-liver axis, while influencing neurotransmitter synthesis through the gut-brain axis to mitigate brain dysfunction. These findings underscore tea's potential in preventive health and disease intervention, offering modern scientific evidence for the health benefits of traditional tea consumption.

Keywords Gut microbiota · Tea · Gut-organ axis · Disease improvement · Molecular mechanisms

11.1 Introduction

In the vast domain of life sciences and health research, the gut microbiome is akin to a mysterious and vibrant "tropical rainforest," quietly governing many critical pathways of human physiological functions and emerging as a highly sought-after and deeply explored frontier treasure in contemporary academia (de Vos et al., 2022; Van Hul et al., 2024). Within this microscopic world, trillions of microorganisms reside

S. Hu (✉)
State Key Laboratory of Tea Plant Germplasm Innovation and Resource Utilization, School of Tea Science, Anhui Agricultural University, Hefei, China
e-mail: Hshan1008@163.com

International Joint Laboratory on Tea Chemistry and Health Effects of Ministry of Education, Anhui Agricultural University, Hefei, China

X. Zhai (ed.), *Tea Chemistry and Health Benefits*,
https://doi.org/10.1007/978-3-032-11828-8_11

in the gut, encoding a number of genes that far exceeds that of the human genome itself. It functions like a hidden "metabolic factory" and "immune fortress," deeply involved in fundamental life processes such as human digestion, immune defense, and neuroendocrine regulation, maintaining the delicate balance of the internal environment. The stability of this microbiome is like the "Sword of Damocles," hanging over the balance between health and disease.

Looking back, our understanding of the gut microbiome has evolved from vague to clear. Early on, limited by technological constraints, we could only glimpse a small part of it. Today, with the advent of high-throughput sequencing and metabolomics, the veil of mystery surrounding the gut microbiome is gradually being lifted. Metabolites such as short-chain fatty acids, bile acid derivatives, and indole compounds have come into view, acting asmess "engers" in the microscopic world. They shuttle between the gut and other organs throughout the body, conveying "secret codes" that regulate health or induce disease. These metabolites are not only defenders of local gut homeostasis but are also closely linked to obesity, diabetes, inflammatory bowel disease, and even neuropsychiatric disorders (Van Hul et al., 2024). They have begun to play a role in disease prediction and risk assessment, using unique metabolic fingerprints to warn of health crises and detect the "latent" presence of diseases years or even decades in advance, thus opening a "window of opportunity" for early and precise disease intervention.

Against this backdrop, tea, an ancient beverage with a history of thousands of years and a cross-continental legacy, rich in cultural heritage and bioactive components, has moved from traditional health anecdotes into the modern scientific spotlight. Natural compounds in tea, such as tea polyphenols, tea polysaccharides, and theanine, act like "molecular keys" that precisely fit the "lock" of gut microbial ecology. They fine-tune community structure and reshape metabolic trajectories, leveraging the regulation of gut microbiota as a "hidden lever" to trigger a "chain reaction" of overall health. From microscopic analysis in the laboratory to clinical empirical exploration, from mechanistic studies in animal models to tracking effects in human cohorts, research on tea's regulation of gut microbiota unfolds like a grand academic scroll. On the "palette" of interdisciplinary integration, it harmonizes the colors of basic science, clinical medicine, and nutrition, promising to outline a new paradigm for preventive health care and disease intervention. It unlocks the health secrets hidden beneath the waves of time for traditional tea drinking and offers a way to solve modern health dilemmas.

11.2 Key Role of Gut Microbiota in Disease Prediction and Improvement

The gut microbiota plays a crucial role in human health. The human gut is home to an enormous number of microorganisms, with a quantity as high as 100 trillion and a weight of approximately 1–2 kg (Carvalho et al., 2012). The common gut bacteria are primarily composed of nine phyla, among which 0.1% are facultative anaerobes, mainly Enterobacteriaceae, while 99.9% are obligate anaerobes, primarily consisting of *Bacillus* and *Bifidobacterium* (de Vos et al., 2022; Lu et al., 2021). The gut microbiota comprises beneficial bacteria, intermediate bacteria or conditional pathogens, and harmful bacteria. The interdependence and antagonism among these three groups maintain a balanced and stable gut environment.

11.2.1 Important Physiological Functions

The gut microbiota possesses a variety of important physiological functions (Fan & Pedersen, 2021; Ross et al., 2024). First, it provides a protective barrier for the host. Normal gut microbiota adheres, colonizes, and proliferates at specific sites, forming a biofilm barrier that inhibits and excludes the invasion, adhesion, and growth of pathogens and conditional pathogens, thereby maintaining the balance between gut microbiota and human physiological functions. Second, normal microbiota stimulates the host's immune and clearance functions. The presence of beneficial, harmful, and conditional pathogenic bacteria in the gut maintains normal immunity when the gut microbiota is balanced. However, when gut microbiota imbalance occurs, with an overgrowth of harmful bacteria, immune dysfunction can lead to various immune-related diseases. Additionally, gut microbiota can break down gut contents into trace nutrients that are more readily absorbed by the human body. Both animal experiments and clinical evidence have shown that gut microbiota imbalance can trigger the onset of metabolic diseases such as obesity, diabetes, and cardiovascular and cerebrovascular diseases. A normal gut microbiota also promotes the normal development of the brain's nervous system. It can metabolize nutrients required by the nervous system, such as gamma-aminobutyric acid (GABA). In contrast, gut microbiota imbalance not only loses this function but also produces large amounts of enterotoxins (e.g., lipopolysaccharides, LPS), which can induce depression, Parkinson's disease, and autism (Tan et al., 2022). Moreover, under balanced conditions, gut microbiota can inhibit tumor development. For example, beneficial bacteria such as Bifidobacterium can degrade nitrosamines into nitrites and amines, exerting anticancer effects. In contrast, Escherichia coli, Clostridium, Enterococcus, and Bacteroides can facilitate the combination of dietary nitrites and amines into highly carcinogenic nitrosamines (Bahuguna & Dubey, 2023; Lindell et al., 2022). Under balanced conditions, gut microbiota can also reduce cholesterol absorption, lower the concentration of endotoxins in the blood, and exert anti-aging effects.

11.2.2 *Early Warning Indicators for Diseases' Onset and Progression*

Changes in the composition and function of gut microbiota can serve as early warning indicators for the onset and progression of diseases. Through analyses of the gut microbiota's metagenomics and metabolomics, specific microbial or metabolic biomarkers associated with disease risk can be identified.

Recent publications in high-impact journals such as Nature have highlighted that the human gut microbiota is a more potent predictor of disease presence than genetic factors and may even forecast an individual's future mortality risk (Salosensaari et al., 2021; Tito et al., 2024). One study, involving a 15-year longitudinal follow-up of a human cohort, established a particularly strong association between Enterobacteriaceae and mortality related to gastrointestinal and respiratory systems. Further analysis using the Kyoto Encyclopedia of Genes and Genomes (KEGG) revealed that microbiome functions associated with mortality involve pathways related to drug biodegradation, carbohydrate metabolism, lipid metabolism, and infectious diseases. Another study investigated the fecal microbiota of 589 patients with varying stages of colorectal cancer. The analysis revealed that *Anaerococcus vaginalis*, *Dialister pneumosintes*, *Parvimonas micra*, *Peptostreptococcus anaerobius*, *Porphyromonas asaccharolytica*, and *Prevotella intermedia* are strongly associated with the development of colorectal cancer and hold promise as potential targets for risk prediction.

Consistent with these findings, the concept of "gut age" was first proposed in Nature Medicine, suggesting that gut microbiota is a significant determinant of human age (Wang et al., 2024a, 2024b). Elderly individuals with healthy gut microbiota tend to be "younger," while unhealthy gut microbiota can make younger individuals "age prematurely." The study also revealed that these "young" microbiota significantly regulate long-term cardiovascular disease in elderly individuals with metabolic disorders.

Risk scores based on host genetics and gut microbiota may improve the prediction of various human diseases. A research team investigated 20 diseases influenced by both genetic and environmental factors, including schizophrenia, Parkinson's disease, hypertension, asthma, diabetes, and obesity (Liu et al., 2024). They found that changes in the gut microbiota are more reflective of an individual's disease susceptibility than genetic factors, with the exception of type 1 diabetes. These findings reiterate that while human diseases are influenced by both genetic and environmental factors, the impact of any environmental factor can ultimately be reflected in the gut microbiota. Scientists from the University of Cambridge and other institutions further suggested that integrating traditional clinical risk factors currently used in clinical practice—such as age, sex, BMI, blood pressure, non-HDL cholesterol, and HbA1c—with emerging quantitative genetic risks (polygenic risk scores) and gut microbiome data could potentially establish a robust and refined predictive framework (Liu et al., 2024). Current research results have partially elucidated disease-associated gut microbiota compositions. For example, in IBD, significant increases are observed in *Enterobacteriaceae, Fusobacterium, Ruminococcus gnavus, Streptococcus anginosus, Enterococcus, Megasphaera, Campylobacter, Gammaproteobacteria,* and *Deltaproteobacteria,* while *Faecalibacterium prausnitzii, Christensenellaceae, Collinsella, Roseburia, Ruminococcus,* and other butyrate-producing bacteria are significantly reduced (Metwaly et al., 2022). Additionally, *Salmonella typhi* and *Helicobacter* are associated with the occurrence of cholangiocarcinoma, and an increase in the abundance of *Bacteroidetes* and a decrease in *Actinobacteria* are commonly related to Alzheimer's disease (Liu & Shah, 2022).

11.2.3 Disease Predictive Indicators

Notably, microbial metabolites can also serve as disease predictive indicators. Elevated levels of conjugated bile acids and ammonia play a key role in hepatic encephalopathy and age-related cognitive impairment (Ren et al., 2024). Increases in taurocholic acid, glycocholic acid, taurolithocholic acid, and taurocholic acid are respectively associated with higher degrees of steatosis, lobular inflammation, portal inflammation, and hepatocellular ballooning (Liu et al., 2022). The increase of the gut commensal *Desulfovibrio* and its product hydrogen sulfide (H_2S) is closely related to the occurrence of metabolic syndrome (obesity) (Qi et al., 2024). In February 2024, a study published in the journal Gut analyzed 411 oral samples and 391 rectal samples, revealing that *Parabacteroides distasonis, Enterocloster bolteae, Dysosmobacter welbionis, Flavonifractor plautii,* and *Lachnospira GAM79* are highly enriched in the gut of patients with acute pancreatitis (Ammer-Herrmenau et al., 2024). The abnormal enrichment of these microbial taxa provides a new perspective for the early diagnosis of acute pancreatitis. The aforementioned changes in gut microbiota and their metabolites offer new insights and approaches for the early diagnosis and prognostic assessment of diseases.

11.2.4 Potential Therapeutic Strategy to Improve Symptoms and Prognosis in Diseases

As the functions of gut microbiota continue to be elucidated, directly intervening in the host's gut microbiota and their metabolites, or supplementing with specific microorganisms, can serve as a potential therapeutic strategy to improve symptoms and prognosis in a variety of diseases. Globally, there has been a significant number of clinical trials related to fecal microbiota transplantation (FMT), covering indications such as IBD, irritable bowel syndrome (IBS), Clostridium difficile infection (CDI), multidrug-resistant bacterial infections, hypertension, type 1 diabetes (T1D), type 2 diabetes (T2D), obesity, graft-versus-host disease (GVHD), anorexia nervosa, liver cirrhosis, insomnia, non-alcoholic fatty liver disease, autism, depression, Parkinson's disease, and cancer (Halkjær et al., 2018; Ooijevaar et al., 2019). In animal studies, the intervention with exogenous beneficial bacteria has yielded remarkable results. Scientists colonized the intestines of high-fat diet-induced obese mice with *Bifidobacterium* and *Lactobacillus* that produce short-chain fatty acids (SCFAs) (Lan et al., 2022; Wang et al., 2020a, 2020b). The experiments revealed that these beneficial bacteria can modulate intestinal energy metabolism and increase the levels of SCFAs such as acetate and propionate. As key metabolites, SCFAs not only activate intestinal endocrine cells, enhance satiety, and reduce food intake but also promote the browning of adipose tissue, increase energy expenditure, and significantly reduce body weight, body fat percentage, and insulin resistance in mice (van der Hee & Wells, 2021).

Similarly, fecal microbiota transplantation from lean mice into obese mice improved the metabolic status of the obese mice, reducing body weight, blood glucose, and insulin resistance. In animal models of IBD, fecal microbiota transplantation has shown significant advantages. Transplanting fecal microbiota from healthy donors into the intestines of DSS-induced colitis mice rapidly restored gut microbiota diversity (Zhang et al., 2024). The restored beneficial bacteria, such as those from the *Clostridium* and *Bacteroides genera*, activated the tryptophan metabolism pathway, normalizing the levels of metabolites like kynurenine and serotonin, exerting anti-inflammatory and mucosal repair effects and alleviating symptoms of intestinal inflammation, diarrhea, and bloody stools in the mice.

With the advancement of clinical medicine, clinical trials have gradually validated the findings from animal experiments and expanded the clinical application prospects of exogenous beneficial bacteria. In the treatment of clostridium difficile infection (CDI), fecal microbiota transplantation is recognized as an effective method. After recurrent CDI patients received fecal microbiota suspensions from healthy relatives, which had been strictly screened and processed, via enema or endoscopic infusion, the beneficial bacteria in the gut quickly regained dominance, inhibiting the growth of *C. difficile*. At the metabolic level, SCFAs such as butyrate were re-enriched, strengthening the intestinal mucosal immune barrier, with a cure rate exceeding 85%, far surpassing traditional antibiotic therapy (Drekonja et al., 2015; Khoruts et al., 2021). Clinical trials for autism spectrum disorder (ASD) have also yielded promising results. After some ASD children orally consumed probiotic preparations containing specific strains, the abundance of *Bifidobacterium* and *Akkermansia muciniphila* in their gut increased (Zhang et al., 2023). These microorganisms influenced neurotransmitter metabolism, increasing the precursors of serotonin, regulating brain neural transmission, and improving social impairments and repetitive behaviors in children, while also significantly alleviating gastrointestinal symptoms such as abdominal pain and constipation.

It is worth noting that many products on the market have highlighted the key role of gut microbiota. Yogurts supplemented with *Lactobacillus acidophilus* and *Lactobacillus bulgaricus* are widely popular (Pannerchelvan et al., 2024). Studies have shown that regular consumption of such yogurt can supplement active beneficial bacteria, which metabolize to produce lactic acid in the gut, lowering the intestinal pH, inhibiting the growth of harmful bacteria, stimulating intestinal motility, improving digestive function, maintaining daily gut microbiota stability, and preventing constipation and diarrhea. In pharmaceutical formulations, *Bacillus coagulans* spore tablets can be used to treat intestinal dysfunction. After entering the human body, the spores germinate and proliferate, secreting antimicrobial peptides to inhibit pathogenic bacteria such as *Escherichia coli* and *Salmonella.* At the same time, they promote the synthesis of various vitamins in the gut, enhancing the absorption efficiency of intestinal cells for nutrients, and alleviating symptoms such as bloating and diarrhea (Gao et al., 2023; Ma et al., 2024a, 2024b). This further demonstrates the core value of microbiota in medical interventions for intestinal diseases.

11.3 Regulatory Effect of Tea on Gut Microbiota

In regulating gut microbiota, tea's most direct function is to modulate its diversity and composition, and stabilize its functions. Gut microbiota, a key intestinal component, is dominated by *Bacteroidetes* and *Firmicutes* (over 90%) in healthy adults (Nishida et al., 2018), followed by *Proteobacteria* and *Actinobacteria.* However, psychological stress, viral or bacterial infections can reduce microbial diversity, increase intestinal permeability, disrupt microbial homeostasis, boost harmful bacteria, and potentially cause metabolic disorders, inflammation, and diseases like gastrointestinal and neurodegenerative diseases (Niño-Narvión et al., 2023; Nishino et al., 2018). Tea intervention can effectively reshape gut microbiota. Studies show that green, oolong, and white teas can increase gut microbial diversity in mice, raise the *Firmicutes/Bacteroidetes* ratio, alleviate dysbiosis, and relieve gut inflammation (Li et al., 2022; Liu et al., 2020). Pu-erh tea can increase the abundance of beneficial *Akkermansia* (Hu et al., 2022). Tea's main natural active components, such as tea polyphenols, theanine, polysaccharides, and pigments, are key to improving gut microbiota (Wei et al., 2023).

11.3.1 Tea Polyphenols

Tea polyphenols can increase the abundance of *Lactobacillus plantarum, Klebsiella, Akkermansia, Bifidobacteria,* and *Bacteroides*, inhibit potential pathogens like *Helicobacter pylori* and *Staphylococcus aureus*, and stabilize the intestinal barrier (Shao et al., 2022; Sun et al., 2021). They also inhibit the Cytc/Cleaved-caspase-3 pathway to regulate mitochondrial dysfunction and reduce colon cell apoptosis (Wei et al., 2023). Moreover, tea polyphenols can reverse the decrease of *Blautia* and *Faecalibaculum* caused by a high-fat diet, activate G-protein-coupled receptors, inhibit histone deacetylases, enhance tight junction expression, and lower intestinal permeability to maintain barrier integrity (Tian et al., 2024).

EGCG, a key tea polyphenol, can improve gut homeostasis and peripheral inflammation by increasing *Firmicutes* and *Anaerobe* abundance, and reduce intestinal permeability and exert anti-anxiety effects by downregulating hippocampal TLR4/MyD88/NF-κB and PI3K/Akt/NFκB phosphorylation and p65 acetylation (Chiou et al., 2012; Yang et al., 2023). Notably, colon-targeted unextractable bound polyphenols can modulate gut microbiota composition to relieve inflammation, with increased *Dubosiella* and *Enterorhabdus* abundance and decreased *Alistipes* and *Romboutsia* abundance being key mechanisms (Yang et al., 2024).

11.3.2 Tea Polysaccharides

In recent years, tea polysaccharides have attracted attention for their ability to modulate gut microbiota and metabolites, improving oxidative stress, inflammation, blood glucose and lipid levels, and preventing and treating cardiovascular diseases. Studies show that tea polysaccharides can reshape gut microbiota structure, increasing beneficial bacteria like *Bifidobacterium*, *Muribaculum*, and *Lactobacillus* (Yang et al., 2021). They activate the PINK1/Parkin pathway, boosting anti-inflammatory metabolites (IL-10, IL-13) in the gut and reducing pro-inflammatory cytokines (TNF-α, IL-6, IL-1β, IFN-γ, LPS) in serum. They also decrease inflammation-related mRNA expression of TJ proteins in colon tissue, enhancing gut barrier function, reducing inflammation, and alleviating brain dysfunction (Xu et al., 2022; Zeng et al., 2022).

As a key regulator of gut homeostasis, tea polysaccharides can lower blood glucose and lipids, and improve obesity by modulating gut microbiota. They significantly reduce the abundance of obesity- and inflammation-related bacteria like *Oscillibacter*, *Bilophila*, and *Clostridium XIVb*, improving metabolic and lipid disorders in type 2 diabetic and obese rats (Chen et al., 2018). Additionally, tea polysaccharides upregulate glucagon-like peptide-1 and increase goblet cell numbers, inhibiting Bax expression to strengthen the gut barrier, balance immunity, suppress inflammation, and improve insulin resistance, offering anti-hyperglycemic and pancreas-protective effects (Ma et al., 2024a, 2024b).

11.3.3 Tea Pigments

Tea pigment, derived from tea polyphenols, are categorized by oxidation degree into theabrownins, theaflavins, and thearubigins. They also modulate gut bacteria to combat obesity, lower inflammation, scavenge free radicals, and enhance cognitive function and circadian rhythm (Huang et al., 2019; Li et al., 2023). Theabrownins notably adjust gut microbiota imbalance, promoting *Lactobacillus*, *Bifidobacterium*, and *Akkermansia* growth, reducing pro-inflammatory cytokines (IL-6, TNF-α), and boosting immunity to ease inflammation and oxidative stress (Li et al., 2023; Yasuhisa et al., 2019). Theaflavin intake restores gut microbiota α-diversity in D-galactose-induced aged mice, increasing *Actinobacteria*, *Bifidobacterium*, and *Clostridium*, while reducing gut oxidative stress and inflammation. Via the "microbiota-gut-brain" axis, it upregulates CREB, BDNF, and TrkB, enhancing cognitive function and relieving anxiety. Tea saponins in tea increase *Bacteroides* and *Prevotella spp.* abundance to combat obesity and modulate glucose transporter and glucagon expression, improving hyperglycemia and hyperlipidemia in type 2 diabetic rats. Theanine reshapes gut microbiota, boosting *Prevotella*, *Lachnospira*, and *Ruminococcus*, stimulating short-chain fatty acid metabolism to enhance intestinal mucosal immunity and the gut immune barrier (He et al., 2021; Wei et al., 2020).

In summary, tea and its bioactive compounds act as prebiotics, reshaping gut microbiota to strengthen the gut barrier and modulate immune function, thereby alleviating inflammation and diseases caused by gut dysbiosis.

11.4 Regulatory Metabolites of Tea on Gut Microbiota

11.4.1 Bile Acid Metabolism

Tea not only reshapes the gut microbiota but also positively regulates microbial metabolites like bile acid, short-chain fatty acid, and tryptophan metabolism. In bile acid metabolism regulation, tea consumption can modulate bile acid synthesis and metabolism. Bile acids, synthesized by hepatocytes, are crucial for digestion, absorption, and metabolism. While the liver produces primary bile acids, secondary bile acids are generated by gut microbiota through chemical reactions, indicating a close link between gut microbiota and bile acid metabolism.

A study found that treating mice with non-alcoholic fatty liver disease with dark tea extract altered the expression of key bile acid synthesis enzymes in the liver, modulated bile acid synthesis and excretion, and changed the abundance of gut microbiota related to bile acid metabolism (Wang et al., 2017, 2019). This suggests that tea consumption can improve bile acid metabolism by modulating gut microbiota. Research indicates that tea components primarily regulate gut microbiota homeostasis via the gut-liver axis, activate FXR and TGR5 signaling pathways, modulate the PI3K/AKT/eNO pathway, and inhibit hepatic NF-κB signaling and NLRP3 inflammasome activation to alleviate liver injury, influencing bile acid synthesis, transport, and metabolism, and maintaining immune balance (Xiao et al., 2023).

Theabrownins in tea can downregulate hepatic TG levels, inhibit microbes associated with bile salt hydrolase, block the FXR-FGF15/19 signaling pathway, and promote ileal bile acid synthesis and secretion to alleviate liver injury and maintain bile acid balance (Somm & Jornayvaz, 2018).

Tea polysaccharides can downregulate genes related to lipid synthesis in the liver, reducing serum triglycerides, LDL-cholesterol, and free fatty acids while increasing HDL-cholesterol, thereby controlling fat accumulation.

EGCG supplementation can alter systemic bile acid metabolism by limiting increases in taurocholate, a serum marker of liver injury, upregulating cholate levels, promoting intestinal bile acid reabsorption, and enhancing deacetylase signaling, EIF2 pathways, and the biological clock to combat fat and protect the liver (Naito et al., 2020).

Caffeine in tea can lower CYP7A1 mRNA and protein expression, affecting bile acid synthesis rate, maintaining appropriate bile acid levels, and stimulating gallbladder contraction and Oddi sphincter relaxation to promote bile excretion. Increased bile excretion delivers more bile acids to the intestine, aiding fat digestion and absorption and maintaining bile acid homeostasis.

Tea and its bioactive components regulate bile acid metabolism not only by modulating bile acid synthesis and metabolism but also by influencing bile acid receptor function. The expression and function of bile acid receptors (FXR, TGR5, PXR, VDR, and RORγt) rely on gut microbiota structure and are negatively regulated by gut inflammation (Fiorucci et al., 2021). These receptors are widely distributed in innate and adaptive immune system cells and play a key role in regulating gut and liver immunity through the gut microbiota—bile acid axis (Biagioli et al., 2021). Tea components can modulate the expression and function of these receptors. Tea consumption activates FXR and TGR5 signaling pathways (Zhou et al., 2023), enhancing gut barrier function and reducing liver steatosis and oxidative stress damage (Ren et al., 2015; Xiang et al., 2023). FXR activation regulates bile acid circulation and excretion in the gut—liver axis by modulating bile acid transporter expression (Fuchs et al., 2017). Recent studies indicate that intestinal FXR activation can regulate pro-inflammatory gene expression by directly interfering with the NLRP3 receptor in a SHP-independent manner (Larabi et al., 2023). TGR5 activation increases gut motility, regulates macrophage M2 polarization, and inhibits inflammation and cholesterol metabolism disorders by activating the cAMP/PKA signaling pathway upon bile acid binding (Lun et al., 2024; Shi et al., 2020). Treating intestinal cells with tea's bioactive components upregulates FXR and TGR5 expression and alters phosphorylation levels of key proteins in downstream signaling pathways. This suggests that tea components directly act on intestinal cells to modulate bile acid receptor function, affecting bile acid metabolism and gut immunity. In summary, tea and its bioactive components regulate bile acid secretion and FXR/TGR5 expression to reshape bile acid metabolism and improve gut inflammation and lipid metabolism disorders.

11.4.2 Short-Chain Fatty Acid Metabolism

As the main metabolic products of distal gut microbiota, short-chain fatty acids (SCFAs) are formed by fermenting undigested food residues in the gut. The six major SCFAs are acetate, propionate, butyrate, isobutyrate, valerate, and isovalerate. They are crucial for maintaining gut homeostasis by enhancing the gut mucosal barrier, modulating immune cell function, and supplying energy to colonocytes (Dąbek-Drobny et al., 2022).

Tea consumption can significantly impact the abundance of SCFAs. Aged white tea increases the fecal abundance of butyrate, isobutyrate, and isovalerate in mice. Pu-erh tea promotes the production of SCFAs, especially isobutyrate, butyrate, isovalerate, and valerate, and activates their receptors to regulate gut nutrient absorption and hormone production, exerting a prebiotic effect (Xie et al., 2023). L-theanine boosts SCFA production by reducing serum diamine oxidase in colon tissue, restoring mucosal barrier expression, and maintaining colonocyte barrier integrity (Wang et al., 2020a, 2020b). Tea polyphenols, by inhibiting α-amylase and α-glucosidase, leave residual carbohydrates in the colon, providing a substrate for SCFA production. Moreover, L-theanine, when combined with a dietary fiber diet, increases the ratio of *Prevotella*, *Lachnospira*, and *Ruminococcus*, as well as the total SCFAs, acetate, propionate, and butyrate content in the feces of SD rats (Xu et al., 2020), thereby promoting SCFA metabolism and influencing intestinal mucosal immunity, including increasing villus height and crypt depth. Tea polyphenols also promote goblet cell growth, which in turn promotes the growth of SCFA-producing bacteria. EGCG supplementation positively affects the abundance of SCFAs in the intestinal contents of mice, increases the expression of gut barrier-related proteins, reduces intestinal permeability, and lowers corticotropin-releasing hormone, adrenocorticotropic hormone, and corticosterone levels via the gut-brain axis, alleviating HPA axis overactivation and cognitive dysfunction, and improving depression and memory impairment (Luo et al., 2022; Van de Wouw et al., 2018).

Butyrate, as an immune regulator in SCFAs, plays a central role in maintaining colonic homeostasis and regulating lipid metabolism. The polyphenols and catechins in tea can increase the abundance of *Bifidobacteria* and *Lactobacilli*, promoting SCFA production (especially butyrate), significantly inhibiting neutrophil production of pro-inflammatory cytokines, chemokines, and calprotectin, increasing tight junction protein expression (occludin, claudin 1, claudin 3, and claudin 4), downregulating the PI3K/AKT/NF-κB pathway-induced intestinal oxidative stress and inflammation, and potentially further improving brain dysfunction via the gut-brain axis (Fachi et al., 2019; Hu et al., 2022, 2023; Yan & Ajuwon, 2017). In summary, tea not only increases the abundance of SCFAs but also enhances their health benefits. SCFAs regulate immune responses by activating free fatty acid receptors (FFARs) or G protein-coupled receptors (GPR41, GPR43, GPR109A) on colonocytes, inhibiting cAMP-dependent pathways like mTOR (Sun et al., 2017).

Tea polyphenols alleviate chronic metabolic disorders in mice on a high-fat diet, increasing SCFA abundance, activating G protein-coupled receptors, suppressing inflammatory cytokines, and inhibiting TLR4 pathway activation to repair gut barriers (Tian et al., 2024). EGCG boosts SCFA levels, activates GPR43 on colonocytes to inhibit Th1 polarization, reduces gut inflammation, and modulates microbiota. It also enhances host anti-infection immunity via leukocyte SCFA receptors. In short, tea regulates microbes to promote SCFA-producing bacteria, increasing SCFAs (especially butyrate) to maintain gut homeostasis, enhance gut barriers, and improve cognition.

11.4.3 Tryptophan Metabolism

As an essential aromatic amino acid, tryptophan plays a crucial role in regulating gut immunity and inflammation (Agus et al., 2018). Tryptophan metabolism includes kynurenine, 5-HT, and indole metabolism, which are vital for gut immunity, barrier function, and neurotransmitter production (Michaudel et al., 2023). Aged Pu-erh tea reshapes the gut microbiota (boosting *Lachnospiraceae_NK4A136_group* and *Akkermansia*), modulates tryptophan and bile acid metabolism, and inhibits gut oxidative stress-mediated inflammatory pathways (TLR4/MyD88/ROS/p38MAPK/NF-κB p65), enhancing gut function and lipid metabolism (Hu et al., 2021, 2022). Fu brick tea's polysaccharides and theaflavins promote *Lactobacillus*, *Parasutterella*, and *Akkermansia* growth, increasing indole-3-aldehyde and indole-3-acetic acid levels (Wuqi et al., 2021, 2023). This activates AhRs, raises IL-22, and promotes ZO-1 and occludin expression to repair gut barriers. In particular, the disorder of tryptophan metabolism and the obstruction of 5-HT metabolic pathway are one of the key factors inducing brain dysfunction. Over 95% of 5-HT is degraded by gut microbiota and TPH1 (Bader, 2020). Post-tea consumption, increased gut tryptophan suppresses the AMPK-sirtuin1 pathway, maintaining autophagy and gut homeostasis (Gao et al., 2022). Tea's bioactives (tea polyphenols, theabrownine, theanine) reshape the microbiota (upregulating *Lactobacillus*, *Bifidobacterium*, *Prevotella*, and downregulating *Helicobacter pylori*, *Staphylococcus aureus*) (Qiaoyu et al., 2021; Wei et al., 2020), stimulating 5-HT synthesis enzymes (TPH, 5-HTPDC), and altering tryptophan metabolism (raising 5-HT/kynurenine ratios). This may underlie tea's improvement of neurodegenerative mental disorders via the gut-brain axis. Components like tea polyphenols and polysaccharides boost indole-producing bacteria (*Lactobacillus reuteri*, *Bifidobacterium*), activating Trpa1 receptors in small intestinal EECs to induce 5-HT secretion and gut neuron activation (Hokanson et al., 2024; Wang et al., 2024a, 2024b).

EGCG and theanine also increase brain 5-HT1A receptor expression. In kynurenine metabolism, tea's theaflavins limit excessive synthesis of metabolites (TDO, IDO, quinolinic acid), lowering kynurenine/tryptophan ratios to reduce neuroinflammation and PHA axis hyperactivation. Tea-induced changes in gut microbiota and SCFA metabolism also alter tryptophan metabolism. Elevated SCFAs repair gut barriers, increase TPH1 mRNA expression in a concentration-dependent manner, and activate ECs to boost 5-HT release (Reigstad et al., 2015). They also bind to colonic GPR43, activating mast cells containing 5-HT, indirectly increasing the relative content of 5-HT (Karaki et al., 2006). In summary, tea reshapes gut microbiota, alters tryptophan metabolism, enhances 5-HT pathways, and inhibits excessive kynurenine metabolite synthesis, thereby modulating gut immunity, promoting 5-HT for gut motility and brain activity via the gut-brain axis.

References

Agus, A., Planchais, J., & Sokol, H. (2018). Gut microbiota regulation of tryptophan metabolism in health and disease. *Cell Host & Microbe, 23*(6), 716–724. https://doi.org/10.1016/j.chom.2018.05.003

Ammer-Herrmenau, C., Antweiler, K. L., Asendorf, T., Beyer, G., Buchholz, S. M., Cameron, S., et al. (2024). Gut microbiota predicts severity and reveals novel metabolic signatures in acute pancreatitis. *Gut, 73*(3), 485–495. https://doi.org/10.1136/gutjnl-2023-330987

Bader, M. (2020). Inhibition of serotonin synthesis: A novel therapeutic paradigm. *Pharmacology & Therapeutics, 205*, 107423. https://doi.org/10.1016/j.pharmthera.2019.107423

Bahuguna, A., & Dubey, S. K. (2023). Overview of the mechanistic potential of probiotics and prebiotics in cancer chemoprevention. *Molecular Nutrition & Food Research, 67*(19), 2300221. https://doi.org/10.1002/mnfr.202300221

Biagioli, M., Marchianò, S., Carino, A., Di Giorgio, C., Santucci, L., Distrutti, E., & Fiorucci, S. (2021). Bile acids activated receptors in inflammatory bowel disease. *Cells, 10*(6). https://doi.org/10.3390/cells10061281

Carvalho, F. A., Aitken, J. D., Vijay-Kumar, M., & Gewirtz, A. T. (2012). Toll-like receptor–gut microbiota interactions: Perturb at your own risk! *Annual Review of Physiology, 74*(2012), 177–198. https://doi.org/10.1146/annurev-physiol-020911-153330

Chen, G., Xie, M., Wan, P., Chen, D., Dai, Z., Ye, H., et al. (2018). Fuzhuan brick tea polysaccharides attenuate metabolic syndrome in high-fat diet induced mice in association with modulation in the gut microbiota. *Journal of Agricultural and Food Chemistry, 66*(11), 2783–2795. https://doi.org/10.1021/acs.jafc.8b00296

Chiou, Y. S., Ma, N. J., Sang, S., Ho, C. T., Wang, Y. J., & Pan, M. H. (2012). Peracetylated (-)-epigallocatechin-3-gallate (AcEGCG) potently suppresses dextran sulfate sodium-induced colitis and colon tumorigenesis in mice. *Journal of Agriculture and Food Chemistry, 60*(13), 3441–3451. https://doi.org/10.1021/jf300441p

Dąbek-Drobny, A., Kaczmarczyk, O., Woźniakiewicz, M., Paśko, P., Dobrowolska-Iwanek, J., Woźniakiewicz, A., et al. (2022). Association between fecal short-chain fatty acid levels, diet, and body mass index in patients with inflammatory bowel disease. *Biology (Basel), 11*(1). https://doi.org/10.3390/biology11010108

de Vos, W. M., Tilg, H., Van Hul, M., & Cani, P. D. (2022). Gut microbiome and health: Mechanistic insights. *Gut, 71*(5), 1020–1032. https://doi.org/10.1136/gutjnl-2021-326789

Drekonja, D., Reich, J., Gezahegn, S., Greer, N., Shaukat, A., MacDonald, R., et al. (2015). Fecal microbiota transplantation for clostridium difficile infection: A systematic review. *Annals of Internal Medicine, 162*(9), 630–638. https://doi.org/10.7326/m14-2693

Fachi, J. L., Felipe, J. S., Pral, L. P., da Silva, B. K., Corrêa, R. O., de Andrade, M. C. P., et al. (2019). Butyrate protects mice from clostridium difficile-induced colitis through an HIF-1-dependent mechanism. *Cell Reports, 27*(3), 750-761.e757. https://doi.org/10.1016/j.celrep.2019.03.054

Fan, Y., & Pedersen, O. (2021). Gut microbiota in human metabolic health and disease. *Nature Reviews Microbiology, 19*(1), 55–71. https://doi.org/10.1038/s41579-020-0433-9

Fiorucci, S., Carino, A., Baldoni, M., Santucci, L., Costanzi, E., Graziosi, L., et al. (2021). Bile acid signaling in inflammatory bowel diseases. *Digestive Diseases and Sciences, 66*(3), 674–693. https://doi.org/10.1007/s10620-020-06715-3

Fuchs, C. D., Paumgartner, G., Wahlström, A., Schwabl, P., Reiberger, T., Leditznig, N., et al. (2017). Metabolic preconditioning protects BSEP/ABCB11(-/-) mice against cholestatic liver injury. *Journal of Hepatology, 66*(1), 95–101. https://doi.org/10.1016/j.jhep.2016.08.017

Gao, N., Yang, Y., Liu, S., Fang, C., Dou, X., Zhang, L., & Shan, A. (2022). Gut-derived metabolites from dietary tryptophan supplementation quench intestinal inflammation through the AMPK-SIRT1-autophagy pathway. *Journal of Agricultural and Food Chemistry, 70*(51), 16080–16095. https://doi.org/10.1021/acs.jafc.2c05381

Gao, X., Yu, J., Chang, L., Wang, Y., Sun, X., Mu, G., & Qian, F. (2023). In vitro antibacterial activity of Bacillus coagulans T242 on Caco-2 cells infected with Salmonella Typhimurium. *Food Bioscience, 52*, 102512. https://doi.org/10.1016/j.fbio.2023.102512

Halkjær, S. I., Christensen, A. H., Lo, B. Z. S., Browne, P. D., Günther, S., Hansen, L. H., & Petersen, A. M. (2018). Faecal microbiota transplantation alters gut microbiota in patients with irritable bowel syndrome: Results from a randomised, double-blind placebo-controlled study. *Gut, 67*(12), 2107. https://doi.org/10.1136/gutjnl-2018-316434

He, J., Chen, J., He, Q., Li, S., Jian, L., Xie, F., et al. (2021). Oral L-theanine administration promotes fat browning and prevents obesity in mice fed high-fat diet associated with the modulation of gut microbiota. *Journal of Functional Foods, 81*, 104476. https://doi.org/10.1016/j.jff.2021.104476

Hokanson, K. C., Hernández, C., Deitzler, G. E., Gaston, J. E., & David, M. M. (2024). Sex shapes gut–microbiota–brain communication and disease. *Trends in Microbiology, 32*(2), 151–161. https://doi.org/10.1016/j.tim.2023.08.013

Hu, S., Li, S., Liu, Y., Sun, K., Luo, L., & Zeng, L. (2021). Aged Ripe Pu-erh tea reduced oxidative stress-mediated inflammation in dextran sulfate sodium-induced colitis mice by regulating intestinal microbes. *Journal of Agriculture and Food Chemistry, 69*(36), 10592–10605. https://doi.org/10.1021/acs.jafc.1c04032

Hu, S., Luo, L., Bian, X., Liu, R. H., Zhao, S., Chen, Y., et al. (2022). Pu-erh tea restored circadian rhythm disruption by regulating tryptophan metabolism. *Journal of Agriculture and Food Chemistry, 70*(18), 5610–5623. https://doi.org/10.1021/acs.jafc.2c01883

Hu, S., Lin, Z., Zhao, S., Zhang, B., Luo, L., & Zeng, L. (2023). Pu-erh tea alleviated colitis-mediated brain dysfunction by promoting butyric acid production. *Food and Chemical Toxicology, 172*, Article 113594. https://doi.org/10.1016/j.fct.2022.113594

Huang, F., Zheng, X., Ma, X., Jiang, R., Zhou, W., Zhou, S., et al. (2019). Theabrownin from Pu-erh tea attenuates hypercholesterolemia via modulation of gut microbiota and bile acid metabolism. *Nature Communications, 10*(1), 4971. https://doi.org/10.1038/s41467-019-12896-x

Karaki, S., Mitsui, R., Hayashi, H., Kato, I., Sugiya, H., Iwanaga, T., et al. (2006). Short-chain fatty acid receptor, GPR43, is expressed by enteroendocrine cells and mucosal mast cells in rat intestine. *Cell and Tissue Research, 324*(3), 353–360. https://doi.org/10.1007/s00441-005-0140-x

Khoruts, A., Staley, C., & Sadowsky, M. J. (2021). Faecal microbiota transplantation for Clostridioides difficile: Mechanisms and pharmacology. *Nature Reviews Gastroenterology & Hepatology, 18*(1), 67–80. https://doi.org/10.1038/s41575-020-0350-4

Lan, Y., Sun, Q., Ma, Z., Peng, J., Zhang, M., Wang, C., et al. (2022). Seabuckthorn polysaccharide ameliorates high-fat diet-induced obesity by gut microbiota-SCFAs-liver axis. *Food & Function, 13*(5), 2925–2937. https://doi.org/10.1039/D1FO03147C

Larabi, A. B., Masson, H. L. P., & Bäumler, A. J. (2023). Bile acids as modulators of gut microbiota composition and function. *Gut Microbes, 15*(1), 2172671. https://doi.org/10.1080/19490976.2023.2172671

Li, A., Wang, J., Kou, R., Chen, M., Zhang, B., Zhang, Y., et al. (2022). Polyphenol-rich oolong tea alleviates obesity and modulates gut microbiota in high-fat diet-fed mice. *Frontiers in Nutrition, 9*, Article 937279. https://doi.org/10.3389/fnut.2022.937279

Li, M., Zhang, C., Xiao, X., Zhu, M., Quan, W., Liu, X., et al.. (2023). Theaflavins in black tea mitigate aging-associated cognitive dysfunction via the microbiota-gut-brain axis. *Journal of Agricultural and Food Chemistry, 71*(5), 2356–2369. https://doi.org/10.1021/acs.jafc.2c06679

Lindell, A. E., Zimmermann-Kogadeeva, M., & Patil, K. R. (2022). Multimodal interactions of drugs, natural compounds and pollutants with the gut microbiota. *Nature Reviews Microbiology, 20*(7), 431–443. https://doi.org/10.1038/s41579-022-00681-5

Liu, Y., Wang, X., Chen, Q., Luo, L., Ma, M., Xiao, B., & Zeng, L. (2020). Camellia sinensis and Litsea coreana ameliorate intestinal inflammation and modulate gut microbiota in dextran sulfate sodium-induced colitis mice. *Molecular Nutrition & Food Research, 64*(6), e1900943. https://doi.org/10.1002/mnfr.201900943

Liu, L., & Shah, K. (2022). The potential of the gut microbiome to reshape the cancer therapy paradigm: A review. *JAMA Oncology, 8*(7), 1059–1067. https://doi.org/10.1001/jamaoncol.2022.0494

Liu, Y., Liu, T., Zhao, X., & Gao, Y. (2022). New insights into the bile acid-based regulatory mechanisms and therapeutic perspectives in alcohol-related liver disease. *Cellular and Molecular Life Sciences, 79*(9), 486. https://doi.org/10.1007/s00018-022-04509-6

Liu, Y., Ritchie, S. C., Teo, S. M., Ruuskanen, M. O., Kambur, O., Zhu, Q., et al. (2024). Integration of polygenic and gut metagenomic risk prediction for common diseases. *Nature Aging, 4*(4), 584–594. https://doi.org/10.1038/s43587-024-00590-7

Lu, W.-W., Chen, X., Ni, J.-L., Zhu, S.-L., Fei, A.-H., & Wang, X.-S. (2021). The role of gut microbiota in the pathogenesis and treatment of acute pancreatitis: a narrative review. *Annals of Palliative Medicine, 10*(3), 3445–3451. https://apm.amegroups.org/article/view/65874

Lun, W., Yan, Q., Guo, X., Zhou, M., Bai, Y., He, J., et al. (2024). Mechanism of action of the bile acid receptor TGR5 in obesity. *Acta Pharm Sin B, 14*(2), 468–491. https://doi.org/10.1016/j.apsb.2023.11.011

Luo, Y. P., Tang, X. F., Zhang, Y. C., Chen, S. M., Wu, Q., & Li, W. J. (2022). Epigallocatechin-3-gallate alleviates galactose-induced aging impairment via gut-brain communication. *Food & Function, 13*(21), 11200–11209. https://doi.org/10.1039/d2fo00994c

Ma, L., Tian, G., Pu, Y., Qin, X., Zhang, Y., Wang, H., et al. (2024a). Bacillus coagulans MF-06 alleviates intestinal mucosal barrier from damage in chicks infected with Salmonella pullorum via activating the Wnt/β-catenin pathway. *Frontiers in Microbiology, 15*. https://doi.org/10.3389/fmicb.2024.1492035

Ma, Y., Xie, H., Xu, N., Li, M., Wang, L., Ge, H., et al. (2024b). Large yellow tea polysaccharide alleviates HFD-induced intestinal homeostasis dysbiosis via modulating gut barrier integrity, immune responses, and the gut microbiome. *Journal of Agriculture and Food Chemistry, 72*(13), 7230–7243. https://doi.org/10.1021/acs.jafc.4c00616

Metwaly, A., Reitmeier, S., & Haller, D. (2022). Microbiome risk profiles as biomarkers for inflammatory and metabolic disorders. *Nature Reviews. Gastroenterology & Hepatology, 19*(6), 383–397. https://doi.org/10.1038/s41575-022-00581-2

Michaudel, C., Danne, C., Agus, A., Magniez, A., Aucouturier, A., Spatz, M., et al. (2023). Rewiring the altered tryptophan metabolism as a novel therapeutic strategy in inflammatory bowel diseases. *Gut, 72*(7), 1296–1307. https://doi.org/10.1136/gutjnl-2022-327337

Naito, Y., Ushiroda, C., Mizushima, K., Inoue, R., Yasukawa, Z., Abe, A., & Takagi, T. (2020). Epigallocatechin-3-gallate (EGCG) attenuates non-alcoholic fatty liver disease via modulating the interaction between gut microbiota and bile acids. *Journal of Clinical Biochemistry and Nutrition, 67*(1), 2–9. https://doi.org/10.3164/jcbn.20-39

Niño-Narvi6n, J., Rojo-L6pez, M. I., Martinez-Santos, P., Rossell, J., Ruiz-Alcaraz, A. J., Alonso, N., et al. (2023). NAD+ precursors and intestinal inflammation: Therapeutic insights involving gut microbiota. *Nutrients, 15*(13). https://doi.org/10.3390/nu15132992

Nishida, A., Inoue, R., Inatomi, O., Bamba, S., Naito, Y., & Andoh, A. (2018). Gut microbiota in the pathogenesis of inflammatory bowel disease. *Clinical Journal of Gastroenterology, 11*(1), 1–10. https://doi.org/10.1007/s12328-017-0813-5

Nishino, K., Nishida, A., Inoue, R., Kawada, Y., Ohno, M., Sakai, S., et al. (2018). Analysis of endoscopic brush samples identified mucosa-associated dysbiosis in inflammatory bowel disease. *Journal of Gastroenterology, 53*(1), 95–106. https://doi.org/10.1007/s00535-017-1384-4

Ooijevaar, R. E., Terveer, E. M., Verspaget, H. W., Kuijper, E. J., & Keller, J. J. (2019). Clinical application and potential of fecal microbiota transplantation. *Annual Review of Medicine, 70*(2019), 335–351. https://doi.org/10.1146/annurev-med-111717-122956

Pannerchelvan, S., Rios-Solis, L., Wasoh, H., Sobri, M. Z. M., Faizal Wong, F. W., Mohamed, M. S., et al. (2024). Functional yogurt: A comprehensive review of its nutritional composition and health benefits. *Food & Function, 15*(22), 10927–10955. https://doi.org/10.1039/D4FO03671A

Qi, Q., Zhang, H., Jin, Z., Wang, C., Xia, M., Chen, B., et al. (2024). Hydrogen sulfide produced by the gut microbiota impairs host metabolism via reducing GLP-1 levels in male mice. *Nature Metabolism, 6*(8), 1601–1615. https://doi.org/10.1038/s42255-024-01068-x

Qiaoyu, S., Cheng, L., Xin, Z., Zufang, W., & Peifang, W. (2021). The interaction between tea polyphenols and host intestinal microorganisms: An effective way to prevent psychiatric disorders. *Food & Function*. https://doi.org/10.1039/d0fo02791j

Reigstad, C. S., Salmonson, C. E., Rainey, J. F., Szurszewski, J. H., Linden, D. R., Sonnenburg, J. L., et al. (2015). Gut microbes promote colonic serotonin production through an effect of short-chain fatty acids on enterochromaffin cells. *The FASEB Journal, 29*(4), 1395–1403. https://doi.org/10.1096/fj.14-259598

Ren, D., Hu, Y., Luo, Y., & Yang, X. (2015). Selenium-containing polysaccharides from Ziyang green tea ameliorate high-fructose diet induced insulin resistance and hepatic oxidative stress in mice. *Food & Function, 6*(10), 3342–3350. https://doi.org/10.1039/c5fo00557d

Ren, Z., Zhao, L., Zhao, M., Bao, T., Chen, T., Zhao, A., et al. (2024). Increased intestinal bile acid absorption contributes to age-related cognitive impairment. *Cell Reports Medicine, 5*(5), Article 101543. https://doi.org/10.1016/j.xcrm.2024.101543

Ross, F. C., Patangia, D., Grimaud, G., Lavelle, A., Dempsey, E. M., Ross, R. P., & Stanton, C. (2024). The interplay between diet and the gut microbiome: Implications for health and disease. *Nature Reviews Microbiology, 22*(11), 671–686. https://doi.org/10.1038/s41579-024-01068-4

Salosensaari, A., Laitinen, V., Havulinna, A. S., Meric, G., Cheng, S., Perola, M., et al. (2021). Taxonomic signatures of cause-specific mortality risk in human gut microbiome. *Nature Communications, 12*(1), 2671. https://doi.org/10.1038/s41467-021-22962-y

Shao, J., Wei, Y., & Wei, X. (2022). A comprehensive review on bioavailability, safety and antidepressant potential of natural bioactive components from tea. *Food Research International, 158*, Article 111540. https://doi.org/10.1016/j.foodres.2022.111540

Shi, Y., Su, W., Zhang, L., Shi, C., Zhou, J., Wang, P., et al. (2020). TGR5 regulates macrophage inflammation in nonalcoholic steatohepatitis by modulating NLRP3 inflammasome activation. *Frontiers in Immunology, 11*, Article 609060. https://doi.org/10.3389/fimmu.2020.609060

Somm, E., & Jornayvaz, F. R. (2018). Fibroblast growth factor 15/19: From basic functions to therapeutic perspectives. *Endocrine Reviews, 39*(6), 960–989. https://doi.org/10.1210/er.2018-00134

Sun, M., Wu, W., Liu, Z., & Cong, Y. (2017). Microbiota metabolite short chain fatty acids, GPCR, and inflammatory bowel diseases. *Journal of Gastroenterology, 52*(1), 1–8. https://doi.org/10.1007/s00535-016-1242-9

Sun, Q., Cheng, L., Zhang, X., Wu, Z., & Weng, P. (2021). The interaction between tea polyphenols and host intestinal microorganisms: An effective way to prevent psychiatric disorders. *Food & Function, 12*(3), 952–962. https://doi.org/10.1039/d0fo02791j

Tan, A. H., Lim, S. Y., & Lang, A. E. (2022). The microbiome–gut–brain axis in Parkinson disease—From basic research to the clinic. *Nature Reviews Neurology, 18*(8), 476–495. https://doi.org/10.1038/s41582-022-00681-2

Tian, B., Huang, P., Pan, Y., Gu, H., Yang, K., Wei, Z., & Zhang, X. (2024). Tea polyphenols reduced obesity by modulating gut microbiota-SCFAs-barrier and inflammation in high-fat diet-induced mice. *Molecular Nutrition & Food Research, 68*(24), Article e2400685. https://doi.org/10.1002/mnfr.202400685

Tito, R. Y., Verbandt, S., Aguirre Vazquez, M., Lahti, L., Verspecht, C., Lloréns-Rico, V., et al. (2024). Microbiome confounders and quantitative profiling challenge predicted microbial targets in colorectal cancer development. *Nature Medicine, 30*(5), 1339–1348. https://doi.org/10.1038/s41591-024-02963-2

Van de Wouw, M., Boehme, M., Lyte, J. M., Wiley, N., Strain, C., O'Sullivan, O., et al. (2018). Short-chain fatty acids: Microbial metabolites that alleviate stress-induced brain-gut axis alterations. *The Journal of Physiology, 596*(20), 4923–4944. https://doi.org/10.1113/jp276431

van der Hee, B., & Wells, J. M. (2021). Microbial regulation of host physiology by short-chain fatty acids. *Trends in Microbiology, 29*(8), 700–712. https://doi.org/10.1016/j.tim.2021.02.001

Van Hul, M., Cani, P. D., Petitfils, C., De Vos, W. M., Tilg, H., & El-Omar, E. M. (2024). What defines a healthy gut microbiome? *Gut, 73*(11), 1893. https://doi.org/10.1136/gutjnl-2024-333378

Wang, D., Gao, Q., Wang, T., Zhao, G., Qian, F., Huang, J., et al. (2017). Green tea infusion protects against alcoholic liver injury by attenuating inflammation and regulating the PI3K/Akt/eNOS pathway in C57BL/6 mice. *Food & Function, 8*(9), 3165–3177. https://doi.org/10.1039/c7fo00791d

Wang, D., Zhang, M., Wang, T., Cai, M., Qian, F., Sun, Y., & Wang, Y. (2019). Green tea polyphenols prevent lipopolysaccharide-induced inflammatory liver injury in mice by inhibiting NLRP3 inflammasome activation. *Food & Function, 10*(7), 3898–3908. https://doi.org/10.1039/c9fo00572b

Wang, D., Cai, M., Wang, T., Liu, T., Huang, J., Wang, Y., & Granato, D. (2020a). Ameliorative effects of L-theanine on dextran sulfate sodium induced colitis in C57BL/6J mice are associated with the inhibition of inflammatory responses and attenuation of intestinal barrier disruption. *Food Research International, 137*, Article 109409. https://doi.org/10.1016/j.foodres.2020.109409

Wang, G., Jiao, T., Xu, Y., Li, D., Si, Q., Hao, J., et al. (2020b). Bifidobacterium adolescentis and Lactobacillus rhamnosus alleviate non-alcoholic fatty liver disease induced by a high-fat, high-cholesterol diet through modulation of different gut microbiota-dependent pathways. *Food & Function, 11*(7), 6115–6127. https://doi.org/10.1039/C9FO02905B

Wang, G., Fan, Y., Zhang, G., Cai, S., Ma, Y., Yang, L., et al. (2024a). Microbiota-derived indoles alleviate intestinal inflammation and modulate microbiome by microbial cross-feeding. *Microbiome, 12*(1), 59. https://doi.org/10.1186/s40168-024-01750-y

Wang, T., Shi, Z., Ren, H., Xu, M., Lu, J., Yang, F., et al. (2024b). Divergent age-associated and metabolism-associated gut microbiome signatures modulate cardiovascular disease risk. *Nature Medicine, 30*(6), 1722–1731. https://doi.org/10.1038/s41591-024-03038-y

Wei, X., Ling, L., An, L., Tuo, Z., Sheng, Z., Yinhua, L., et al. (2020). L-Theanine affects intestinal mucosal immunity by regulating short-chain fatty acid metabolism under dietary fiber feeding. *Food & Function*. https://doi.org/10.1039/d0fo01069c

Wei, K., Wei, Y., Wang, Y., & Wei, X. (2023). Amelioration effects and regulatory mechanisms of different tea active ingredients on DSS-induced colitis. *Journal of Agriculture and Food Chemistry, 71*(44), 16604–16617. https://doi.org/10.1021/acs.jafc.3c04524

Wuqi, Y., Daoyuan, R., Yan, Z., Lei, L., & Xingbin, Y. (2021). Fuzhuan brick tea polysaccharide improved ulcerative colitis in association with gut microbiota-derived tryptophan metabolism. *Journal of Agricultural and Food Chemistry*. https://doi.org/10.1021/acs.jafc.1c02774

Wuqi, Y., Haiping, D., Hongjun, S., Xiangnan, Z., Ting, L., Li, Z., et al. (2023). Theabrownin from Fu brick tea improves ulcerative colitis by shaping the gut microbiota and modulating the tryptophan metabolism. *Journal of Agricultural and Food Chemistry*. https://doi.org/10.1021/acs.jafc.2c06821

Xiang, G., Sun, H., Chen, Y., Guo, H., Liu, Y., Li, Y., et al. (2023). Antioxidant and hypoglycemic activity of tea polysaccharides with different degrees of fermentation. *International Journal of Biological Macromolecules, 228*, 224–233. https://doi.org/10.1016/j.ijbiomac.2022.12.114

Xiao, J., Dong, L. W., Liu, S., Meng, F. H., Xie, C., Lu, X. Y., et al. (2023). Bile acids-mediated intracellular cholesterol transport promotes intestinal cholesterol absorption and NPC1L1 recycling. *Nature Communications, 14*(1), 6469. https://doi.org/10.1038/s41467-023-42179-5

Xie, Z., Zeng, Z., Chen, G., Dong, W., Peng, Y., Xu, W., et al. (2023). Intracellular Polysaccharides of Aspergillus cristatus from Fuzhuan brick tea leverage the gut microbiota and repair the intestinal barrier to ameliorate DSS-induced colitis in mice. *Journal of Agriculture and Food Chemistry, 71*(21), 8023–8037. https://doi.org/10.1021/acs.jafc.3c00611

Xu, W., Lin, L., Liu, A., Zhang, T., Zhang, S., Li, Y., et al. (2020). L-Theanine affects intestinal mucosal immunity by regulating short-chain fatty acid metabolism under dietary fiber feeding. *Food & Function, 11*(9), 8369–8379. https://doi.org/10.1039/d0fo01069c

Xu, A., Zhao, Y., Shi, Y., Zuo, X., Yang, Y., Wang, Y., & Xu, P. (2022). Effects of oxidation-based tea processing on the characteristics of the derived polysaccharide conjugates and their regulation of intestinal homeostasis in DSS-induced colitis mice. *International Journal of Biological Macromolecules, 214*, 402–413. https://doi.org/10.1016/j.ijbiomac.2022.06.115

Yan, H., & Ajuwon, K. M. (2017). Butyrate modifies intestinal barrier function in IPEC-J2 cells through a selective upregulation of tight junction proteins and activation of the Akt signaling pathway. *Plos One, 12*(6), Article e0179586. https://doi.org/10.1371/journal.pone.0179586

Yang, W., Ren, D., Zhao, Y., Liu, L., & Yang, X. (2021). Fuzhuan brick tea polysaccharide improved ulcerative colitis in association with gut microbiota-derived tryptophan metabolism. *Journal of Agricultural and Food Chemistry, 69*(30), 8448–8459. https://doi.org/10.1021/acs.jafc.1c02774

Yang, J. Z., Zhang, K. K., Liu, Y., Li, X. W., Chen, L. J., Liu, J. L., et al. (2023). Epigallocatechin-3-gallate ameliorates polystyrene microplastics-induced anxiety-like behavior in mice by modulating gut microbe homeostasis. *Science of the Total Environment, 892*, Article 164619. https://doi.org/10.1016/j.scitotenv.2023.164619

Yang, C., Yang, W., Wang, Y., Du, Y., Zhao, T., Shao, H., et al. (2024). Nonextractable polyphenols from Fu brick tea alleviates ulcerative colitis by controlling colon microbiota-targeted release to inhibit intestinal inflammation in mice. *Journal of Agriculture and Food Chemistry, 72*(13), 7397–7410. https://doi.org/10.1021/acs.jafc.3c06883

Yasuhisa, A., Rena, O., Masahiro, K., Yoshimasa, T., & Keiji, K. (2019). Theaflavins improve memory impairment and depression-like behavior by regulating microglial activation. *Molecules*. https://doi.org/10.3390/molecules24030467

Zhang, J., Zhu, G., Wan, L., Liang, Y., Liu, X., Yan, H., et al. (2023). Effect of fecal microbiota transplantation in children with autism spectrum disorder: A systematic review. *Front Psychiatry, 14*, 1123658. https://doi.org/10.3389/fpsyt.2023.1123658

Zhang, Q.-W., Yang, M.-J., Liao, C.-Y., Taha, R., Li, Q.-Y., Abdelmotalab, M. I., et al. (2024). Atractylodes macrocephala Koidz polysaccharide ameliorates DSS-induced colitis in mice by regulating the gut microbiota and tryptophan metabolism. *British Journal of Pharmacology, n/a*(n/a). https://doi.org/10.1111/bph.17409

Zeng, Z., Xie, Z., Chen, G., Sun, Y., Zeng, X., & Liu, Z. (2022). Anti-inflammatory and gut microbiota modulatory effects of polysaccharides from Fuzhuan brick tea on colitis in mice induced by dextran sulfate sodium. *Food & Function, 13*(2), 649–663. https://doi.org/10.1039/d1fo02702f

Zhou, S., Yang, L., Hu, L., Qin, W., Cao, Y., Tang, Z., et al. (2023). Blueberry extract alleviated lipopolysaccharide-induced inflammation responses in mice through activating the FXR/TGR5 signaling pathway and regulating gut microbiota. *Journal of the Science of Food and Agriculture, 103*(9), 4638–4648. https://doi.org/10.1002/jsfa.12560

Epilogue

Great efforts have been made toward revealing tea flavor and bio-active compounds. It is anticipated that in future research, the improvement of analytical techniques is highly-required to meet the growing needs for accuracy, precision, and sensitivity, while also being efficient, portable, and suitable for real-time quality monitoring during tea production. For sensory analysis, it's crucial to develop a more effective integration of analytical-instrumental data with sensory evaluation results, alongside innovative computational approaches in flavor chemistry, to advance tea flavor analysis and obtain comprehensive flavor profiles.

Regarding tea's health benefits, although numerous studies and experiments have demonstrated tea's efficacy in alleviating various diseases and its potential in preventing chronic conditions, researchers must continue to investigate several critical questions in future studies. These questions include understanding the key mechanisms by which tea acts in *vivo* (particularly in human) to combat diseases, identifying specific essential components directly responsible for the health benefits, and developing practical techniques to translate laboratory findings of tea's bio-active components into clinical applications.

In summary, tea holds a significant position in modern life due to its long-standing medicinal use, rising global consumption, and strong connection to health benefits like antioxidation and anti-inflammation. As a culturally rich and healthy-beneficial beverage, tea is historically important and increasingly relevant today. Future research should deeply explore tea's health benefits, clarify its mechanisms, and find better ways to use tea for human health. Also, attention should be paid to the rationality and safety of tea consumption to ensure its scientific and healthy enjoyment.

X. Zhai (ed.), *Tea Chemistry and Health Benefits*,
https://doi.org/10.1007/978-3-032-11828-8

Acknowledgment

This book of Tea Chemistry and Health Benefits: Flavour and Processing Dynamics compiles eleven chapters written by a young board of tea researchers from China including Dr. Xiaoting Zhai from Anhui Agricultural University, Dr. Shanshan Hu from Anhui Agricultural University, Dr. Zisheng Han from Anhui Agricultural University, Dr. Yujie Wang from Zhejiang University, Dr. Piaopiao Long from Anhui Agricultural University, Ph.D candidate Mr. Dongzhu Huang from Anhui Agricultural University, and Ph.D candidate Miss. Jing Wang from Hunan Agricultural University. The authors would like to express their sincere thanks to the support of Natural Science Foundation of China (32302605 and 32402071), National Key Research and Development Program of China (2021YFD1601100), the earmarked fund CARS-19, and Anhui Key Research and Development Plan (2022h11020022).

MIX
Papier aus verantwortungsvollen Quellen
Paper from responsible sources
FSC® C105338

If you have any concerns about our products,
you can contact us on
ProductSafety@springernature.com

In case Publisher is established outside the EU,
the EU authorized representative is:
Springer Nature Customer Service Center GmbH
Europaplatz 3, 69115 Heidelberg, Germany

Printed by Libri Plureos GmbH
in Hamburg, Germany